T0260018

Corporate Financial Distress, Restructuring, and Bankruptcy

Corporate Financial Distress, Restructuring, and Bankruptcy

Analyze Leveraged Finance, Distressed Debt, and Bankruptcy

Fourth Edition

EDWARD I. ALTMAN
EDITH HOTCHKISS
WEI WANG

WILEY

Published by John Wiley & Sons, Inc., Hoboken, New Jersey.
Published simultaneously in Canada.

For general information on our other products and services or for technical support, please contact our Customer Care Department within the United States at (800) 762-2974, outside the United States at (317) 572-3993, or fax (317) 572-4002.

Wiley publishes in a variety of print and electronic formats and by print-on-demand. Some material included with standard print versions of this book may not be included in e-books or in print-on-demand. If this book refers to media such as a CD or DVD that is not included in the version you purchased, you may download this material at http://booksupport.wiley.com. For more information about Wiley products, visit www.wiley.com.

Library of Congress Cataloging-in-Publication Data is available.

ISBN 978-1-119-48180-5 (Hardcover)
ISBN 978-1-119-48181-2 (ePDF)
ISBN 978-1-119-48185-0 (ePub)

Cover Design: Wiley
Cover Image: © Ps Pong / Shutterstock

Printed and bound by CPI Group (UK) Ltd, Croydon CR0 4YY

C9781119481805_270724

Ed Altman dedicates this book to his wife and partner for over 50 years, Dr. Elaine Altman, whose support and advice have sustained him and helped in crafting these four editions over 35 years.

Edie Hotchkiss dedicates this book to her husband Steven and daughter Jenny for their constant support.

Wei Wang dedicates this book to his wife Ella and children Andrea, Mia, Julia, and Ethan for their endless love and inspiration.

Contents

PART TWO

High-Yield Debt, Prediction of Corporate Distress, and Distress Investing

About the Authors

Edward Altman is the Max L. Heine Professor of Finance Emeritus at New York University, Stern School of Business, and Director of the Credit and Fixed Income Research Program at the NYU Salomon Center.

Dr. Altman has an international reputation as an expert on corporate bankruptcy, high-yield bonds, distressed debt, and credit risk analysis. He is the creator of the world-famous Altman Z-Score models for bankruptcy prediction of firms globally. He was named Laureate 1984 by the Hautes Études Commerciales Foundation in Paris for his accumulated works on corporate distress prediction models and procedures for firm financial rehabilitation and awarded the Graham & Dodd Scroll for 1985 by the Financial Analysts Federation for his work on Default Rates and High Yield Corporate Debt. He was a Founding Executive Editor of the *Journal of Banking & Finance* and serves on the editorial boards of several other scholarly finance journals.

Professor Altman was inducted into the Fixed Income Analysts Society Hall of Fame in 2001 and elected President of the Financial Management Association (2003) and a Fellow of the FMA in 2004, and was among the inaugural inductees into the Turnaround Management Association's Hall of Fame in 2008.

In 2005, Dr. Altman was named one of the "100 Most Influential People in Finance" by *Treasury & Risk Management* magazine and is frequently quoted in the popular press and on network TV.

Dr. Altman has been an advisor to many financial institutions including Merrill Lynch, Salomon Brothers, Citigroup, Concordia Advisors, Investcorp, Paulson & Co., S&P Global Market Intelligence and the RiskMetrics Group (MSCI, Inc.). He is currently (2018) Advisor to Golub Capital, Classis Capital (Italy), Wiserfunding in London, Clearing Bid, Inc., S-Cube Capital (Singapore), ESG Portfolio Management (Frankfurt) and AlphaFixe (Montreal). He serves on the Board of Franklin Mutual Series and Alternative Investments Funds. He is also Chairman of the Academic Advisory Council of the Turnaround Management Association. Dr. Altman was a Founding Trustee of the Museum of American Finance and was Chairman of the Board of the International Schools Orchestras of New York.

Edith S. Hotchkiss is a Professor of Finance at the Carroll School of Management at Boston College, where she teaches courses in corporate finance, valuation, and restructuring. She received her AB in engineering and economics summa cum laude from Dartmouth College and her PhD in finance from NYU's Stern School of Business. Prior to entering academics, she worked in consulting and for the Financial Institutions Group of Standard & Poor's Corporation.

Professor Hotchkiss's research covers topics including: corporate financial distress and restructuring; the efficiency of Chapter 11 bankruptcy; and trading in corporate debt markets. Her work has been published in leading finance journals including the *Journal of Finance, Journal of Financial Economics*, and *Review of Financial Studies*. She has served on the national board of the Turnaround Management Association, and as a consultant to FINRA on fixed income markets. She has also served as a consultant for several recent Chapter 11 cases.

Wei Wang is an Associate Professor and RBC Fellow of Finance and Director of Master of Finance–Beijing program at the Smith School of Business at Queen's University, Canada. His research interests are in bankruptcy restructuring, distressed investing, and corporate governance. His work has been published in leading academic journals including the *Journal of Finance* and *Journal of Financial Economics*, and featured in the *Wall Street Journal* and other media. He has published a number of Harvard Business School finance cases. He worked in commodities derivative trading and financial engineering prior to entering academics.

Dr. Wang has taught corporate restructuring and distressed investing at the Wharton School's undergraduate, MBA, and EMBA programs as a visiting professor. He is in active collaboration with the Aresty Institute for Executive Education at the Wharton School to deliver lectures and workshops on bankruptcy restructuring, leveraged loans, and distressed M&A to banking executives. Dr. Wang has also taught corporate restructuring and fixed income securities with an Asian market perspective at Hong Kong University of Science and Technology Business School. He is retained as a foreign expert at the Mingde Center for Corporate Acquisition and Restructuring Research at Shanghai University of Finance and Economics in China.

Acknowledgments

We would like to acknowledge an impressive group of practitioners and academics who have assisted us in the researching and writing of this book. We are enormously grateful to all of these persons for helping us to shape our analysis and commentary in our writings and in our classes at the New York University Stern School of Business, Boston College, Queens University Smith School of Business, and the Wharton School.

Among the many practitioners who have helped out over the many years in the writing of this volume, Ed Altman would like to single out Robert Benhenni, Allan Brown, Martin Fridson, Michael Gordy, Tony Kao, Stuart Kovensky, and James Peck. Edie Hotchkiss and Wei Wang further thank Brian Benvenisty, Michael Epstein, Elliot Ganz, Joseph Guzinski, David Keisman, Bridget Marsh, Abid Qureshi, Ted Osborn, and Robert Stark for the many hours of conversations and comments on our work.

We also would like to sincerely thank the many academic colleagues who helped to enrich the content of this book. Our academic colleagues include Yakov Amihud, Alessandro Danovi, Sanjiv Das, Jarred Elias, Malgorzata Iwanicz-Drozdowski, Erkki Laitinen, Frederik Lundtote, Herbert Rijken, and Arto Suvas. Ed would also like to sincerely acknowledge the great assistance of the staff at the NYU Salomon Center, including Brenda Kuehne, Mary Jaffier, Robyn Vanterpool and last, but not least, Lourdes Tanglao.

To his family, especially his wife Elaine and son Gregory, Professor Altman has only sincere words of gratitude for their endless support. To his colleagues and co authors Edith Hotchkiss and Wei Wang, for their amazing collegiality and great efforts in making this volume a reality. Edie Hotchkiss would like to thank Ed for first introducing her to this field as her PhD dissertation adviser, and for his guidance and friendship through many years of research in this area. Wei Wang sincerely thanks Ed and Edie for inviting him to work on this volume. He would also like to thank his students at both the Smith School of Business and the Wharton School, including Aneesh Chona and Xiaobing Ma, for spending many hours reading the manuscript and providing valuable feedback.

Preface

In looking back over the first three editions of *Corporate Financial Distress and Bankruptcy* (1983, 1993, 2006), we note that with each publication, the incidence and importance of corporate bankruptcy in the United States had risen to ever more prominence. The number of professionals dealing with the uniqueness of corporate death in this country was increasing so much that it could have perhaps been called a "bankruptcy industry." There is absolutely no question in 2019 that we can now call it an industry. The field has become even more significant in the past 15 years, accompanied by an increase of academics specializing in the area of corporate financial distress. Indeed, there is nothing more important in attracting rigorous and thoughtful research than data! With this increased theoretical and especially empirical interest, Wei Wang, has joined the original author (Altman) of the first three editions and Edith Hotchkiss (from the third edition) to produce this volume.

It is now quite obvious that the bankruptcy business is big-business. While no one has done an extensive analysis of the number of people who deal with corporate distress on a regular basis, we would venture a guess that it is at least 45,000 globally, with the vast majority in the United States but a growing number abroad. We include turnaround managers (mostly consultants); bankruptcy and restructuring lawyers; bankruptcy judges and other court personnel; accountants, bankers, and other financial advisers who specialize in working with distressed debtors; distressed debt investors, sometimes referred to as "vultures"; and, of course, researchers. Indeed, the prestigious Turnaround Management Association (www.tumaround.org) total members numbered more than 9,000 in 2018.

The reason for the large number of professionals working with organizations in various stages of financial distress is the increasing number of large and complex bankruptcy cases. Despite the fact that the United States has been in a benign credit cycle since 2010, during the six-year period of 2012–2018, 130 companies with liabilities greater than $1 billion filed for protection under Chapter 11 of the Bankruptcy Code. Over the past 47 years (1971–2018), there have been at least 450 of these large, mega-bankruptcies in the United States. Just before finishing our first draft of this book, one of the nation's largest retailers, Sears, Roebuck and Company, filed for Chapter 11 with over $11 billion of liabilities! And the number of mega-bankruptcies, as well as the total of all filings, will spike dramatically when the next financial crisis hits, especially due to the enormous build-up of corporate debt in recent years.

This book is a completely updated volume that includes updated key statistics and surveys the most recent academic studies. Newly added chapters include those on leveraged finance, out-of-court restructurings, and international insolvency codes, as well as a review of the Altman Z-Score family of models and their applications to celebrate the 50th birthday of the original Z-Score model. The 16 chapters in this new edition cover the most important aspects of leveraged finance, high-yield markets, corporate restructuring, bankruptcy, and credit risk modeling.

Starting with Chapter 1, we define corporate distress and present the statistical background for corporate defaults and bankruptcies over the past few decades. The chapter also discusses the common reasons for corporate failures and presents the organization theory that guides practice. In addition, the chapter introduces the key industry players in distressed restructuring and investing.

The leveraged finance market experienced an unprecedented boom in the past two decades, the total issuance of leveraged loans and high yield bonds reaching close to $1 trillion in 2017. The markets have been quite creative at producing new financial instruments (e.g. second-lien, covenant-lite) as the markets have grown. These instruments are attractive to not only traditional commercial lenders but also alternative investors due to their high yield and high fee structure. Chapter 2 provides an overview of the two major categories of debt instruments in this space and discusses typical features of these instruments, lender protections, default and remedies, as well as debt subordination issues. This material is particularly necessary to understanding the priority of debt claims and their relative bargaining position in distressed restructuring.

Chapter 3 provides an overview of the U.S. bankruptcy system. We begin by briefly illustrating the evolution of the U.S. bankruptcy law since the equity receiverships of 1898. We provide a primer on Chapter 11 by introducing the key provisions of the U.S. Bankruptcy Code after the Bankruptcy Abuse Prevention and Consumer Protection Act of 2005 (BAPCPA). Our summary and interpretation of important sections of the Code is written to be accessible to students and practitioners in finance as well as a legal audience. We review the many relevant academic studies, and also provide examples from recent cases to help readers gain an in-depth understanding of the bankruptcy process. The conclusion to this chapter summarizes the ABI Commission Report of 2014 advocating revisions to the existing code.

Firms suffer large costs of financial distress, and bankruptcy restructuring can be even costlier. These costs include not only direct costs such as out-of-pocket expenses for lawyers and finance professionals, but also a wide range of opportunity costs known as indirect costs. Firms generally have strong incentives to avoid these costs by conducting private negotiations and restructuring out-of-court. When and how can firms successfully restructuring out of court? Why do others restructure in court? Chapter 4 attempts to answer these questions.

Chapter 5 explores the analytics and process for distressed firm valuation. We provide a careful discussion of valuation models, and consider why we observe wide disagreements over firm value between different stakeholders in the negotiation process. We describe in depth best practices in valuation methods, using the example of Cumulus Media which filed for Chapter 11 in November 2017.

Virtually every aspect of a firm's governance can change in some way when a firm becomes financially distressed. Management turnover increases, board size declines, and boards often change in their entirety at reorganization. A substantial number of restructurings lead to a change in control of the company. Chapter 6 discusses key corporate governance issues for distressed firms, including fiduciary duties of managers and boards, complexities in providing compensation, and the value of creditor control rights. We wrap up the chapter by discussing managerial labor markets and labor issues.

In Chapter 7, we explore the success of the bankruptcy reorganization process, especially with respect to the postbankruptcy performance of firms emerging from Chapter 11. In numerous instances, emerging firms suffer from continued operating and financial problems, sometimes resulting in a second filing, unofficially called a Chapter 22. Indeed, we are aware of at least 290 of these two-time filers over the period 1984-2017 (see Chapter 1), and 18 three-time filers (Chapter 33s). There are even three Chapter 44s and at least one Chapter 55! Despite the numbers of bankruptcy repeaters, there are also some spectacular success stories upon emergence from bankruptcy, at least from the perspective of equity holders in the reorganized company.

Chapter 8 provides a brief summary on international insolvency regimes, paying particular attention to countries including France, Germany, Japan, Sweden, UK, China, and India. We focus on these representative countries because of the distinct nature in their legal procedures, significant growth in their restructuring industry in recent years, and the availability of related empirical academic research. Our brief discussions for these countries highlight the most important features of their legal systems for restructuring as well as ongoing issues and reforms.

The second part of the book provides comprehensive coverage of high yield bond markets, default prediction models and their applications, and distressed investing. We explore in depth the estimation of default probabilities for issuers in the United States (Chapter 10) and for sovereign issuers (Chapter 13) and the loss given default or recovery rates (Chapter 16). Chapters 11 and 12 demonstrate applications of these models for many different scenarios, including credit risk management, distressed debt investing, turnaround management and other advisory capacities, and legal issues. Chapters 14 and 15 go on to examine the size and development of the distressed and defaulted debt market.

Chapter 9 explores risk-return aspects of the U.S. high-yield bond and bank loan markets, most important to highly levered and distressed firms. Since high-yield or "junk" bonds are the raw material for future possible distress situations, it is important to investigate their properties. Among the most relevant statistics to investors in this market are default rates, as well as recovery rates for firms that default. The U.S. high-yield corporate bond market reached more than

$1.6 trillion outstanding in 2017, a 60% increase since the year of publication of the previous edition of this book. Globally, the high-yield bond market reached approximately $2.5 trillion.

It has been 50 years since the seminal work by Professor Altman developing the first family of default prediction models. With advancements in financial research such as the Black-Scholes-Merton framework, we have gained substantially greater understanding of methods for predicting and pricing default risk. Yet the Altman Z-score remains one of the most popular models in this domain, due to not only its high predictive power but also its simplicity. In Chapter 10, Professor Altman provides a 50-year retrospective on the evaluation and applications of the Z-score family of models and other credit risk models. The three chapters that follow present applications of the Z-score models. Chapter 11 focuses on applications performed by analysts who are external to the distressed firm in order to improve their position or to exploit profitable opportunities presented by distressed firms and their securities. With respect to the turnaround management arena, Chapter 12 further explores the possibility of using distressed firm predictive models to assist in the management of the distressed firm itself in order to manage a return to financial health. We illustrate this via an actual case study - the GTI Corporation -and its rise from near extinction. Finally, in Chapter 13, we apply our updated distress prediction model, called Z-Metrics, in a bottom-up analysis of the default assessment of sovereign debt.

The distressed and defaulted debt market has grown tremendously over the past two decades. As of 2017, the publically traded and private issued market was about $747 billion (face value) and $414 billion (market value). This substantial market is poised to grow considerably when the next credit crisis hits. Debt market investors, particularly hedge funds, recognize distressed debt as an important and unique asset class. In Chapters 14 and 15, we explore the size, growth, risk-reward dimensions, and investment strategies in distressed debt. Last, in Chapter 16, we provide a comprehensive survey of studies devoted to modeling and estimating debt recovery rates.

As the restructuring industry and the high yield and distressed markets continue to evolve, we hope readers will find this book a valuable reference to understanding the state of the market and prepare for the next downturn. We hope that the framework, methodologies, research findings, and statistics we present will be useful to *practitioners* who seek a deep understanding of the practice and the state-of-the-art academic research, *academic researchers* who continue to explore and create knowledge to guide restructuring practices, *policy makers* who pay close attention to the design of bankruptcy law and market regulation, and *students* who aspire to learn about exciting opportunities in the world of distress!

Edward I. Altman
Edith Hotchkiss
Wei Wang

Corporate Financial Distress, Restructuring, and Bankruptcy

The Economic and Legal Framework of Corporate Restructuring and Bankruptcy

Corporate Financial Distress

Introduction and Statistical Background

Corporate financial distress, and the legal processes of corporate bankruptcy reorganization (Chapter 11 of the Bankruptcy Code) and liquidation (Chapter 7 of the Bankruptcy Code), has become a familiar economic reality to many U.S. corporations. The business failure phenomenon received some exposure during the 1970s, more during the recession years of 1980–1982 and 1989–1991, heightened attention during the explosion of defaults and large firm bankruptcies in the 2001–2003 post-dotcom period, and unprecedented interest in the 2008–2009 financial and economic crisis period. Between 1989 and 1991, 34 corporations with liabilities greater than $1 billion filed for protection under Chapter 11 of the Bankruptcy Code; in the three-year period from 2001 to 2003, 102 of these "billion-dollar-babies" with liabilities totaling $580 billion filed for bankruptcy protection; and from 2008 to 2009, 74 such companies filed for bankruptcy with an unprecedented amount of liabilities totaling over $1.2 trillion.

The line-up of major corporate bankruptcies was capped by the mammoth filings of Lehman Brothers ($613 billion in liabilities), General Motors ($173 billion in liabilities), CIT Group ($65 billion in liabilities), and Chrysler ($55 billion in liabilities) during the 2008–2009 financial crisis. In fact, the total amount of liabilities of these four mega cases accounted for 75% of the liabilities of all billion-dollar firms filing for bankruptcy from 2008 to 2009. Three other mega cases from the 2001–2003 period also make the list of the top 10 largest filings, including Conseco ($56.6 billion in liabilities), WorldCom ($46.0 billion) and Enron ($31.2 billion—or, almost double this amount if one adds in Enron's enormous off-balance liabilities, making it the fourth "largest" bankruptcy in the United States). We note that it is most relevant to discuss the size of bankruptcies in terms of liabilities at filing rather than assets. For example, WorldCom had approximately $104 billion in book value of assets, but its market value at the

3

time of filing was probably less than one fifth of that number. General Motors had $91 billion in book value of assets, but liabilities amounting to $172 billion. It is the claims against the bankruptcy estate, as well as the going-concern value of the assets, that are most relevant in a bankrupt company. Firm size is no longer a proxy for corporate health and safety. Figure 1.1 shows the number of Chapter 11

Year	Number of Filings	Prepetition Liabilities ($ millions)	Number of Filings ≥ $1B	≥$1B/Total Filings (%)
1989	23	34,516	10	43
1990	35	41,115	10	29
1991	53	82,424	12	23
1992	38	64,677	14	37
1993	37	17,701	5	14
1994	24	8,396	1	4
1995	32	27,153	7	22
1996	33	11,949	1	3
1997	36	18,866	5	14
1998	55	31,913	6	11
1999	107	70,516	19	18
2000	137	99,091	23	17
2001	170	229,861	39	23
2002	136	338,176	41	30
2003	102	115,172	26	25
2004	45	40,100	11	24
2005	36	142,950	11	31
2006	34	22,775	4	12
2007	38	72,338	8	21
2008	146	724,222	24	16
2009	233	603,120	49	21
2010	114	56,835	14	12
2011	84	109,119	7	8
2012	69	71,613	14	20
2013	66	39,480	11	17
2014	59	91,992	14	24
2015	70	79,841	19	27
2016	98	125,305	37	38
2017	91	121,079	24	26
Mean No. of Filings, 1989–2017	76		16	21
Median No. of Filings, 1989–2017	59		12	21
Median No. of Filings, 1998–2017	88		17	
Mean Liabilities, 1989–2017		$120,424		
Median Liabilities, 1989–2017		$71,613		

FIGURE 1.1   Chapter 11 Filing Statistics (1989–2017)
Source: Altman and Kuehne (2018b) and Salomon Center.

filings and prepetition liabilities of firms with at least $100 million in liabilities from 1989 to 2017 (the mega cases). Figure 1.2 lists the top 40 largest bankruptcy filings of all time by the total amount of liabilities. Figure 1.3 lists the top 40 largest bankruptcy filings of all time by Consumer Price Index adjusted total amount of liabilities (in constant 2017 dollars).

Company	Liabilities	Filing Date
Lehman Brothers Holdings, Inc.	613,000	9/15/2008
General Motors Corp.	172,810	6/1/2009
CIT Group, Inc.	64,901	11/1/2009
Conseco, Inc.	56,639	12/2/2002
Chrysler, LLC	55,200	4/30/2009
Energy Future Holdings Corp.	49,701	4/29/2014
WorldCom, Inc.	45,984	7/21/2002
MF Global Holdings Ltd.	39,684	10/31/2011
Refco, Inc.	33,300	10/5/2005
Enron Corp.	31,237	12/2/2001
AMR Corp.	29,552	11/29/2011
Delta Air Lines, Inc.	28,546	9/5/2005
General Growth Properties, Inc.	27,294	4/22/2009
Pacific Gas & Electric Co.	25,717	4/6/2001
Thornburg Mortgage, Inc.	24,700	5/1/2009
Charter Communications, Inc.	24,186	3/27/2009
Calpine Corp.	23,358	12/5/2005
New Century Financial Corp.	23,000	4/2/2007
UAL Corp.	22,164	12/2/2002
Texaco, Inc.	21,603	4/1/1987
Capmark Financial Group, Inc.	21,000	10/25/2009
Delphi Corp.	20,903	10/5/2005
Conseco Finance Corp.	20,279	12/2/2002
Caesars Entertainment Operating Co., Inc.	19,869	1/15/2015
Olympia & York Realty Corp.	19,800	5/15/1992
Lyondell Chemical Co.	19,337	1/6/2009
American Home Mortgage Investment Corp.	19,330	8/6/2007
Adelphia Communications Corp.	18,605	6/1/2002
Northwest Airlines Corp.	17,915	9/5/2005
Mirant Corp.	16,460	7/14/2003
SunEdison, Inc.	16,141	4/21/2016
Residential Capital, LLC	15,276	5/14/2012
Walter Investment Management Corp.	15,216	11/30/2017
Global Crossing, Ltd.	14,639	1/28/2002
Executive Life Insurance Co.	14,577	4/1/1991
NTL, Inc.	14,134	5/2/2002
Mutual Benefit Life Insurance Co.	13,500	7/1/1991
Tribune Co.	12,973	12/8/2008
Reliance Group Holdings, Inc.	12,877	6/12/2001
R.H. Donnelley Corp.	12,374	5/28/2009

FIGURE 1.2   List of 40 Largest Bankruptcy Filings of All Time

Company	Liabilities in 2017 $	Filing Date
Lehman Brothers Holdings, Inc.	697,846	9/15/2008
General Motors Corp.	197,426	6/1/2009
Conseco, Inc.	77,167	12/2/2002
CIT Group, Inc.	74,146	11/1/2009
Chrysler, LLC	63,063	4/30/2009
WorldCom, Inc.	62,650	7/21/2002
Energy Future Holdings Corp.	51,456	4/29/2014
Texaco, Inc. (incl. subsidiaries)	46,610	4/1/1987
MF Global Holdings Ltd.	43,241	10/31/2011
Enron Corp.	43,231	12/2/2001
Refco, Inc.	41,791	10/5/2005
Delta Air Lines, Inc.	35,825	9/5/2005
Pacific Gas & Electric Co.	35,591	4/6/2001
Olympia & York Realty Corp.	34,590	5/15/1992
AMR Corp.	32,201	11/29/2011
General Growth Properties, Inc.	31,182	4/22/2009
UAL Corp.	30,197	12/2/2002
Calpine Corp.	29,314	12/5/2005
Thornburg Mortgage, Inc.	28,218	5/1/2009
Charter Communications, Inc.	27,631	3/27/2009
Conseco Finance Corp.	27,628	12/2/2002
New Century Financial Corp.	27,189	4/2/2007
Delphi Corp.	26,233	10/5/2005
Executive Life Insurance Co.	26,232	4/1/1991
Adelphia Communications Corp.	25,348	6/1/2002
Mutual Benefit Life Insurance Co.	24,294	7/1/1991
Capmark Financial Group, Inc.	23,991	10/25/2009
American Home Mortgage Investment Corp.	22,851	8/6/2007
Northwest Airlines Corp.	22,483	9/5/2005
Baldwin United Corp.	22,148	9/1/1983
Lyondell Chemical Co.	22,091	1/6/2009
Mirant Corp.	21,926	7/14/2003
Penn Central Transportation	20,846	6/1/1970
Caesars Entertainment Operating Co., Inc.	20,546	1/15/2015
Global Crossing, Ltd.	19,945	1/28/2002
Southeast Banking Corp.	19,574	9/20/1991
NTL, Inc.	19,256	5/2/2002
Campeau Corp. (Allied & Federated)	18,653	1/1/1990
Reliance Group Holdings, Inc.	17,822	6/12/2001
First City Banc. of Texas	16,830	10/31/1992

FIGURE 1.3   List of 40 Largest Bankruptcy Filings of All Time in 2017 Dollars

A variety of terms are used in practice to depict the condition and formal process confronting the distressed firm and characterize the economic problem involved. Four generic terms commonly found in the literature are *failure, insolvency, default,* and *bankruptcy.* Although these terms are sometimes used interchangeably, they are distinctly different in their meanings and formal usage.

Failure, in an economic sense, means that the realized rate of return on invested capital, with allowances for risk consideration, is significantly lower than prevailing rates on similar investments. Somewhat different economic criteria have also been used, including insufficient revenues to cover costs, or an average return on investment that is continually below the firm's cost of capital. These definitions make no statement about whether to discontinue operations. The term *business failure* was adopted by *Dun & Bradstreet* (D&B), which for many years provided statistics on various business conditions, including exits. D&B defined business failures to include "businesses that cease operation following assignment or bankruptcy; those that cease with loss to creditors after such actions or execution, foreclosure, or attachment; those that voluntarily withdraw, leaving unpaid obligations, or those that have been involved in court actions such as receivership, bankruptcy reorganization, or arrangement; and those that voluntarily compromise with creditors."

Insolvency is another term depicting negative firm performance and is generally used in a more technical fashion. Technical insolvency exists when a firm is unable to meet its debts as they come due. This may, however, be a symptom of a cash flow or liquidity shortfall, which may be viewed as a temporary, rather than a chronic, condition. Balance sheet insolvency is especially critical and refers to when total liabilities exceed a fair valuation of total assets. The real net worth of the firm is, therefore, negative. This condition has implications for how and whether the firm will restructure, and requires a comprehensive analysis of both a going concern and liquidation value. In some countries (but not the United States), a determination of insolvency may be needed for a court to commence formal bankruptcy proceedings.

Default refers to a borrower violating an agreement with a creditor, as specified in the contract with the lender. Technical defaults take place when the firm violates a provision other than a scheduled payment, for example, by violating a covenant such as maintaining a specified minimum current ratio or maximum debt ratio. Violating a loan covenant frequently leads to renegotiation rather than immediate demand for repayment of the loan, and typically signals deteriorating firm performance. When a firm misses a required interest or principal payment, a more formal default occurs. If the problem is not "cured" within a grace period, usually 30 days, the security is declared "in default." After this period, the creditor can exercise its contractually available remedies, such as declaring the full amount of the debt immediately due. Often, an impending payment default triggers a restructuring of debt payments or a formal bankruptcy filing.

Defaults on publicly held indebtedness peaked in the two most recent recession periods, 2001–2002 and 2008–2009. Indeed, in 2001 and 2002, over $160 billion of publicly held corporate bonds defaulted. In 2009, defaults soared to an unbelievable level of over $120 billion in a single year! Figure 1.4 shows the history of U.S. public bond defaults from 1971 to 2017, including the dollar amounts and the amounts as a percentage of total high-yield bonds outstanding—the so-called "junk bond default rate." Default rates climbed to over 10% in only four years in history (1990, 2001, 2002, and 2009).

Finally, a firm is sometimes referred to as *bankrupt* when, as described above, its liabilities exceed the going concern value of its assets. Until a firm declares bankruptcy in a federal bankruptcy court, accompanied by a petition either to liquidate its assets (Chapter 7) or to reorganize (Chapter 11), it is difficult to discern if a firm is bankrupt. In this book, we refer to firms as bankrupt when they enter court supervised proceedings. In Chapter 3 herein, we study in depth the process and evolution of bankruptcy laws for the United States.

REASONS FOR CORPORATE FAILURES

Corporate failures and bankruptcy filings are a result of financial and/or economic distress. A firm in financial distress experiences a shortfall in cash flow needed to meet its debt obligations. Its business model does not necessarily have fundamental problems and its products are often attractive. In contrast, firms in economic distress have unsustainable business models and will not be viable without asset restructuring. In practice, many distressed firms suffer from a combination of the two. Many factors contribute to the high number of corporate failures. We list the most common reasons below.

1. *Poor operating performance and high financial leverage*

    A firm's poor operating performance may result from many factors, such as poorly executed acquisitions, international competition (e.g., steel, textiles), overcapacity, new channels of competition within an industry (e.g., retail), commodity price shocks (e.g., energy), and cyclical industries (e.g., airlines). High financial leverage exacerbates the effect of poor operating performance on the likelihood of corporate failure.

2. *Lack of technological innovation*

    Technological innovation creates negative shocks to firms that do not innovate. The arrival of a new technology often threatens the survival of firms that possess related, yet less competitive, technologies. For example, when digital recording eventually took over dry-film technologies in the 2000s, firms focusing on the older technologies were driven out of business.

3. *Liquidity and funding shock*

    A potential funding risk known as rollover risk received heightened attention from both academics and practitioners after the 2008–2009 financial crisis. In periods of weak credit supply, some firms are unable to roll over maturing debt because of illiquidity in credit markets. This concern was particularly acute following the onset of the 2008–2009 financial crisis.

Year	Par Value Outstanding[a]	Par Value Defaults	Default Rates
2017	$1,622,365	$29,301	1.806%
2016	$1,656,176	$68,066	4.110%
2015	$1,595,839	$45,122	2.827%
2014	$1,496,814	$31,589	2.110%
2013	$1,392,212	$14,539	1.044%
2012	$1,212,362	$19,647	1.621%
2011	$1,354,649	$17,963	1.326%
2010	$1,221,569	$13,809	1.130%
2009	$1,152,952	$123,878	10.744%
2008	$1,091,000	$50,763	4.653%
2007	$1,075,400	$5,473	0.509%
2006	$993,600	$7,559	0.761%
2005	$1,073,000	$36,209	3.375%
2004	$933,100	$11,657	1.249%
2003	$825,000	$38,451	4.661%
2002	$757,000	$96,858	12.795%
2001	$649,000	$63,609	9.801%
2000	$597,200	$30,295	5.073%
1999	$567,400	$23,532	4.147%
1998	$465,500	$7,464	1.603%
1997	$335,400	$4,200	1.252%
1996	$271,000	$3,336	1.231%
1995	$240,000	$4,551	1.896%
1994	$235,000	$3,418	1.454%
1993	$206,907	$2,287	1.105%
1992	$163,000	$5,545	3.402%
1991	$183,600	$18,862	10.273%
1990	$181,000	$18,354	10.140%
1989	$189,258	$8,110	4.285%
1988	$148,187	$3,944	2.662%
1987	$129,557	$7,486	5.778%
1986	$90,243	$3,156	3.497%
1985	$58,088	$992	1.708%
1984	$40,939	$344	0.840%
1983	$27,492	$301	1.095%
1982	$18,109	$577	3.186%
1981	$17,115	$27	0.158%
1980	$14,935	$224	1.500%
1979	$10,356	$20	0.193%
1978	$8,946	$119	1.330%
1977	$8,157	$381	4.671%
1976	$7,735	$30	0.388%
1975	$7,471	$204	2.731%
1974	$10,894	$123	1.129%
1973	$7,824	$49	0.626%
1972	$6,928	$193	2.786%
1971	$6,602	$82	1.242%

FIGURE 1.4 Historical Default Rates—Straight Bonds Only (Excluding Defaulted Issues from Par Value Outstanding), 1971–2017 ($ Millions)
Source: Salomon Center at New York University Stern School of Business.

			Standard Deviation
Arithmetic Average Default Rate	1971 to 2017	3.104%	3.006%
	1978 to 2017	3.347%	3.191%
	1985 to 2017	3.759%	3.312%
Weighted Average Default Rate[b]	1971 to 2017	3.378%	
	1978 to 2017	3.381%	
	1985 to 2017	3.394%	
Median Annual Default Rate	1971 to 2017	1.906%	

[a] As of mid-year.
[b] Weighted by par value of amount outstanding for each year.

FIGURE 1.4   (*Continued*)

4. *Relatively high new business formation rates in certain periods*

New business formation is usually based on optimism about the future. But new businesses fail with far greater frequency than do more seasoned entities, and the failure rate can be expected to increase in the years immediately following a surge in new business activity.

5. *Deregulation of key industries*

Deregulation removes the protective cover of a regulated industry (e.g., airlines, financial services, healthcare, energy) and fosters larger numbers of entering and exiting firms. Competition is far greater in a deregulated environment. For example, after the airline industry was deregulated at the end of the 1970s, airline failures multiplied in the 1980s and have continued since then.

6. *Unexpected liabilities*

Firms may fail because off-balance sheet contingent liabilities suddenly become material on-balance sheet liabilities. For example, a number of U.S. firms failed due to litigation related to asbestos, tobacco, and silicone breast implants. Firms may also inherit uncertain liabilities through acquisitions. Energy firms and mining firms may inherit unanticipated environmental obligations via asset purchases. Financial institutions, such as Washington Mutual, inherited liabilities related to subprime mortgage related litigation in the aftermath of the 2008–2009 financial crisis.

These factors play heavily in the prediction and avoidance of financial distress and bankruptcy. Fifty years after its introduction, the Altman Z-score remains one of the most widely used credit scoring models used by practitioners and academics to indicate the probability of default. Part Two of this book is devoted to default and bankruptcy prediction models, including the Altman Z-score and its derivatives.

BANKRUPTCY AND REORGANIZATION THEORY

The continuous entrance and exit of productive entities are natural components of any economic system. The phrase "creative destruction," referring to the ongoing process by which innovation leads new producers to replace outdated ones, was coined by Joseph Schumpeter (1942), who described it as an "essential fact about capitalism."

Because of the inherent costs to society of the failure of business enterprises, laws and procedures have been established (1) to protect the contractual rights of interested parties, (2) to orderly liquidate unproductive assets, and (3) when deemed desirable, to provide for a moratorium on certain claims to give the debtor time to become rehabilitated and to emerge from the process as a continuing entity. Both liquidation and reorganization are available courses of action in many countries of the world and are based on the following premise: If an entity's intrinsic or going-concern value is greater than its current liquidation value, then the firm should be permitted to attempt to reorganize and continue. If, however, the firm's assets are worth more "dead than alive" – that is, if liquidation value exceeds the economic going-concern value – liquidation is the preferred alternative. In the end, the efficiency of any bankruptcy system can be judged by its ability to appropriately identify and provide for the restructuring of firms that arguably should be able to survive.

There are, however, challenges to reach an economically efficient outcome. These include, for example, conflicting incentives of differing priority claimants regarding the liquidation versus continuation decision; incentives of one set of claimants to accelerate its claims to the detriment of the firm value as a whole, known as the "collective action" problem; and inability to reach agreement among dispersed claimants. Perhaps one of the largest challenges to the process is that the going concern and liquidation values are not objective and observable. Such challenges often make a less costly out-of-court solution impossible and necessitate a formal legal framework for restructuring or liquidating a firm under court supervision. In Chapters 3 and 4 of this book, we explore the various options, both in and out of court, for restructuring distressed firms.

The primary benefit of a reorganization-based system is to enable economically productive assets to continue to contribute to society's supply of goods and services, to say nothing of preserving the jobs of the firm's employees, revenues of its suppliers, and tax payments. However, these benefits need to be weighed against the costs of bankruptcy to the firm and to society.

DISTRESSED RESTRUCTURING IN A NUTSHELL

Distressed restructuring is all about fixing failed firms. The general goal is to restructure either the left-hand side of the balance sheet, known as *asset restructuring*, and/or the right-hand side of the balance sheet, known as *financial restructuring*. The motivation for asset restructuring is to improve operations and thus cash flows and redeploy underperforming or unexploited assets to more efficient users. One common way to achieve this is to install new managers, often with the help of turnaround specialists, with a focus on maximizing the size of the company value "pie." The motivation for financial restructuring is to make the firm's cost of capital cheaper. Firms with an "expensive" capital structure need financial restructuring to deleverage the firm to a level that is sustainable in the long-term.

There are many restructuring options available to a distressed firm. In out-of-court restructurings, firms bargain with creditors and other stakeholders in private negotiations. Such restructurings typically result in senior debt claims being exchanged for new debt claims, either senior or junior, and junior debt claims being exchanged for equity claims, with equity holders taking significant dilution. The success of such debt-for-equity swaps depends largely on whether creditors can effectively coordinate their votes on the distressed exchange proposal and whether they fare better in an out-of-court restructuring than an in-court restructuring. The in-court option refers to restructuring under the supervision of the bankruptcy court. The major benefits for the formal bankruptcy proceedings are that the Bankruptcy Code equips the debtor with many valuable options for restructuring debt claims and assets and resolves the coordination problems of bargaining by debtholders. However, the disadvantage is that they are lengthier and thus more expensive than the out-of-court option. We explore the outcomes and costs of distressed restructurings in Chapter 4 herein.

THE DISTRESSED RESTRUCTURING INDUSTRY PLAYERS

The fact that corporate distress and bankruptcy in the United States is a major industry can be demonstrated by the size and scope of activities associated with this field. The bankruptcy "space" today attracts a record number of practitioners and researchers. One reason is the size of the entities that found it necessary to file for bankruptcy during and after the 2008–2009 financial crisis. A list of the major "players" in the bankruptcy "game" and the related distressed firm industry are:

- Bankrupt firms (debtors)
- Bankruptcy legal system (judges, trustees, etc.)
- Creditors and committees
- Bankruptcy law specialists
- Bankruptcy insolvency accountants and tax specialists
- Distressed turnaround specialists
- Financial restructuring advisors
- Distressed securities traders and analysts
- Bankruptcy and workout publications and data providers

To a large extent, the 1978 Bankruptcy Act provides that management of the bankrupt firm, known as the "debtor in possession," retains significant influence, if not control, over the process. This in turn affects, ex-ante, the firm's ability to renegotiate claims in advance of or to avoid a filing.

As of 2016, there were 349 bankruptcy judgeships nationwide authorized to guide the debtors and their various creditors through the bankruptcy process.[1] These are federal judges who serve in 94 judicial districts encompassing the 50 states, the District of Columbia, Puerto Rico, Guam, and the Northern Mariana Islands. Bankruptcy judges are assisted by the U.S. Trustees Program, a component of the Department of Justice, which plays a major role in administering the huge flow of cases in the system. Among other responsibilities, the U.S. Trustee appoints a committee to represent unsecured creditors, and other committees as justified for a particular case. A trustee oversees the liquidation and distributions in a Chapter 7 case; in Chapter 11, a trustee is more rarely appointed, generally to replace management of the bankrupt debtor in cases of mismanagement or fraud.

The nation's large core of bankruptcy lawyers make up an important constituency in the bankruptcy process. These lawyer-consultants represent the many stakeholders in the process, including the debtor, creditors, equity holders, employees, and even tax authorities. Martindale lists more than 110,000 bankruptcy lawyers in 2017 (see www.martindale.com). The New York area alone has more than 3,000 bankruptcy lawyers listed. Some of the large law firms with specialization in the bankruptcy area include Kirkland & Ellis, Weil Gotshal & Manges; Akin Gump Strauss Hauer & Feld; Jones Day; Skadden, Arps, Slate, Meagher & Flom; Milbank, Tweed, Hadley & McCloy; Paul, Weiss, Rifkind, Wharton & Garrison; and Davis, Polk, & Wardell, among many others.[2]

There are two groups of restructuring advisory firms in the industry. The first group focuses on asset restructuring, helping troubled companies improve operations, often to avoid a bankruptcy filing. These firms are known to house and provide *turnaround specialists* to distressed firms. Well-known players in the field include AlixPartners, Alvarez & Marsal, and FTI. The other group focuses on financial restructuring, managing and advising a company's capital structure renegotiations. Well-known players in the field include Lazard Freres, PJT Partners (formerly the Blackstone Group), Miller Buckfire, N. M. Rothschild & Sons, Evercore, and Greenhill, although there are also several smaller successful operations. On the creditor advisory side, the largest advisers are Houlihan Lokey Howard & Zukin; Jefferies; Chanin; FTI; and Giuliani Partners. The last two are carve-outs or sales of divisions from accounting firms.

The nature of the firm's claims, and the identity of the owners of those claims once a firm is distressed, have an important effect on the dynamics of the renegotiation process, whether in or out of court. In many larger cases, original bank lenders may have sold their position to specialized investors as the firm's performance notably declines. Similarly, private-equity-like investors may have replaced original purchasers of the firm's bonds or notes, or even claims of trade creditors.

Lastly, just as important to strategists and researchers is the availability of data on distressed firms from many sources, as noted throughout this text.

BANKRUPTCY FILINGS

The two broad categories of bankruptcy filings are business (Chapter 7, Chapter 11, Chapter 12, and Chapter 13 of the US Bankruptcy Code) and consumer filings (Chapter 7, Chapter 11, and Chapter 13 of the US Bankruptcy Code). A third and rarely observed category is bankruptcy filings by municipalities, such as the city of Detroit, Michigan (Chapter 9). Figure 1.5 lists the bankruptcy filings for business and nonbusiness entities from 1985–2017, while Figure 1.6 lists bankruptcy filings by the bankruptcy Chapter from 1985 to 2017. Although the vast majority are consumer bankruptcies, with as much as 97% of the total filings in recent years, this book focuses exclusively on large business filings, primarily Chapter 11, and filings by public companies. Figure 1.7 plots the number of filings and prepetition liabilities of companies with a minimum of $100 million in liabilities from 1989 to 2017. Examining Figure 1.5 reveals some observations worth mentioning.

First, the incredible increase in nonbusiness (consumer) bankruptcies before 2005 and again from 2008–2011 is apparent, reflecting the huge increase in personal indebtedness in the United States during the periods. The number of personal bankruptcies increased almost fivefold from 1985 to 2005. With the tougher conditions for consumers filing for bankruptcy under the Bankruptcy Abuse Prevention and Consumer Protection Act of 2005 (BAPCPA), the number of nonbusiness bankruptcies declined sharply after 2005. Interestingly, the large increase in nonbusiness bankruptcy filings from 2004 to 2005 and the large decline in the year after may reflect that consumers strategically timed their filings before the new law was enacted in October 2005.

Second, the absolute number of business filings has been trending downwards in the past three decades. The number of filings decreased to a record low of less than 20,000 in 2006 and tripled in the "heady" years of 2008/2009, before falling to historically low levels starting in 2014.

Third, despite the decrease in the number of filings since the early 1990s, total liabilities of the larger business bankruptcies swelled to record levels in the 2008–2009 period. This trend has fed the distressed debt investment sector and has given unprecedented importance to this "new" alternative asset class (see our discussion in Chapters 14–15 herein).

From 2011–2017, the average annual number of filings with liabilities greater than $100 million, both public and private (77), has been in line with the historical average (76) over the 38-year period (1980–2017). Figure 1.7 shows a declining trend in the number of public filings starting from 2010 to 2015. The 98 filings in 2016 and 91 filings in 2017 are both higher than the historical average from 1989–2017 (76) and the median (59), for the same period. Particularly, energy

Year	Business	Nonbusiness	Total	Business Percent of Total
1985	71,242	341,189	412,431	17.28%
1986	80,879	449,129	530,008	15.26%
1987	81,999	492,850	574,849	14.26%
1988	63,775	549,831	613,606	10.39%
1989	63,227	616,753	679,980	9.30%
1990	64,853	718,107	782,960	8.28%
1991	71,549	872,438	943,987	7.58%
1992	70,643	900,874	971,517	7.27%
1993	62,304	812,898	875,202	7.12%
1994	52,374	780,455	832,829	6.29%
1995	51,959	874,642	926,601	5.61%
1996	53,549	1,125,006	1,178,555	4.54%
1997	54,027	1,350,118	1,404,145	3.85%
1998	44,367	1,398,182	1,442,549	3.08%
1999	37,884	1,281,581	1,319,465	2.87%
2000	35,472	1,217,972	1,253,444	2.83%
2001	40,099	1,452,030	1,492,129	2.69%
2002	38,540	1,539,111	1,577,651	2.44%
2003	35,037	1,625,208	1,660,245	2.11%
2004	34,317	1,563,145	1,597,462	2.15%
2005	39,401	2,039,214	2,078,415	1.90%
2006	19,695	597,965	617,660	3.19%
2007	28,322	822,590	850,912	3.33%
2008	43,546	1,074,225	1,117,771	3.90%
2009	60,837	1,412,838	1,473,675	4.13%
2010	56,282	1,536,799	1,593,081	3.53%
2011	47,806	1,362,847	1,410,653	3.39%
2012	40,075	1,181,016	1,221,091	3.28%
2013	33,212	1,038,720	1,071,932	3.10%
2014	26,983	909,812	936,795	2.88%
2015	24,735	819,760	844,495	2.93%
2016	24,114	770,846	794,960	3.03%
2017	23,157	765,863	789,020	2.93%
Total	1,576,261	34,294,014	35,081,055	4.49%

FIGURE 1.5 Bankruptcy Filings by Type, 1985–2017
Source: The Bankruptcy Yearbook & Almanac and United States Courts Form F-2 (http://www.uscourts.gov/).

companies prominently populated defaults and bankruptcies from 2015 to 2017. The 98 defaults and bankruptcies in the energy sector in the period from January 2015 through June 2017 accounted for 47% of all defaults in that sector over the 47-year time series from 1970 to 2017. The number of mega-bankruptcies with liabilities greater than $1 billion in 2017 (24) was about 1.5 times greater than the historical average over the 38-year period (1980–2017) of 16.

Year	Chapter 7	Chapter 9	Chapter 11	Chapter 12	Chapter 13	Chapter 15
1985	280,986	N/A	23,374	N/A	108,059	–
1986	374,452	N/A	24,740	601	130,200	–
1987	406,761	N/A	19,901	6,078	142,065	–
1988	437,882	5	17,690	2,034	155,969	–
1989	476,993	9	18,281	1,440	183,228	–
1990	543,334	13	20,783	1,346	217,468	–
1991	656,460	18	23,989	1,495	262,006	–
1992	681,663	14	22,634	1,608	265,577	–
1993	602,980	12	19,174	1,243	251,773	–
1994	567,240	16	14,773	900	249,877	–
1995	626,150	10	12,904	926	286,588	–
1996	810,400	8	11,911	1,083	355,123	–
1997	989,372	10	10,765	949	403,025	–
1998	1,035,696	3	8,386	807	397,619	–
1999	927,074	5	9,315	834	382,214	–
2000	859,220	11	9,884	407	383,894	–
2001	1,054,927	8	11,424	383	425,292	–
2002	1,109,923	7	11,270	485	455,877	–
2003	1,176,905	6	9,404	712	473,137	–
2004	1,137,958	6	10,132	108	449,129	–
2005	1,659,017	11	6,800	380	412,130	6
2006	360,890	5	5,163	348	251,179	75
2007	519,364	6	6,352	376	324,771	42
2008	744,424	4	10,160	345	362,762	76
2009	1,050,832	12	15,189	544	406,962	136
2010	1,139,601	7	13,713	723	438,913	124
2011	992,332	13	11,529	637	406,084	58
2012	843,545	20	10,361	512	366,532	121
2013	728,833	9	8,980	395	333,326	88
2014	619,069	12	7,234	361	310,061	58
2015	535,047	4	7,241	407	301,705	91
2016	490,365	8	7,292	461	296,655	179
2017	486,347	7	7,442	501	294,637	86
Total	24,926,042	279	428,190	29,429	10,483,837	1,140

FIGURE 1.6   Bankruptcy Filings by Bankruptcy Chapter, 1985–2017
Source: The Bankruptcy Yearbook & Almanac and United States Courts Form F-2 (http://www.uscourts.gov/).

Trends in bankruptcy filings and their impact on the entire corporate bankruptcy system is more complicated than simply the number and dollar value of filings. For example, the time spent in reorganization from filing to emergence, the number of out of court exchanges or "prepackaged" Chapter 11 filings, the success of the reorganization, and the roles of creditors, are all evolving factors that we discuss in this book.

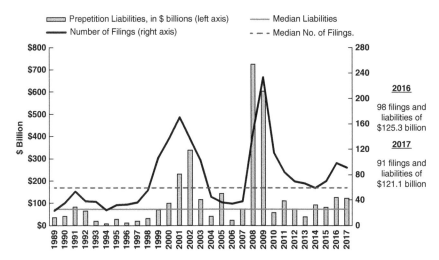

FIGURE 1.7 Number of Filings and Prepetition Liabilities of Public Companies, 1989–2017

Note: Minimum $100 million in liabilities.

Sources: NYU Salomon Center Bankruptcy Filings Database.

CHAPTER 22 DEBTORS AND BANKRUPTCY SUCCESS

The primary goal of the reorganization process is to relieve the burden of the debtor's liabilities and restructure the firm's assets and capital structure so that financial and operating problems will not recur in the foreseeable future.

The bankruptcy reorganization process is, unfortunately, not always successful even if the firm emerges as a continuing entity. It is certainly possible for the emerged firm to fail again and file a second time (or even a third time, etc.) for protection under the Code. We first coined the term "Chapter 22" in the second edition of this book to illustrate those companies that have filed twice. These Chapter 22s were saddled with too much debt and/or the business outlook was overly optimistic at the time of emergence the first time. There have even been Chapter 44 cases; one famous example is Trump Entertainment Resorts (formerly known as Trump Hotels and Casino Resorts and Trump Plaza), which filed for bankruptcy in 1991, 1992, 2004, and 2009; another is Global Aviation Holdings (formerly known as ATA Holdings) that filed in 2004, 2008, 2012, and 2013.

In Chapter 7 of this book, we explore outcomes of bankruptcy cases [as well as post-emergence performance]. Figure 1.8 lists the estimated number of Chapter 22s, 33s, 44s, and even 55s each year since 1984. During this period, 290 firms have filed twice (informally known as Chapter 22), 18 firms have filed three times (informally known as Chapter 33), three firms filed four times (informally known as Chapter 44), and Trump's Casinos and Resorts have had five different filings!

Year	Chapter 22s	Chapter 33s	Chapter 44s	Chapter 55s	Total Bankruptcy Filings[a]	% Multiple Filers
1984–1989	18	0	0	0	788	2.28
1990	10	0	0	0	115	8.7
1991	9	0	0	0	123	7.32
1992	6	0	0	0	91	6.59
1993	8	0	0	0	86	9.3
1994	5	0	0	0	70	7.14
1995	9	0	0	0	85	10.59
1996	12	2	0	0	86	16.28
1997	5	0	0	0	83	6.02
1998	2	1	0	0	122	2.46
1999	10	0	0	0	145	6.9
2000	12	1	0	0	187	6.95
2001	17	2	0	0	265	7.17
2002	11	0	1	0	229	5.24
2003	17	1	0	0	176	10.23
2004	6	0	0	0	93	6.45
2005	9	1	0	0	86	11.63
2006	4	0	0	0	66	6.06
2007	8	1	0	0	78	11.54
2008	19	0	0	0	138	13.77
2009	18	1	1	0	211	9.48
2010	10	1	0	0	106	10.38
2011	5	2	0	0	86	8.14
2012	12	1	0	0	87	14.94
2013	11	2	1	0	71	19.72
2014	7	0	0	1	54	14.81
2015	8	0	0	0	79	10.13
2016	13	2	0	0	99	15.15
2017	9	0	0	0	71	12.68
Totals	290	18	3	1	3,976	
Average, Annual						9.59
Average, Overall						7.85

[a]Must have been a public company at the time of one of the filings.

FIGURE 1.8 Chapter 22s, 33s, 44s and 55s in the United States (1984–2017)
Sources: The Bankruptcy Almanac, annually (Boston: New Generation Research); and
Altman and Hotchkiss, *Corporate Financial Distress and Bankruptcy*, 3rd ed. (Hoboken,
NJ: John Wiley & Sons, 2006).

# of Chapter 22, 33, 44, and 55 Filings (1984–2017)	312
# of Emergences and Acquired (1981–2014)	1,525
% Refiled after Emergence Only	20.46%

FIGURE 1.9 Percent of Chapter 11 Public Company Emergences that Later Result in a Repeat Filing (1984–2017)
Sources: The Bankruptcy Almanac, annually (Boston: New Generation Research); and Altman and Hotchkiss, *Corporate Financial Distress and Bankruptcy* 3rd ed. (Hoboken, NJ: John Wiley & Sons, 2006).

Through 2017 (Figure 1.9), there were 312 multiple filings, almost 8% of total bankruptcy filings for the same period. Importantly, an estimated 20% of all firms emerging from the bankruptcy process as a "going concern" have subsequently refiled. As one can observe, the totals are nontrivial and are interpreted by some observers as indicating problems in our distressed restructuring system.

DISTRESS INVESTING

With the fast development and maturing of the leveraged finance markets, and the significant increase in both the quantity and size of entities that filed for bankruptcy in past two decades, distressed claims has emerged as an important asset class that has become more widespread in the investment community.

Recent industry reports show that distressed investing is regarded as one of the most profitable strategies implemented by alterative investing funds, outperforming many other common hedge fund strategies.[3] There are a few reasons why distressed investments may offer attractive risk-adjusted rate of returns. First, distressed debt is often purchased at large discounts from lenders who lend at par. For example, a bank might sell due to regulatory concerns and unwillingness to get involved in the restructuring process. Further, high-yield mutual funds might unload positions at a discount (so called "fire sales") when they experience shocks to fund flows. Second, there are steep barriers to entry for investing in this market due to required experience, expertise, transaction costs, illiquidity, and scale of funding needed in the restructuring process. These barriers have resulted in a smaller group of sophisticated distressed debt investors.

There are generally two major types of distressed investors. The first group focuses on distress-for-control investing. These investors are typically private equity firms, who pursue the "loan-to-own" strategy through which they identify and purchase the "fulcrum" security with the goal of converting it to majority equity ownership in the emerged entity. These investors do not sell out the equity stake immediately after restructuring and typically have a three- to five-year investment horizon. They proactively get involved in corporate governance such as management and board selection and the business operations of the firm in reorganization and after.

The second group of investors are typically hedge funds with expertise in trading distressed claims and managing the bankruptcy process. They do not aim for a majority equity stake but often seek profits through identifying underpriced claims, though they sometimes also adopt strategies to influence the reorganization process. Some investors within this group focus on purchasing and consolidating trade or other claims, and gain from resolving the coordination problems among dispersed creditors.

In practice, while we have presented the two types of distressed investors here as distinct, the line can blur with hedge funds sometimes going for control and private equity firms sometimes focusing more on trading profits. Part Two of this book provides a comprehensive overview of the strategies employed by distressed investors and the returns and risk profiles of distressed debt.

NOTES

1. See http://www.uscourts.gov/statistics-reports/status-bankruptcy-judgeships-judicial-business-2016.
2. Vault releases an annual list of best law firms for restructuring and bankruptcy. For the most up-to-date list, see http://www.vault.com/company-rankings/law/.
3. Credit Suisse compiles hedge fund index returns for various trading strategies and releases periodic reports on their performance.

CHAPTER 2

An Introduction to Leveraged Finance

The leveraged loan markets and high-yield bond markets play a critical role in helping finance speculative-grade borrowers. These debt contracts have a variety of special features that make them unique in the financial markets. They are key to the restructuring of distressed firms because leveraged structures are more likely to become distressed than nonleveraged structures. Investment banks, hedge funds, and private equity funds (PEs) pay close attention to this special segment of the financial markets.

Both the leveraged loan markets and high-yield bond markets in the United States have experienced fast growth since the end of the 2008–2009 financial crisis. Both markets experienced a record amount of new issuance in 2013 – over $600 billion of leveraged loans and over $330 billion of high-yield bonds issued in total (based on data from Standard & Poor's Leveraged Commentary & Data (S&P LCD) and SIFMA). Total issuance in both markets declined shortly after 2013, when concerns about deteriorated underwriting standards led the Federal Reserve, Federal Deposit Insurance Corporation (FDIC), and the Office of the Comptroller of the Currency to issue new leveraged lending guidelines.[1] By 2017, however, issuance had returned to the peak level reached earlier ($651 billion in leveraged loan issuance and $284 in high-yield bond issuance).

In this chapter, we provide a brief introduction to these two major instruments for speculative-grade financing. We touch on important features of credit agreements and bond indentures that govern the rights and responsibilities of the creditor and the borrower. We also discuss how lenders design contracts to protect their claims. Finally, this chapter introduces debt subordination, which is important to understanding creditor rights in bankruptcy and performing the reorganization and waterfall analysis (discussed in Chapter 3, Chapter 5, and Chapter 6 herein).

LEVERAGED LOANS: OVERVIEW

Leveraged loans are a segment of the syndicated loan market focusing on lower credit borrowers. They emerged in the mid-1990s as a new asset class. The expanded investor base resulted in significant growth in demand for this loan segment. The fast growth of this market and the need for uniform market practices and standardized trading documentation prompted the formation of the Loan Syndications and Trading Association (LSTA) and the Loan Pricing Corporation (LPC). The standardization of loan documents contributed to the increase in market liquidity and efficiency, which, in turn contributed to the growth of a robust, liquid secondary market.

There are various definitions of leveraged loans. Many practitioners use a yield-spread cutoff. For example, loans with a spread over LIBOR above 150 basis points would be referred to as leveraged loans. Practitioners also refer to loans that are rated below investment grade, which refers to credit ratings at Ba1 or below by Moody's and at BB+ or below by S&P or Fitch as leveraged. Compared to investment-grade loans, leveraged loans typically carry higher interest rates, larger fees, more stringent collateral requirements, and more and tighter covenants. Leveraged loans are issued for purposes ranging from refinancing, financing mergers and acquisitions (including leveraged buyouts (LBOs)), and recapitalizations (e.g., dividend recaps) to general corporate purposes.

Like other syndicated loans, leveraged loans are structured, arranged, and administered by commercial and investment banks. These banks, known as arrangers and agents, charge a fee for their investment banking and administration services. The *lead arranger* structures the loan and credit agreement and solicits interest from potential lenders. It is also referred to as the *book runner*, acting as the "top dog" in a syndication. The *agent* bank manages servicing and administration of the loan after syndication. There are many other titles assigned to participants performing various roles in the lending process (e.g., administrative agent, syndication agent, documentation agent, etc.).

Figure 2.1 shows the amount of annual leveraged loan issuance. The *pro rata* portion consists of revolving credit and amortizing loans, which are typically issued to and held by banks, while the *institutional* portion consists of nonamortizing term loans, which are typically held by institutional investors. The figure shows that revolving credit and amortizing loans accounted for about half of the total issuance in the early 2000s but only about one-third in recent years, suggesting an increasing involvement by nonbank institutional investors in the leveraged financing markets. The main lenders in this market segment include collateralized loan obligations (CLOs), loan mutual funds, high-yield funds and hedge funds, and finance and insurance companies. According to S&P LCD, as of the end of 2017, CLOs accounted for 64% of the institutional market share, and hedge funds, high-yield funds, and loan mutual funds account for approximately

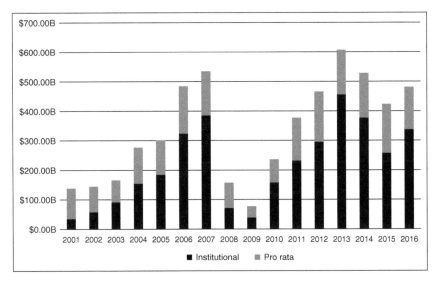

FIGURE 2.1 Leveraged Loan Annual Issuance, 2001–2016
Source: Image created based on data from S&P Global.

30% of the institutional loan market share, while insurance companies and finance companies account for the rest.[2]

The shift in the identity of the lenders from banks to nonbank institutions occurs not only at the stage of loan origination but also through turnover that occurs in the secondary market. In secondary market trading, institutional investors take part in a syndicated leveraged loan through either an *assignment* or *participation*. Through assignment, the buyer of the loan becomes the direct lender of record and receives interest and principal payments directly from the agent bank. The buyer is entitled to all voting privileges of the other lenders of record. In contrast, through participation, the buyer takes a share of an existing lender's loan. The buyer does not generally have full voting rights, except for material changes in the loan document such as interest rate, maturity, and collateral. Assignments may be subject to borrower consent, while participations are granted without consent unless they appear on the disqualified institutions (DQ) list, which comprises entities identified by the borrower as not permitted to own its debt.[3]

While the precise identity of nonbank institutional lenders at origination is not publicly observable, a number of academic studies suggest that their entry into this market expanded the supply of capital and led to a lower cost of capital for borrowing firms. Ivashina and Sun (2011) show that corporate loan spreads tend to fall during times when institutional loans are syndicated more quickly, and that syndication speed depends on flows of capital to large institutional investors. Nadauld and Weisbach (2012) focus on the role of securitization and show that loans which were more likely to be purchased by CLOs carried lower interest rate spreads, suggesting that securitization was associated with a lower cost of borrowing. Benmelech, Dlugosz, and Ivashina (2012) show that borrowers whose loans were securitized did not experience adverse outcomes (declines in credit quality) compared to borrowers whose loans were not securitized.

Figures 2.2a and 2.2b present the S&P/LSTA 12-month default rate (both by issuer count and dollar-denominated amount) on leveraged loans from 1998 to 2017. In most months since March 2013, the dollar-denominated loan default rate exceeded the issuer-denominated rate due to defaults by only a few relatively large

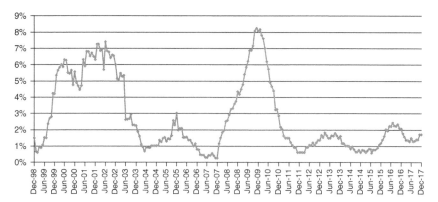

FIGURE 2.2a   S&P Leverage Loan Index 12-Month Moving Average Default Rate (by Issuer Count), 1998–2017
Source: Altman and Kuehne (2018b).

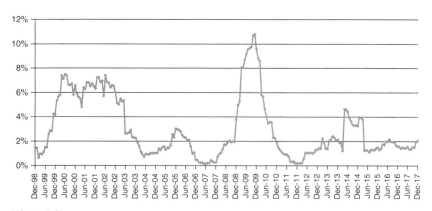

FIGURE 2.2b   S&P Leverage Loan Index 12-Month Moving Average Default Rate (by Dollar-Denominated Amount), 1998–2017
Source: Altman and Kuehne (2018b).

issuers. Note that the average default rates of leveraged loans are generally lower than those for bonds (Figure 1.4).

Conditional upon a default, the identity of the holders of the loans impacts a firm's ability to renegotiate its debt. A longstanding belief, with both theoretical and empirical basis, is that debt which is more dispersedly held is subject to hold-out problems and therefore more difficult to renegotiate (Gilson, John, & Lang, 1990; Asquith, Gertner, and Scharfstein, 1994). The most recent study to examine this question is by James and Demiroglu (2015), who study 344 debt restructurings from 2000 to 2012. Confirming the earlier studies, these authors find that firms are most likely to be able to restructure out of court when the firm relies on a single bank. Firms that rely on institutional loans, in particular those which have been securitized, are more likely to restructure in bankruptcy. An important additional finding is that firms that rely more on securitized loans are more likely to use prepackaged bankruptcies. They further find that firm level recovery rates and the likelihood of emerging from bankruptcy are not related to the lender identity.

There are two major instrument types within the leveraged loan universe: *revolving credit facilities* (*revolvers*) and *term loans*.

Revolvers

A *revolving credit facility*, like a corporate credit card, allows the borrower to draw down, repay, and reborrow up to a specified credit limit over the life of the loan. The facility is primarily used to meet temporary working capital needs. Lenders, typically commercial banks, are committed to providing funds if conditions for lending are met. Revolvers typically have a 364-day maturity (the 364-day facility) due to banks' concern about the regulatory capital requirement for issuing loans with a maturity over one year to speculative-grade borrowers.

Interest rates are quoted as a base rate or London Interbank Offered Rate (LIBOR) plus an applicable margin.[4] Common base rates are the prime rate and federal funds rate. A floor is often imposed on LIBOR or the base rate. For example, a LIBOR floor of 1% suggests that 1% will be used as the base rate if LIBOR is below that threshold. The applicable margin is typically a spread of 150–400 basis points. Lenders typically adopt performance-based pricing for the revolving credit facility. For example, the applicable margin or the spread can be tied to specific financial ratios (e.g., interest coverage ratio or leverage ratio) or the credit rating of the borrower.

Banks impose various types of fees on the revolver. These typically include an *upfront fee*, paid at loan closing, with the largest share going to the lead arranger; a *commitment fee*, charged on the daily average undrawn balance; a *facility fee*, charged on the entire amount of the facility; an *administrative fee*, paid annually to the administrative agent for its services; and others. The lender often imposes extra interest (typically 2% above the applicable rate) if the borrower defaults.[5] These fees can at times add up to over 500 bps, imposing additional costs on the borrower over and above the interest rate.

For speculative-grade companies, borrowing under a revolver is almost always secured by collateral, which can be substantially all assets of the borrower. The amount of credit available is determined from the *borrowing base*, defined as the value of specified assets of the borrower. Since most revolvers are short term, the most common types of assets in the borrowing base are current assets, including cash and marketable securities, accounts receivable, and inventory. Commodity and energy producers typically pledge reserves as the borrowing base. The credit available is generally calculated as the product of the advance rate (the maximum percentage of the borrowing base the lender is willing to extend for the loan) and the value of those assets.

Figure 2.3 shows the percentage of revolvers using accounts receivable and inventory as the borrowing base, based on 10,061 revolving facilities initiated by US corporations (both investment-grade and high yield) between 1996 and 2016, drawn from Thomson Reuters' LPC Dealscan database. The figure shows that the typical advance rates for accounts receivable are between 75% and 85%, while they range from 50% to 65% for facilities relying on inventory. This is intuitive, because receivables are considered a more liquid and safer current asset than inventory, which is subject to a large potential discount if it becomes necessary to sell in order to raise immediate cash.

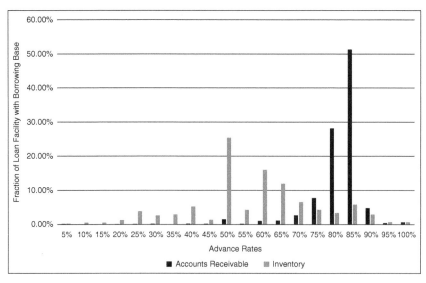

FIGURE 2.3   Advance Rates Based on Various Borrowing Bases
Source: Authors' compilation based on Thomson Reuters' LPC Dealscan.

Term Loans

Term loans are installment loans that have scheduled interest and principal payments. They typically have a longer maturity than revolving facilities. The most common maturities for term loans are five to eight years. There are two types of term loans, and they differ in whether a progressive prepayment of principal is required. An *amortizing term loan* (Term Loan A, or TLA), normally syndicated to banks along with the revolving facility, sets up a payment schedule for the borrower to pay down the principal over time, before maturity. In contrast, a *nonamortizing loan* (Term Loan B, Term Loan C, etc.), normally syndicated to institutional investors, often has debt structured with a balloon principal payment at maturity.

LENDER PROTECTION

Classic finance theories suggest that severe incentive conflicts arise between debtholders and shareholders. These conflicts of interest are manifested in *distributions to shareholders*, whereby shareholders are paid a large amount of cash dividends; *claim dilution*, whereby shareholders have a preference for issuing new debt with equal or greater priority to existing debt; *asset substitution and over-investment*, whereby shareholders prefer risky projects even if they carry negative NPVs; and *underinvestment*, whereby shareholders pass up positive NPV projects. Lenders take actions by imposing various protection clauses in the creditor agreement for their own protection. We provide a brief overview of the standard clauses for lender protection specified in a typical credit agreement below.

Conditions Precedent

Conditions precedent are conditions that must be met for a borrower to receive its funds. There are initial conditions, which must be met at deal closing, and ongoing conditions, which must be satisfied at each borrowing (e.g., under a revolver). A lender that believes a condition has not been satisfied is legally entitled to decline to lend. Some examples of conditions precedent are legal opinions from a borrower's legal counsel; no material adverse change (MAC) of a business, condition, assets, and operations; and environmental due diligence.

Representations and Warranties

Representations and warranties are given by a borrower to assure the bank that the borrower has provided accurate facts and information. Representations and warranties, particularly those that relate to legal condition, are often covered by the opinion of the borrower's counsel. If a representation is inaccurate, which may be considered a default, lenders have the legal right to stop lending and even demand repayment of loans.

Mandatory Prepayments

This section ensures that company funds are not used to benefit other parties at a lender's expense. Mandatory prepayments are typically tied to asset sales, debt issuance (*debt issuance sweep*), equity issuance (*equity issuance sweep*), excess cash flow (*excess cash flow sweep*), and change of control.

Prepayments tied to asset sales require a borrower to use a fraction, or all, of the proceeds from the sale to pay down debt. Similarly, a debt issuance sweep and an equity issuance sweep require the borrower to make prepayments using the proceeds of debt issuance or equity issuance, respectively. Excess cash flow sweep requires that some percentage of the borrower's excess cash flows is applied to prepay outstanding loans. Excess cash flows are typically defined as the excess of EBITDA over the sum of (i) capital expenditures (ii) the aggregate amount of debt service (iii) the tax payment amount, and (iv) any decreases in working capital, for an agreed-upon accounting period. However, the actual definition highly depends on what is specified in the contract and can be complicated. In fact, just the definition of EBITDA may take up to a full page of the credit agreement. Finally, the loan contract often requires the firm to retire the debt upon a change of control, unless not applicable because of a "debt portability" carve-out in the loan documents. An increasing number of portability clauses have been used in institutional term loan contracts, enabling the firm to avoid having to refinance its debt upon a change of control.

There are many other types of prepayment requirements, most of which are specific to the revolving facilities. For example, a "revolving clean downs" clause requires a borrower to have no outstanding amount drawn under the revolving facility for a period of 30 days during each calendar year. For asset-based lending (ABLs), prepayments are required when there are negative shocks to the value of the underlying collateral. ABLs are commonly used by commodity producers, such as exploration and production companies, in the energy sector. These companies typically pledge their underlying reserves as the borrowing base for ABLs. The advance rates depend on the nature and the value of the reserve (rates being higher for proved reserves than for probable or possible reserves). This creates potential problems when underlying commodity prices experience large swings. When the price of oil plummeted, for example, not only did commodity producers lose their revenue streams, but the value of their collateral also shrank as a result, resulting in lenders accelerating the loan payments. This can further exacerbate the financial constraints faced by these companies.

Covenants

There are three primary types of loan covenants: affirmative covenants, negative covenants, and financial covenants.

Affirmative covenants impose conditions and actions that a borrower must take. There are generally three categories: disclosure covenants, standard covenants, and other covenants. Disclosure covenants focus on information disclosure and delivery, such as borrowers submitting timely financial statements, compliance certificates, notices of material events, and so on. Standard covenants require borrowers' proper maintenance of books, records, and property; compliance with law; tax payments; and so on. Other affirmative covenants range from insurance and inspection rights to the use of proceeds and *pari passu* ranking, that is, of equal seniority ranking.

Negative covenants, also known as *incurrence covenants*, require borrowers to either meet certain requirements before taking actions or refrain from taking certain actions that may negatively affect a borrower's ability to repay a loan or the value of the collateral. For example, lenders may restrict borrowers from issuing additional debt, regardless of seniority, to avoid claim dilution. Lenders also often prohibit borrowers from using assets as collateral for another loan – a "negative pledge." Borrowers may be restricted from making large distributions to shareholders via dividend payments or stock repurchases. All of these actions provide lenders with a firm grip on corporate actions that may negatively affect their stake.

Financial covenants, also known as *maintenance covenants*, require a borrower to achieve a prespecified level of financial performance throughout the life of a loan. Financial covenants can be date-specific, performance based, or a hybrid type. Date-specific covenants test a borrower's financial condition on a specific date. Typical covenants include requirements for a borrower's tangible net worth, debt-to-equity ratio, current ratio, working capital level, and so on. They are mostly based on a firm's balance sheet information at a specific time. Performance-based financial covenants, such as coverage ratios (interest coverage ratio, debt service coverage ratio, fixed charge coverage ratio), capital expenditure, and lease payments, tend to be "flow"-based and rely on information from income statements and cash flow statements. The hybrid type, such as the debt to EBITDA ratio, relies on information from both the balance sheet and income statements, and can be a rolling measure over several quarters.[6]

Figure 2.4a plots the histograms of interest coverage and Figure 2.4b plots debt-to-EBITDA covenants for all leveraged loans listed in Thomson Reuters' LPC Dealscan from 1996 to 2016. The figure shows that interest coverage, when required by the lender, tends to concentrate in the 1.5–3 range. The debt-to-EBITDA covenant typically requires the ratio to have the maximum value of 3–5.

Some leveraged loans carry affirmative and incurrence covenants, but not traditional maintenance covenants. These loans increasingly appear in periods of easier credit, such as prior to the 2008–2009 financial crisis. These "covenant-lite" loans contain many borrower friendly terms. For example, they may allow an unlimited amount of debt so long as the borrower meets the incurrence test, or even allow unlimited dividends, subject to satisfying a leverage test. According to S&P LCD, these loans made up 75% of new institutional loans in 2017 and accounted for about 80% of the U.S. institutional leveraged loans

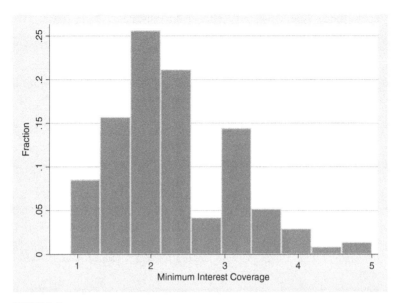

FIGURE 2.4a   Interest Coverage Covenant Distribution (U.S. Leveraged Loans, 1996–2016)

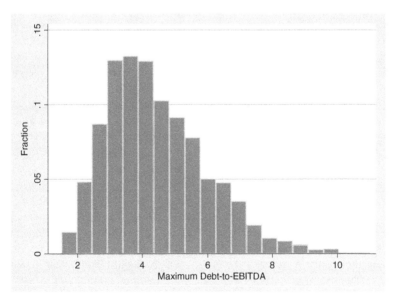

FIGURE 2.4b   Debt-EBITDA Coverage Covenant Distribution (U.S. Leveraged Loans, 1996–2016)
Source: Author's compilation based on Thomson Reuters LPC Dealscan.

as of February 2018. This represented a dramatic increase from 2010, when covenant-lite loans constituted less than 10% of new institutional loans. The total outstanding amount of covenant-lite loans stood at a record high of over $600 billion as of mid-2017. Regulators showed concerns over the fast rise of covenant-lite loans in the leveraged loan market from 2010 to 2014, which was perceived as a reflection of weakened bank lending standards.

Several academic studies examine the economic rationales for the fast development of covenant-lite loans. Becker and Ivashina (2016) find that covenant-lite loans are overwhelmingly used when the institutional ownership is high. They suggest that covenant-lite loans mitigate bargaining frictions in loan syndicates by reducing the likelihood of ex post bargaining that is triggered by covenant violations. Further, Berlin, Nini, and Yu (2016) find that covenant-lite loans are not used in isolation, and are usually structured together with a revolving facility that is not covenant-lite. They refer to the structure with concentrated control rights assigned to revolving lenders as "split control rights." Yet another positive view of covenant-lite loans is that they are more likely to be used when moral hazard problems are less severe – empirically, Billett, Elkamhi, Popov, and Pungaliya (2016) show that they are more often used by firms that are larger and that have lower leverage, higher profitability, and better liquidity. Still, a primary concern with covenant-lite loans is that a future performance decline might not trigger intervention by lenders to protect their interests. If the firm value were to fall significantly more prior to a default, lenders may not be adequately compensated for the risk of lower future recoveries.

Loan Contract Renegotiations

Loan contracts are designed in a state-contingent manner to grant lenders strong control rights through ex post renegotiations. These contract terms are often renegotiated during the life of the loan contract. Amendments to the contract range from revisions to one single covenant to the change of the whole collateral package.

A growing number of academic studies (including Roberts and Sufi, 2009a; Denis and Wang, 2014; and Roberts, 2015) demonstrate that loan contracts are frequently renegotiated before their originally stated maturity. The timing of the renegotiation coincides with the arrival of new information regarding the borrower's credit quality, investment opportunities and collateral, and macroeconomic conditions. Besides modifications to pricing-related items, such as interest rate, amount, and maturity, covenants are most often modified. This is intuitive because, at times, companies find their investments and financing strategies are constrained by existing covenants. These firms can either retire the old debt and issue new debt or negotiate with lenders to make amendments to covenants and grant waivers.

Default and Remedies

The precise definition of *default* is generally laid out in the "Events of Default" section of a credit agreement. These events typically include failure to pay principal or interest on time (payment default), false representation (defined in the "Representations and Warranties" section in a credit agreement), default on other debt (within cross-default or cross-acceleration clauses), insolvency or bankruptcy, and other events (e.g., invalidity of liens or guarantees, ERISA events, etc.).

Events of default are meaningless unless they are associated with remedies or actions that lenders can act on. They typically include stopping lending, terminating commitments, "accelerating" payments, demanding "cover" (cash collateral) for letters of credit, foreclosing on collateral, instituting suit for past-due payments, and demanding payment from guarantors. In the event of default, lenders are *not required* to follow through on these actions, but they serve as an important consideration for staking out their bargaining positions against the borrower, shareholders, and junior creditors.

Covenant Violations

Covenant violations are triggered when a borrower violates one or more covenants (financial covenants, typically) laid out in the credit agreement. Covenant violations do not trigger actual default but are trip wires that allow the lender to gain control of the borrower. Covenant violations are often referred to as *technical default*.

Recent academic studies, including Dichev and Skinner (2002), Chava and Roberts (2008), Roberts and Sufi (2009a), Nini, Smith, and Sufi (2012), and a long list of follow-up studies, empirically examine loan covenant violations and their effects on corporate investment and financing policies. These studies document that, because lenders set tight covenants initially, covenant violations occur relatively often. Such violations are not necessarily associated with financial distress. Rather, lenders use such defaults to gain indirect control of a borrower's major corporate policies by making amendments to the credit agreement. The borrower's capital investment, net debt issuance, and shareholder payout decline sharply after covenant violations. Interestingly, borrowers rarely switch lenders following a violation.

HIGH-YIELD BONDS

High-yield bonds are composed of bonds issued by two types of companies. The first type, referred to as "fallen angels," are securities issued by companies that at one time (usually at issuance) were investment-grade but, like most of us, get uglier as they age and "migrate" down to noninvestment-grade or "junk" level status. When the modern age of the high-yield market began in the late 1970s, just about 100% of this very small market was made up of these fallen angels. Fallen angels may be the result of consumer preference changes or technological transformation. For example, large retailers such as JC Penney and Sears used to be rated investment grade, but they were downgraded to junk due to changes in consumer spending patterns. The other source of high-yield bonds is original-issue securities, which receive a noninvestment grade rating at birth due to their aggressive capital structure and high-risk nature. Examples are debt issued by LBOs, startups, and firms emerging from bankruptcy.

High-yield bond offerings are not typically registered with the Securities and Exchange Commission (SEC). Most come to the market under the exception of Rule 144A – companies place bonds with investors privately and then register them with the SEC a few years later (or never register them with the SEC). High-yield bond issuance is cyclical. However, as Figure 2.5 shows, the high-yield markets have experienced an unprecedented boom since the 2008–2009 financial crisis (based on data from SIFMA), and the high-yield market share of the total US corporate debt markets has remained around 20% to 30% in recent years. Major investors in this market are mutual funds, pension funds, insurance companies, and collateralized debt obligations (CDOs).

Two special types of high-yield bonds are commonly issued by non-investment-grade companies: discount (zero coupon) notes and pay-in-kind (PIK)

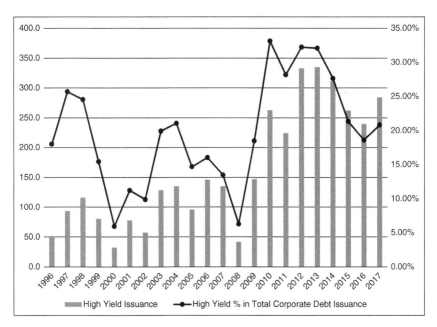

FIGURE 2.5   High-Yield Bond Annual Issuance
Source: Based on data from SIFMA.

notes. Discount notes do not pay cash coupons and are issued at a discount to their face value. The value of a discount note increases as it moves closer to maturity. PIK notes allow speculative-grade issuers to pay interest by issuing additional bonds (hence "in kind") instead of cash. After the "interest" is paid, the next interest payment will be made on the original bonds plus those issued for the PIK payment. "Toggle" bonds are special types of PIK notes; they allow companies the option to either pay the notes in kind or pay cash. Toggle bonds have gained in popularity against traditional PIK notes in recent years. See Chapter 9 for a comprehensive discussion of the high-yield markets.

DEBT SUBORDINATION

The priority of debt in the capital structure is determined by various factors, including seniority status, security and lien ranking, corporate organization structure, and guarantees. These factors could allow a class of seemingly senior debt to be ranked lower in priority than a class of junior debt in bankruptcy. In this section, we briefly describe the three major types of debt subordination.

Contractual Subordination

This refers to a situation where creditors of a particular class agree to contractually subordinate themselves in priority to another creditor group. Since bank loans are typically senior and secured, *contractual subordination* often refers to the inclusion of a subordination clause in a bond indenture.

The subordination clause typically specifies that one set of (subordinated) unsecured noteholders or bondholders agrees to subordinate its cash flow claims to another set of creditors. A standard subordination clause for a senior subordinated note, for example, would include the following provisions:

1. The payment of obligations owing in respect to the notes is *subordinated in right of payment* to all existing and future senior indebtedness of the issuer;
2. The notes are senior in right of payment *to all existing and future* subordinated indebtedness of the issuer;
3. The notes in all respects rank *pari passu* in right of payment with all existing and future senior subordinated indebtedness of the issuer, and they will be senior in right of payment to all existing and future subordinated indebtedness of the issuer.

The description of the subordination is usually followed by an extensive set of definitions. For example, the subordination section would clearly state that, in the case of insolvency, bankruptcy, or liquidation, all senior indebtedness must be paid in full (most likely in the form of cash or equivalents) before subordinated obligations are paid. With a subordination clause in place, subordinated noteholders are contractually subordinated to senior debtholders.

Lien Subordination

Secured Lending and Lien Perfection Debt can be secured against assets of companies through the granting of collateral. Secured lenders are given security interests and legal rights to repossess and realize the collateral to satisfy their claims in the case of bankruptcy. However, because secured lenders are "stayed" from foreclosure in bankruptcy, they cannot repossess the collateral during bankruptcy reorganization; however, they receive adequate protections for their claims and enjoy the highest priority in the capital structure.

Pledgeable assets that can be used as collateral range from current assets to long-term assets. They can be both tangible (such as real estate and equipment) and intangible (such as rights, patents, and other intellectual properties).[7] For most secured debt, companies pledge a specific type of asset for secured borrowing. At times, companies may pledge *substantially all* their assets as collateral. Typically, a borrower and all lenders need to contractually agree (typically through a Guarantee and Collateral Agreement) that the security interest has been created and granted.

To properly secure an asset, lenders must "perfect" a *lien* – an interest in property to secure payment of a debt or performance of an obligation against the asset. Lenders need to verify that a borrower has publicly filed and stated that the beneficiary of the collateral has prior rights to the pledged collateral. Lenders must make a public announcement and record the pledge in a public registry. In practice, since secured transactions are governed by state law and the interpretation of Article 9 of the Uniform Commercial Code (UCC), a lien is deemed to be "perfected" once a UCC financing statement has been filed in the office of the secretary of state in the state where the borrower is incorporated and a public announcement is made.

The timing of lien perfection matters to how debt claims are ranked in bankruptcy. If multiple creditor classes are using the same collateral as security, the date of perfection can be used for seniority ranking. Lenders should be concerned that perfection can lapse after several years or if the borrower changes its name or status.

Second- (Junior-) Lien Loans The market for second-lien loans, one of the most significant innovations in the U.S. leveraged loan market, has grown rapidly in the past two decades, from less than $1 billion annual issuance in the late 1990s to $40 billion in 2014. The popularity of second-lien loans can be attributed to their advantages for the borrower compared to the issuance of unsecured notes. Second-lien loans have better pricing, provide better access to investors who invest in secured loans (e.g., CLOs), and are easier to renegotiate (given their more concentrated ownership structure) than high-yield bond debt. CLOs had been the predominant investors in second-lien loans in the 2000s; however, over the past several years, the major investors have shifted to high-yield funds and distressed hedge funds, which, as an investor base, account for more than 70% of annual issuance as of 2015.[8]

Second-lien lenders typically have their rights restricted by first lien lenders through an intercreditor agreement. The agreement prohibits second-lien lenders from exercising any rights regarding the collateral until the first lien lenders are paid in full. In nature, they are subordinated in lien to first-lien lenders. Other salient features of the intercreditor agreement include provisions that require second-lien lenders to waive rights (e.g., waivers to adequate protection) and vote along with the first lien lenders. Some of these waivers pertain only to the "standstill" period (typically 90–180 days after default). The primary goal of including these provisions is that the first lien lenders want the second-lien lenders to be silent. Thus, second-lien lenders are often referred to as the "silent seconds."

Some of the provisions and restrictions for second-lien lenders are specific to postbankruptcy issues. For example, first-lien lenders may require second-lien lenders' preconsent for firms to use cash collateral. They also want second-lien lenders to remain silent to the lifting of automatic stay and the sale of collateral. The agreement may preclude the second-lien creditors from providing DIP financing absent the first lien holders' approval. Because of the silent second provisions, second-lien lenders effectively lose their separate vote for the credit agreement – they must support the agreement put forward by the first-lien lenders. However, this may depend on whether the first-lien and the second-lien lenders are documented in one creditor agreement or separate agreements. Being treated as two lending facilities and drafted in separate contracts increases the likelihood that they will be treated as separate classes by a bankruptcy judge, allowing the second-lien holders to have great opportunities to play an active role in the bankruptcy reorganization process.[9] There are at times cases for a crossing lien and split collateral structure, where the first-lien and the second-lien lenders are structured as two separate credit facilities.[10]

Structural Subordination

Corporate organizational structure matters to ranking the priority of debt and, thus, the potential debt recovery. If a borrower were a holding company with all real assets and operations at the subsidiary level, any debt incurred at the holding company level would be subordinated to debt incurred at the subsidiary level, regardless. This is referred to as *structural subordination*. In general, lenders want to be close to the real assets of a company to avoid structural subordination. We use the corporate organization of the company shown in Figure 2.6 to illustrate how structural subordination works.

Suppose Company X is a corporation with three levels of entities in its corporate reorganization structure. TopCo is the parent company that directly wholly owns HoldCo, an intermediate holding company that owns three subsidiaries, two domestic and one in a foreign country. Neither TopCo nor HoldCo owns real assets or operations, except for their equity ownership of the owned entities.

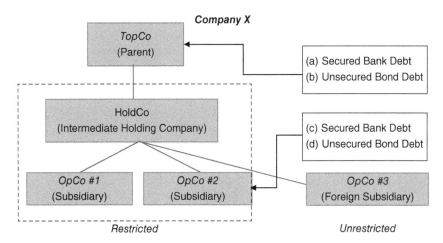

FIGURE 2.6 Organization Structure and Debt Structure of Company X

The three subsidiaries own and control all assets and operations. Suppose there are four classes of debt issued by various entities of the company as a whole. TopCo has secured bank borrowing (a), with all its assets used as collateral, and an unsecured debt (b) outstanding. Similarly, OpCo #2 has a secured loan facility (c) and an unsecured debt (d). How do we rank the seniority of the four classes of debt?

The general principal is to follow how close the debt is to operations and assets – the closer, the better. The secured bank debt (c) is secured by the operations and assets of OpCo #2 and is clearly ranked highest in seniority. Debt (d), while unsecured, is still at the operating company level and has access to cash flows; therefore, it is ranked second. Bank debt (a), which is essentially secured by TopCo's equity ownership of the three OpCos through the intermediate holding company HoldCo, acts as a residual claim to debt classes (c) and (d) and, therefore, is ranked third in the capital structure. Finally, unsecured debt (b) is the last in line. Secured bank debt (a) and unsecured debt (b) are said to be *structurally subordinated* by debt (c) and debt (d), while debt (d) would be *contractually subordinated* to (c), based on existing security agreements.

Note that the credit agreement for the loan (c) or indenture of bond debt (d) may contain restrictions and covenants that apply to all other subsidiaries and even HoldCo. These entities, shown in the dashed box, are known as the *restricted* subsidiaries. Those that are outside the box, often foreign subsidiaries, are *unrestricted* subsidiaries and are not bound by the credit agreement.[11]

Debt guarantees may complicate the priority ranking of debt. There are, in general, three types of guarantees: *upstream* (subsidiary guaranteeing debt that is issued by the parent or holding company), *downstream* (the holding or parent company guarantees debt that is issued by a subsidiary), and *cross-stream* (one subsidiary guarantees debt that is issued by another subsidiary). Suppose there is an upstream guarantee from OpCo #2 to the debt issued by TopCo. If both OpCo #1 and OpCo #2 were insolvent, the unsecured debt obligations of OpCo #2 and debt obligations of TopCo would share the same level of priority and rank *pari passu*. Therefore, an upstream guarantee effectively improves the recovery of debt of TopCo.

NOTES

1. These government agencies observed the tremendous growth in the leveraged lending market after the financial crisis, and in particular cited concerns with the fast growth in loans with "light" covenants and weak lender protections. A joint press release from the regulators is available at https://www.federalreserve.gov/newsevents/pressreleases/bcreg20130321a.htm.
2. This is in contrast with years right after the financial crisis (2009–2011) when CLO issuance was dramatically reduced, and hedge funds, high-yield funds, and loan mutual funds accounted for 40–50% of the institutional loan market share.
3. For a detailed description and discussion of the leveraged loan markets, interested readers can refer to "A Guide to the U.S. Loan Market" by S&P LCD, September 2012, and "The U.S. Leveraged Loan Market: A Primer" by Milken Institute, October 2004.
4. See articles posted at LSTA.org for discussion of the uncertain fate of LIBOR.
5. There are two different approaches to calculating default interest. One approach is that the extra default interest is charged only on amounts that are to be paid when overdue (e.g. installment amount), while the other is charged on the entire amount of the loan outstanding.
6. For detailed descriptions of various lender protections included in a typical credit agreement, readers can refer to Bellucci and McCluskey (2016).
7. Hochberg, Serrano, and Ziedonis (2016) and Mann (2016) document a rising trend in patents used as collateral for secured borrowing.
8. See "Second Lien Loans: A Complex Yet Potentially Compelling Investment Opportunity" by DDJ Capital Management, October 2016.
9. For example, in the 2009 bankruptcy case of Trump Entertainment Resorts, the second-lien note holders, led by Marc Lasry of Avenue Capital, formed an ad hoc creditors committee and played a showdown with the first-lien lenders. They eventually "crammed up" the first-lien lenders and obtained a super-majority control of the business after it emerged from bankruptcy.
10. An interesting case that demonstrates particular complexities in a capital structure is RadioShack's 2015 bankruptcy. Separate from the single (unitranche) credit agreement, two groups of lenders entered an intercreditor agreement that specified their liens on certain assets; the ABL lenders had a first lien on current assets, including accounts receivable and inventory, and a second lien on long-term assets such as intellectual properties and long-term tangible assets. The term loan lenders had a first lien on long-term assets and a second lien on current assets.
11. Borrower-friendly lending markets have led to an increase in the use of domestic entities as unrestricted subsidiaries. Taking advantage of this structure, borrowers have transferred assets from restricted entities to an unrestricted subsidiary, then using the transferred assets as collateral for an entirely new loan (e.g., J. Crew).

CHAPTER 3

An Overview of the U.S. Bankruptcy Process

To best understand the U.S. bankruptcy process, it is helpful to briefly review the previous statutes and codes that have helped to form the present system. In this chapter, we trace the developments leading to the current U.S. Bankruptcy Code. We then provide a comprehensive summary of the most important features of Chapter 11 of the Code, which are key to enabling the reorganization and survival of firms via court supervised bankruptcy proceedings.

EVOLUTION OF THE U.S. BANKRUPTCY PROCESS

The Constitution empowers the U.S. Congress to establish uniform laws regulating bankruptcy. By virtue of this authority, various acts and amendments have been passed, starting with the Bankruptcy Act of 1800. A number of revisions were enacted and led to the establishment of the Bankruptcy Act of 1898, known as the Nelson Act. These early acts in the nineteenth century established the modern concepts of debtor-creditor relations. The three most important Bankruptcy Acts subsequently passed since the beginning of the twentieth century are: in 1938, the Chandler Act, replacing the inadequate earlier statute; in 1978, the Bankruptcy Reform Act of 1978, providing the standard until recently; and in 2005, the Bankruptcy Abuse Prevention and Consumer Protection Act.

Equity Receiverships of 1898

The Bankruptcy Act of 1898 provided only for a company's liquidation and contained no provisions allowing corporations to reorganize and thereby remain in existence postbankruptcy. Reorganization could be effected through *equity receiverships*, which were developed to prevent disruptive seizures of property by dissatisfied creditors who were able to obtain liens on specific properties. Receivership in equity is not the same as receivership in bankruptcy. In the latter case, a receiver is a court agency that administers the bankrupt's assets until a trustee is appointed.

Equity receivership was extremely time-consuming and costly, as well as susceptible to severe injustices. Courts had little control over the reorganization plan, and the committees set up to protect security holders were usually made up of powerful corporate insiders who used the process to further their own interests. The initiative for equity receivership was usually taken by the company in conjunction with a friendly creditor. There was no provision made for an independent objective review of a plan. Since ratification required majority creditor support, it usually meant that companies offered cash payoffs to powerful dissenters to gain their support. The procedure led to long delays and charges of unfairness, and essentially was replaced by provisions of the temporary Bankruptcy Acts of 1933 and 1934.

The Chandler Act of 1938

In 1933, a new bankruptcy act was hastily drawn up and enacted during the great depression. This act was short-lived; in 1938 it underwent a comprehensive revision and was thereafter known as the Chandler Act. For our purposes, the two most relevant chapters of the Chandler Act were those related to corporate bankruptcy and to attempts at reorganization, namely Chapter XI and Chapter X.

A Chapter XI arrangement was a voluntary proceeding and applied only to the unsecured creditors of corporations. The debtor's petition for reorganization usually contained a preliminary plan for financial relief. The court had the power to appoint an independent trustee or receiver to manage the corporate property or, in many instances, to permit the old management team to continue its control throughout the proceedings. During the proceedings, a referee would call the creditors together to go over the proposed plan and any new amendments that had been proposed. If a majority in number and amount of each class of unsecured creditors consented to the plan, the court could confirm the arrangement and make it binding on all creditors. The plan typically provided for a scaled-down creditor claim, a composition of claims, and/or an extension of payment over time. New financial instruments could be issued to creditors in lieu of their old claims.

The prospect of continued management control and reduced financial obligations made Chapter XI particularly attractive to incumbent management. Further, the debtor could borrow new funds that had preference over all unsecured indebtedness (essentially debtor-in-possession financing – see the discussion later in this chapter). When successful, Chapter XI arrangements were of relatively shorter duration and thus less costly than a more complex Chapter X proceeding.

Chapter X proceedings applied to publicly held corporations, except railroads, and to those that had both secured and unsecured creditors. The process could be initiated voluntarily by the debtor, or involuntarily by three or more creditors with total claims of $5,000 or more. In most cases, a Chapter XI was preferred by the debtor because Chapter X automatically provided for the appointment of an independent, disinterested trustee. The bankruptcy petition for a Chapter X had to contain a statement explaining why adequate relief could not be obtained under Chapter XI. The aim of this requirement was to make a Chapter X proceeding unavailable to corporations having simple debt and capital structures.

The independent trustee was charged with developing and submitting a reorganization plan that was "fair and feasible" to all the parties involved, including creditors as well as preferred and common stockholders. In addition, the trustee was charged with the day-to-day management responsibilities, but usually delegated such tasks to the management (old or new). In most Chapter X proceedings, the trustee was aided by various experts, as well as by committees representing the creditors and stockholders, to develop and present a reorganization plan. In the case of railroad bankruptcies, the Interstate Commerce Commission (ICC) was charged with this task.

Another important participant in Chapter X proceedings was the SEC. The SEC was charged with rendering an advisory report if the debtor's liabilities exceeded $3 million, but the court could ask for SEC assistance regardless of liability size. The advisory report usually took the form of a critical evaluation of the reorganization plan submitted by the trustee and an opinion as to the fairness and feasibility of the plan. Preparing the report involved comprehensively valuing the debtor's existing assets and comparing the value with the various claims against the assets. Ultimately, the decision whether the firm was permitted to reorganize and submit the plan for final acceptance rested with the federal judge.

The Chandler Act provided that the Chapter X reorganization plan, after approval by the court, be submitted to each class of creditors and stockholders for final approval. Final ratification required approval of two-thirds in dollar amount and one-half in number (or a majority in number, in the case of Chapter XI) of each class of creditors and stockholders (if total liabilities were less than total asset value). If the plan, as accepted by the court, completely eliminated a particular class – such as the common stockholders – the excluded group had no vote in the final ratification, though it could always file a suit on its own behalf. Common stockholders were eliminated when the firm was deemed insolvent in a bankruptcy sense, that is, when the liabilities exceeded a fair valuation of the assets.

Up until 1978, the Chandler Act continued to provide for the orderly liquidation of insolvent debtors under court supervision. Regardless of who filed the petition, liquidations were handled by referees who oversaw the operation until a trustee was appointed. Payments of receipts usually followed the so-called absolute priority doctrine, under which claims with higher priority had to be paid in full before lower priority, or subordinated, claims could receive any funds at all. The liquidation fate was primarily observed for small firms. However, reorganization attempts of some larger firms were not successful, and liquidation often eventually occurred. A glaring example of a failure to successfully reorganize was the billion-dollar W.T. Grant case. The firm voluntarily filed under Chapter XI in 1975 and attempted to reorganize rather than waited for any of its creditors to compel it to file under Chapter X. However, the company was forced to liquidate several months later in 1976. In contrast, several other large reorganizations of that era were successful, including the billion-dollar (in assets) United Merchants & Manufacturing Chapter XI proceeding in July 1977.

The Bankruptcy Reform Act of 1978

Forty years after the passage of the Chandler Act, Congress passed the Bankruptcy Reform Act of 1978, which revised the administrative and to some extent the procedural, legal, and economic aspects of corporate and personal bankruptcy in the United States. The decision to reform reflected a near doubling in the annual rate of bankruptcy filings over the prior 10 years, the largest increase being in personal bankruptcies. Further, referees under the Chandler Act often had problems administering their duties and made suggestions for substantial improvement of the Act.

The 1978 Act created a U.S. Bankruptcy Court in each of 94 districts where there was a U.S. Federal District Court. The new court system was not fully established until April 1, 1984. The bankruptcy court is given exclusive jurisdiction over the property, wherever the debtor is located. All cases under the code and all civil actions and proceedings arising from its enforcement are held before the bankruptcy judge unless he or she decides to abstain from hearing a particular proceeding that is already pending in the state court or in another court that is believed to be more appropriate. Bankruptcy judges were initially appointed by the President and confirmed by the Senate. Since the Bankruptcy Amendments and Federal Judgeship Act of 1984, the court of appeals for the circuit in which the district is located has been responsible for the appointment of bankruptcy judges for that district. The term of appointment for bankruptcy judges is 14 years. Also, under the 1978 Code, the role of the SEC as the public's representative has greatly diminished.

The new act went into effect on October 1, 1979, and is divided into four titles, with Title I containing the substantive and much of the procedural law of bankruptcy. This part, known as the "Code," is divided into eight chapters: Chapter 1 (General Provisions), Chapter 3 (Case Administration), and Chapter 5 (Creditors, the Debtor and the Estate), which apply generally to all cases; and Chapter 7 (Liquidation), Chapter 9 (Adjustment of Municipality Debt), Chapter 11 (Reorganization), Chapter 13 (Adjustment of Debts of Individuals with Regular Income), and Chapter 15 (U.S. Trustee Program), which apply to specific debtors and procedures.

Chapter 7: Liquidation   Chapter 7 provides for an orderly liquidation supervised by a court appointed trustee. Chapter 11 enables corporations to reorganize under bankruptcy court supervision, and is the main focus of our text. Economically, liquidation (reorganization) is justified when the value of the assets sold individually is greater than (less than) the going concern value of the business.

The filing is commenced either by firms filing a voluntary petition or by creditors filing an involuntary petition against the debtor in the bankruptcy court in the judicial district where the company is incorporated, resides, or has its principal place of business or assets. Filing for Chapter 7 triggers an automatic stay, which freezes all collections against the firm. A trustee is then appointed to replace the company's management. The trustee operates the debtor's business for a limited period if doing so is in the best interests of the bankruptcy estate and is consistent with its orderly liquidation. Proceeds from the liquidation are distributed to creditors according to the absolute priority rule (APR). After liquidation, the firm ceases to exist.

Chapter 7 has been more often used by small businesses than Chapter 11. Figure 3.1 shows that there were twice as many Chapter 7 filings as Chapter 11 filings by U.S. businesses between 1990 and 2017. This result is intuitive as many small businesses have no hope for rehabilitation, and perhaps no ability to overcome the relatively larger costs of a Chapter 11 proceeding (see Chapter 4 herein). In contrast, Chapter 7 has rarely been filed by large U.S. public firms, which account for less than 10% of such bankruptcy filings (Source: bankruptcydata .com). Chapter 7 is sometimes used by larger corporations that initially enter Chapter 11 but are unable to reach a viable reorganization plan, or that need to

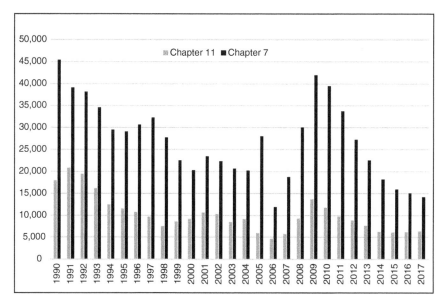

FIGURE 3.1 Number of Business Chapter 11 and Chapter 7 Filings
Source: United States Courts Form F-2 (http://www.uscourts.gov/).

distribute cash proceeds and any remaining assets to claimants after the firm's assets have been substantially sold in Chapter 11.

Chapter 11: Reorganization   An extremely important change in the 1978 Code involved the new Chapter 11 for business rehabilitation, consolidating much of the older Chapters X and XI. Under Chapter 11, the management as "debtor-in-possession" continues to operate the business unless the court orders a disinterested trustee for cause, or it is determined to be in the best interest of the creditors and/or the owners. Cause includes fraud, dishonesty, incompetence, or gross mismanagement, either before or after commencement of the case.

U.S. law imposes no obligation on a company's board to commence court proceedings upon insolvency (see Chapter 8 herein). When a firm does file under Chapter 11, there is no required insolvency test, either on a cash flow basis or balance sheet basis. The vast majority of filing debtors, however, are insolvent in some sense. In some cases, contingent events, which could cause insolvency, have been argued successfully as reasons to seek protection under the Bankruptcy Code. Examples of these contingent events include a number of Chapter 11 filings due to asbestos litigations and tort claims in the early 2000s.[1]

The 1978 Code contains a number of provisions important in enabling the debtor to reorganize, including but not limited to: an automatic stay; an exclusive period during which only the debtor can propose a plan of reorganization; the ability to use cash collateral and/or obtain postpetition financing; the ability to assume or reject leases and other executory contracts; the ability to sell assets free and clear of liens; the ability to retain and compensate key employees; and the ability to reject or renegotiate labor contracts and pension benefits. These specific features are explained in detail in the second portion of this chapter.

Most firms that file for Chapter 11 are not ultimately reorganized. In some cases, while at the time of filing it is estimated that the going concern value of a firm exceeds its liquidation value, the business might continue to deteriorate to such an extent that liquidation become inevitable. Many small firms are unable to successfully reorganize under Chapter 11. A firm can confirm a liquidating Chapter 11 plan, or convert the case to a Chapter 7 liquidation (Chapter 7 herein discusses evidence regarding bankruptcy case outcomes).

Bankruptcy Abuse Prevention and Consumer Protection Act of 2005

The U.S. Congress enacted a revised Bankruptcy Act on April 20, 2005, called the Bankruptcy Abuse Prevention and Consumer Protection Act of 2005 (BAPCPA). Although the most controversial and publicized aspects of the Act impact consumer (personal) bankruptcies by making the laws significantly less debtor friendly, the new Act has many important provisions that impact corporate bankruptcies. Examples include amendments to provisions for the debtor's exclusivity period for proposing a reorganization plan, the composition of creditors' committees, treatment of vendors, treatment of commercial leases, the use of key employee retention plans, and employee retirement benefits.[2] In general, as with consumers, the new Act is viewed as more favorable to creditors. Most of these provisions went into effect on October 20, 2005, six months after the Act was passed.

In addition to major changes to specific sections of the 1978 Code, the BAPCPA enacted a new Chapter 15 on cross-border insolvencies. Chapter 15 is the successor to Section 304 of the 1978 Bankruptcy Code, and is based on the United Nations Commission on International Trade Law (UNCITRAL) "Model Law" which provides a legal framework for cross-border cooperation and communication between courts (see Chapter 8 for a detailed discussion of cross-border insolvency).

A PRIMER ON CHAPTER 11

In this section, we explain some of the most important issues in Chapter 11 reorganization today. While we do not intend to provide a full legal reference, we describe the most up-to-date practice after the enactment of the 1978 Code, with amendments made by BAPCPA.

Venue Choice

A firm may voluntarily file for bankruptcy as long as it does so for legitimate business reasons. However, any three or more creditors holding unsecured claims aggregating to at least $15,775 (as of 2017) may submit an involuntary petition against the borrower under Chapter 7 or Chapter 11. Firms may choose to make a voluntary filing in a different bankruptcy court after creditors file an involuntary petition. For example, in the Caesars Entertainment Operating Company case, an involuntary petition was filed by three major creditors – Appaloosa Investment, OCM Opportunities Fund, and Special Value Expansion Fund – on January 12, 2015, in the District of Delaware. The company filed a voluntary petition on January 15 in the Northern District of Illinois. The Delaware court transferred the involuntary case to the Illinois court two weeks later.

Where is a company required to file its bankruptcy petition? The U.S. system provides a degree of latitude to some debtors in choosing the location or "venue" for the case. According to the U.S. Code Title 28 Chapter 87 Section 1408, a U.S. business can file for bankruptcy in one of the following four locations: (1) the place of domicile or residence, which can be the state of incorporation; (2) the principal place of business, which tends to be the corporate headquarters; (3) the principal place of assets; or (4) any district where a bankruptcy case is pending against the firm's affiliate.

Larger firms, which typically conduct business nationwide and have assets located in multiple states, therefore have some ability to select the court for their filing. For example, on the morning of June 1, 2009, Chevrolet-Saturn of Harlem, a dealership in Manhattan owned by General Motors itself, filed for bankruptcy protection in the U.S. Bankruptcy Court for the Southern District of New York, followed by General Motors Corporation (the main GM in Detroit) and other GM subsidiaries filing in the same court. Some scholars (e.g., LoPucki, 2005) argue that this practice of "forum shopping" enables firms to choose debtor-friendly venues, leading to inefficient reorganizations in some cases. Others argue that sophisticated debtors choose courts that have more expertise in dealing with complex cases.

In the two and one half decades between 1990 and 2017, more than half of U.S. public firms with at least $50 million in book assets filed their petitions in only two courts: the U.S. Bankruptcy courts for the District of Delaware and Southern District of New York (source: bankruptcydata.com). The topic continues to be debated, to the point where two U.S. senators introduced a bill in January 2018, the Bankruptcy Venue Reform Act, to prevent forum shopping.[3]

Judge Assignment

After a petition is filed, a judge is assigned to the case within hours. The majority of bankruptcy courts state that they use a blind rotation system to randomly assign cases to one of the active bankruptcy judges in their divisional office. However, courts that at times have a large caseload relative to their capacity have recruited district judges, visiting bankruptcy judges from other districts, or retired bankruptcy judges to oversee cases. For example, seeing a sharp rise in the number of bankruptcy filings from the late 1990s to the early 2000s, Delaware employed all three types of judges.

Although some people are skeptical that judges are truly randomly assigned, a few studies provide empirical evidence that bankruptcy case characteristics are independent of judge characteristics (e.g., Bernstein, Colonnelli, and Iverson, 2018; Chang and Schoar, 2013; and Iverson, Madsen, Wang, and Xu, 2018). These studies also examine the effect of judges' experiences and preferences on bankruptcy outcomes. They find that judges' on-the-bench experiences have a strong impact on bankruptcy duration, the probability of emergence, and the debtor's post-emergence performance, and that their pro-debtor preferences strongly explain the propensity to grant or deny certain motions.

Automatic Stay

An important aspect of reorganization law is to resolve the "collective action" problem by preventing individual creditors from taking actions that would jeopardize reorganization efforts and allowing the debtor to continue running the business without direct interference from creditors (see, for example, Baird, 2010). This is accomplished with Section 362 of the Bankruptcy Code, which governs the automatic stay in bankruptcy. Specifically, no claimant can foreclose on collateral, collect payment, enforce a judgment from a previous lawsuit, create or perfect a lien on property, sue to recover past claims (including taxes owed against the debtor), or refuse to pay an amount owed to the debtor once the firm is under court supervision in bankruptcy.

There are some limited instances where a judge may grant lifting the stay.[4] Under the Code (Section 361), the debtor is required to ensure that the value of the secured creditors' claims are "adequately protected" through the reorganization process; the judge may grant relief from the automatic stay upon observing evidence of "cause, including the lack of adequate protection of an interest in property of such party in interest." The adequate protection provision requires that if using, leasing, or selling a secured lender's collateral results in a decrease in its value, or if the debtor grants a security interest in the collateral to a new lender, the debtor must compensate the secured lender with a cash payment equal to the decrease in the collateral value, an additional or a replacement lien on other assets, or the "indubitable equivalent" of the lender's interest in the collateral.

Operating in Bankruptcy: The Concept of the Debtor in Possession (DIP)

The United States Trustee Program, a division of the U.S. Department of Justice, supervises the administration of a bankruptcy case and serves as the "watchdog over the bankruptcy process."[5] The United States Trustee is appointed and supervised by the Attorney General and charged with the responsibility of supervising the administration of a bankruptcy case. Specific responsibilities of the U.S. Trustee include the following, each of which we describe in further detail in this chapter: appointment of a trustee in Chapter 7 cases and some Chapter 11 cases; taking legal action as needed to enforce the requirements of the Bankruptcy Code; reviewing professional fees; appointing an unsecured creditors' and other committees for the case; and reviewing disclosure statements. Their primary objective is to make sure that the debtor is operating in good faith and in conformity with the Bankruptcy Code and other laws.

Despite this oversight, Chapter 11 of the Bankruptcy Code presumes that managers will remain in possession and control of a corporate debtor, and continue to operate the business. In doing so, the "debtor in possession" or "DIP" has a fiduciary duty to protect and preserve the assets of the bankruptcy estate and to administer them in the best interest of the creditors. Operating the ongoing business requires the use of the debtor's assets, sales and leasing of property, and so on. The debtor is able to conduct these activities so long as they are "in the ordinary course of business." Transactions beyond the reasonable day-to-day business or customary practices, such as the sale of an essential business division, require approval of the court and potential intervention by the judge and/or creditors.

"First Day Motions" and Events Early in the Case

Chapter 11 provides certain immediate protections for the debtor, such as the automatic stay. Further, the debtor's counsel typically seeks a number of court orders in the first days of the case to facilitate the transition into bankruptcy and minimize disruption to operations. Typical motions are to maintain bank accounts and cash management procedures, to approve payments for prepetition wages and to critical vendors (suppliers), to retain advisors and professionals, and for the use of cash collateral and approval of postpetition financing. Other important decisions, including the appointment of committees and rejection or assumption of unexpired contracts such as leases, are made within a short time of the filing.

Treatment of Vendors   A vendor that has not received payment for goods delivered to an insolvent customer has rights to get back or "reclaim" those goods under the Uniform Commercial Code's reclamation statute (UCC Section 2-702), as well as provided for under the BAPCPA (Section 546(c)). If the customer has filed for bankruptcy, vendors enjoy a reclamation period for goods delivered up to 45 days after shipment or up to 20 days after the customer's bankruptcy filing (if the 45-day period ended after a customer's bankruptcy filing). If the vendor fails to provide notice of its reclamation claim, it can still assert an administrative priority claim for the value of goods received by the debtor within 20 days prior to bankruptcy filing.

Given these complex rules, and the uncertainty of reclaiming goods or being repaid prior to the resolution of the bankruptcy case, vendors are very likely to tighten their trade credit policy as a customer becomes financially distressed and its chances of a bankruptcy filing rise. This can exacerbate a distressed firm's financial condition, if not triggering the filing itself. For example, when a news story published in September 2017 reported that Toys 'R' Us was considering a Chapter 11 filing, nearly 40% of vendors refused to ship goods without cash payment, precipitating a liquidity crisis for the firm.[6]

Once the firm has filed for bankruptcy, vendors may refuse to continue to provide goods unless their prepetition invoices are paid in full. Therefore, critical vendors are given preferential treatment for their claims over other unsecured creditors. DIP financing can provide the funding to make payments to vendors, and so to some extent alleviate vendor concerns for shipping goods after the filing date. However, motions to approve these payments are often subject to objections filed by the unsecured creditors that are effectively *subordinated*. For example, in the Kmart 2002 bankruptcy case (Gilson and Abbott, 2009), the company sought to pay prepetition claims to more than 2,000 vendors, worth more than $300 million. The judge's order was challenged by some of Kmart's unsecured creditors, leading to the ultimate rejection of the critical vendor payment by the Seventh Circuit Court of Appeals.

Utilities that provide essential services to the debtor are treated differently than typical commercial vendors. Under the BAPCPA, the debtor must provide "adequate assurance of payment" to these providers in the form of a cash deposit, letter of credit, prepayment, or surety bond within 30 days of a bankruptcy filing (Section 366).

Retaining Professionals   Financial and legal professionals provide a number of services for the debtor. These include (but are not limited to) negotiating with creditors, building financial models and valuation analyses, performing liquidation analyses, capital raising efforts, strategic business analyses, and employee retention efforts. See also Figure 4.1, describing the roles of professionals often retained in bankruptcy cases.

Professional fees and expenses incurred after the filing of the case are granted the legal status of an administrative expense (Section 503(b)). These fees have priority in payment over any other unsecured claims. For fees incurred prior to filing, professionals can waive prepetition claims and apply to be paid as an administrative expense. Attorneys and financial advisors are typically compensated on a time basis. However, financial advisors' compensation structure may include incentive bonuses, known as a restructuring fee or success fee, that are paid out to advisors upon the completion of a case, such as consummation of a restructuring or a sale. These fees, at times, add up to a large amount (see Chapter 4 herein for a detailed discussion on bankruptcy costs).

Under the old 1978 Bankruptcy Code, most of the large investment banks that participated in continuous underwritings of stocks and bonds for corporations were essentially prohibited from advising debtors who file for bankruptcy protection. This restriction was due to the "disinterestedness role" regarding any prior activities for three years before the filing for the debtor, regardless of whether such securities were still outstanding. The disinterestedness standard was reduced after BAPCPA because the old rule was repealed. Under Section 327 of the current Code, professionals can be retained so long as they are not found to be conflicted based on past activities. For example, an individual elected to the board or who held a prior management position with the firm cannot be retained as a professional. More broadly, an entity found, by reason of any relationship to or interest in the debtor, to have an interest adverse to the interests of the estate or any class of creditors or equity holders cannot be retained.

Financing in Chapter 11: Cash Collateral and DIP Financing   Making payments for postpetition and even some prepetition expenses (for wages or to vendors, for example) is critical to debtor's ongoing business and successful reorganization in bankruptcy. The debtor also needs sufficient liquidity to fund working capital, maintenance and repair expenses, and payments for professionals during the case. Such payments require use of the debtor's cash, but that cash may already serve as collateral for secured claims. One important source of funds may be to access cash that already serves as collateral for secured claims. Section 363(c) of the Code specifies that the debtor may not use, sell, or lease cash collateral unless "each entity that has an interest in such cash collateral consents" or "the court, after notice and a hearing, authorizes such use, sale, or lease." A second important source of funds is DIP financing.

Section 364 of the Code grants the debtor and the trustee the power to obtain postpetition credit with court approval. Specifically, the debtor is encouraged to obtain unsecured financing as an administrative expense. If the debtor is unable to obtain unsecured credit, the court may authorize the debtor to obtain a special form of postpetition credit known as DIP financing. A DIP loan is credit (1) with priority over any or all administrative expenses—a "super-priority" administrative claim, (2) secured by a lien on property that is not otherwise subject to a lien, or (3) secured by a junior lien on property that is subject to a lien. The court can also authorize credit to be secured by a senior or equal lien on property that is already subject to a lien – the "priming lien" provision – only if credit would be unavailable otherwise and the prepetition lien that is subordinated (or primed) by the DIP loan is adequately protected.

The proportion of large U.S. companies in Chapter 11 that took DIP loans grew from less than half in the 1990s (Bharath, Panchapegesan, and Werner, 2010; Dahiya, John, K., Puri, and Ramırez, 2003) to two thirds after the turn of the century (Gilson, Hotchkiss, and Osborn, 2016; Li and Wang, 2016). There has been tremendous interest from both practitioners and academics regarding various aspects of these loan contracts.[7] Companies that need immediate access to this type of loan typically file a DIP financing motion to the court on the day of, or shortly after, the bankruptcy petition.

Eckbo, Li, and Wang (2018) provide a comprehensive study of specific features of 407 loan facilities obtained by large public U.S. firms from 2002 to 2014. DIP loans usually take the form of revolving lines of credit, accompanied by a term loan and/or a letter of credit. The tenure of the loan is short term, typically ranging between one month and two years, suggesting that the funds are intended to match cash flow cycles and provide working capital, as well as to pay for operating expenses. In addition to interest costs, lenders can charge various fees. The total cost of the loan (spread and fees included) can be as high as 20%. In addition, the DIP contract may include restrictive covenants and required payments based on "milestones" such as filing a plan and disclosure statement or completing a sale of assets by a certain date. Thus, while DIP loans can be an important source of financing during the restructuring process, they may also serve as a power tool for senior lenders to exercise control over bankrupt companies.

Who provides the DIP financing? Prepetition lenders may gain an advantage by leading the arrangement of DIP financing as they can control collateral and be in a better position to recover their prebankruptcy exposure. Some DIP contracts may require the borrower to use part of the proceeds to repay the prepetition loan. The ability to "roll-up" some or all of prepetition loans into the DIP loan can be attractive to prepetition lenders. Lenders that specialize in distressed debt financing includes commercial banks, asset-backed lenders, investment banks, and alternative investors such as hedge funds and PE firms (Li and Wang, 2016).[8] Some specialized investors provide DIP financing with the goal of positioning themselves as the future owners of the company, by converting the loan into a majority equity stake when the firm is restructured – a so called "loan-to-own" strategy.

While lending to a firm in bankruptcy requires specialized expertise, DIP lending is remarkably *not* a very risky endeavor given the minimal loss rates and the large spreads and fees charged (see Moody's, 2008). The most publicized loss to a lender on a DIP loan occurred in the 2001 WinStar bankruptcy, with DIP lenders' recovery in the range of 20–30%. Since then, according to Moody's, there has been no major default on a DIP loan except in in the case of ATP Oil & Gas Corp., where the DIP facility was deemed impaired.[9] Even in this case, the eventual recovery to the DIP lenders was close to par. DIP loans have become increasingly attractive of late, with the active involvement of both traditional financial institutions and alternative investors. For example, in the Westinghouse (2017) case, there were 32 potentially interested financing parties and 14 of them submitted binding offers for the DIP loan.[10]

What are the effects of DIP financing on Chapter 11 outcomes? Conceptually, DIP financing enables the debtor to remain liquid during the most difficult days after filing and to invest in positive NPV projects that would not have been possible without the additional credit. Detractors of this type of financing might argue that it leads to overinvestment, giving managers the incentive and means to accept extremely risky or negative NPV projects. Empirical evidence from Dahiya, John, Puri, and Ramirez (2003) shows that DIP financing has a positive correlation with the eventual likelihood of reorganization, and that firms that obtain DIP financing from existing lenders tend to spend less time in bankruptcy.

Appointment of Committees The U.S. Trustee is responsible for the appointment of an official committee to represent the interests of unsecured creditors (the "UCC"). Importantly, the UCC's costs for professional advisors, such as attorneys or accountants, are paid by the debtor. The UCC ordinarily consists of unsecured creditors holding the seven largest unsecured claims who are willing to serve. The UCC does not control the business and assets of the debtor, but rather consults with the debtor on administration of the case and monitors the business operation. Specifically, the Code (Section 1103) permits the UCC to: review and discuss the progress of the case with the management team, investigate the financial condition and operation of the business, participate in the formulation of a reorganization or liquidation plan, ask the court to appoint a trustee or an examiner, and request that the court dismiss a case or convert a case to Chapter 7. The UCC typically files objections with the court for actions taken by the debtor they deem to negatively affect enterprise value and debt recovery, for example for approval of executive retention and incentive bonuses or exclusivity extensions. At times, ad hoc committees of creditors are formed when there is no official committee appointed, or when dissident debt holders do not want to join the official committee. However, the cost of professionals hired to assist an ad hoc committee is typically borne by the committee itself.

Academic studies show that the UCC is formed in less than half of Chapter 11 cases with over $10 million in assets (Waldock, 2017). The members of the UCC are mostly trade creditors, though hedge funds, private equity funds, and other activist investors also often seek representation. Their presence occurs in over 30% of the largest Chapter 11 cases filed from the late 1990s to early 2010s (Goyal and Wang, 2017; Jiang, Li, and Wang, 2012).

Because Chapter 11 affects the interest of claimants besides unsecured creditors, the U.S. Trustee or the court has the discretion to appoint additional committees as appropriate. If such a committee is appointed, the bankruptcy estate pays its fees and expenses. Most common is an official equity committee, but even this occurs in less than 10% of Chapter 11 cases of large U.S. public firms in the past two decades (Goyal and Wang, 2017). Appointment of an equity committee considers several factors including whether there is any substantial likelihood of a meaningful distribution to equity holders, which may be unlikely for a more severely insolvent debtor.

Replacing Management with an Independent Trustee It is possible under the Bankruptcy Code (Section 1104) that management is replaced by a court-appointed independent trustee (not to be confused with the U.S. Trustee's office) if there are "reasonable grounds to suspect" that the debtor's current CEO, CFO, or board members participated in "fraud, dishonesty, or criminal conduct in the management of the debtor or the debtor's public financial reporting either before or after the commencement of the case." When appointed by the court, the trustee takes control of the business, displacing the DIP. While required in a Chapter 7 liquidation, the appointment of a trustee in Chapter 11 cases is not common, and will occur only by the court on its own initiative or after a motion brought by the US Trustee or a party of interest. In many cases, the board of directors will have already removed officers of the company prior to entering Chapter 11, avoiding the disruption of a trustee being appointed postpetition. It is also possible that the U.S. Trustee recommends a court appointed examiner to investigate specific issues of fraud or mismanagement. While the use of an examiner is rare, it has been important in some large and significant Chapter 11 cases, including Enron and Lehman Brothers.

Rejection/Assumption of Leases and Other Executory Contracts An executory contract is a contract between the debtor and another party under which both sides still have important performance remaining. Common examples include real estate leases, equipment leases, development contracts, and licenses to intellectual property. Section 365 allows debtors to choose whether to keep ("assume"), abandon ("reject"), or transfer ("assume and assign") their executory contracts, within time limits provided by the Bankruptcy Code.

A controversial change under the 2005 BAPCPA shortened the time for a debtor to reject a lease to within 120 days of filing, with the bankruptcy court permitting only one 90-day extension beyond that (without landlord consent). A contract is deemed rejected if the debtor does not assume the contract by the deadline. Critics have argued that the accelerated timeframe for these decisions does not provide debtors with sufficient time to analyze performance and make adequate business decisions regarding these contracts. This and other changes under BAPCPA have been blamed for contributing to the failure of reorganization attempts of several large retailers, such as Circuit City in 2010, and other firms that have entered bankruptcy with a large number of commercial leases.

If a lease is rejected, the debtor can walk away from its obligations under the contract and the lessor will be entitled to an unsecured prepetition claim against the debtor, in an amount equal to the greater of one year's rent or 15% of unpaid rent on the unexpired lease (not to exceed three years' rent). Such terms start on the earlier of (1) the petition date or (2) the date on which the lessor repossesses the leased property, or the lessee surrenders it. In contrast, all prepetition and postpetition obligations for those leases that the debtor assumes become administrative expense priority claims. The debtor also has the right to assume and assign a lease to a third party, commonly a buyer of some assets, as long as the buyer is expected to be able to perform under the contract.

The debtor has the right to assign a lease to a third party willing to pay more than contracted rent, without permission from the lessor, and garner any revenues generated. Sometimes firms sell property designation rights—which involves transferring the debtor's right to decide which contracts to assume, to whom they will be assigned, and under what terms—to a third party, for a price. The purchaser, often property management companies, are willing to pay cash to the debtor and assume the responsibility to market the leases. The leases can, therefore, become a valuable asset, especially those with below-market rates. However, property owners may file objections on reassignment of leases to third parties, based on business or moral grounds.

Generally, a debtor will assume or assign leases with good terms and renegotiate or reject those with unfavorable terms. In fact, the threat of rejecting bad leases often gives the debtor considerable negotiating leverage over lessors to modify terms. Rejection and assignment of leases have saved many large retail and airline firms in bankruptcy millions of dollars in operating expenses.[11] Lemmon, Ma, and Tashjian (2009) show that rejecting lease contracts in bankruptcy constitutes a large portion of asset restructurings. With the accelerated timetable of a maximum

210 days to decide, the debtor is not given sufficient time to analyze performance and decide whether to assume or reject leases in bankruptcy and therefore often makes decisions even before filing bankruptcy.

The debtor firm is also equipped with rights to assume and reject supply contracts via Section 365. The debtor can use these rights as a threat to renegotiate terms of their prepetition contracts, potentially saving millions of dollars for the debtor but squeezing margin for their suppliers. Academic studies (e.g., Hertzel, Li, Officer, and Rodgers 2008; Kolay, Lemmon, and Tashjian 2016) find that customer bankruptcy is generally bad news for suppliers, which experience negative equity returns when customers file for bankruptcy, particularly when the costs of replacing those customers is high.

Developing and Confirming a Plan of Reorganization

Unless the case is dismissed or converted to Chapter 7, for the debtor to be able to emerge from bankruptcy, the court must confirm a plan of reorganization. The plan specifies who the claimants are, what those claimants will receive in the restructuring (and therefore what the new capital structure will look like), and other aspects regarding the disposition of assets. Therefore, it is important to review how the plan is developed, what the plan must achieve in order to be approved, and the process for parties to agree to a plan. The plan must be accompanied by a "disclosure statement" that contains sufficient information about the debtor to enable claimants to make an informed judgment whether to accept the plan.

Exclusivity Period The 1978 Code gives the debtor an exclusivity period of 120 days after the petition to file a plan. Other parties, such as the trustee, a creditors' committee, an equity security holders' committee, a creditor, an equity security holder, or any indenture trustee, may file a plan only if a trustee has been appointed, if the debtor has not filed a plan within 120 days of filing, or if the debtor's plan is not accepted within 180 days of filing. However, the frequency with which courts granted extensions to this period prompted the writers of the 2005 Act to limit exclusivity periods to a maximum of 18 months (plus two months to permit solicitation and plan confirmation).

One interesting statistical trend of the past four decades is the time spent in bankruptcy from filing to emergence. Prior to the passage of the BAPCPA in late 2005, the median time in default was about 24 months. Since 2006, the median time has decreased by more than half to 11 months, due in part to limits on extending exclusivity and to a large increase in prepackaged Chapter 11 filings (see Altman and Benhenni, 2017; Altman and Kuehne, 2017).

Absolute Priority and Classifying Claims To receive any distribution under a reorganization or liquidation plan, a creditor must file a proof of claim and an equity holder must file a proof of interest. Under a plan, claims are grouped into classes of "substantially similar" claims and ranked according to their seniority. These classes typically include the following, though more complex plans can specify a number of separate classes within these broader groupings.

Secured creditor claims, that is, those that have specific assets as collateral. Secured claims have priority over all other claims to the extent that the liquidation value of the collateral is greater than the amount of the claim. If the collateral value is sufficient, postpetition interest can accrue as part of the secured claim. The debtor may be able to utilize the collateral for its benefit during reorganization to the extent the secured claims are "adequately protected" (see above). If the collateral value is insufficient to cover the full amount of the secured claim, the balance is considered as an unsecured claim.

Priority administrative expense claims, which include trustee fees, legal fees, and other service fees. Although ranked lower than secured claims, the court requires these fees be paid in order for a plan to be confirmed. Other administrative claims include (with exceptions as specified in Section 503) necessary costs accrued in the ordinary course of the debtor's business or financial affairs after the commencement of the case; for example, for payments to suppliers as described above. In contrast, in a Chapter 7 liquidation, the only administrative claims allowed are trustee, legal, and professional fees, which are paid from the liquidation proceeds before secured claims.

Other priority claims, including prepetition employee wages, unpaid contributions to employee benefit plans, customer claims (money in connection with the future use of goods or services from the debtor), claims of governmental units such as taxes and penalties, and employee injury claims. Prepetition employee wages are allowed with a limited dollar amount from up to 180 days before the filing (or the cessation of the debtor's business, whichever occurs first).

Unsecured claims, which include unsecured bonds, trade claims, legal claims, insurance claims, and so on. Senior debt has priority over all debt that is specified as subordinated to that debt, but has equal priority with all other unsecured debt. The terms of most loan agreements spell out these priorities. In addition, the organizational structure of the company matters to the priority ranking of the debt claims (see Chapter 2 herein).

Prepetition equity holder claims – preferred and common stockholders, in that order.

The proposed plan lays out the distributions to be made to each class of claimants. Distributions are made after the plan has been confirmed in cash or in the form of claims on the reorganized firm. The lowest priority class or classes that are entitled to receive any distribution typically receive shares of stock in the reorganized firm.

Most reorganization plans are guided by the "absolute priority rule" or "APR" (Section 507), which means that higher priority claims must be paid in full before a less senior claim can receive anything. In practice, plans are a negotiated agreement involving each class that is entitled to vote on the plan. Thus, a plan can violate absolute priority, such that a more junior claim receives some payment while more senior claims are not paid in full. This arrangement is often expedient and permits compromise with creditors who would otherwise vote against a plan.

Deviations of APR were common in the 1980s and 1990s but have become less frequent since the 2000s (see Chapter 7 for empirical evidence). APR deviations are guided by the "best interest test" (see below) whereby no impaired creditor class receives less value than would have been received in a liquidation. In essence, absolute priority serves more as a guideline for a Chapter 11 plan than as a requirement.

Reorganization Valuation The centerpiece of a reorganization plan is a valuation of the debtor as a continuing entity. Valuation involves a forecast of expected postbankruptcy "free cash flows" based on the underlying business plan for the restructured firm. The specific valuation methodologies applied to these forecasts are discussed in Chapter 5. The total enterprise value can incorporate other assets such as excess working capital, tax loss carry-forwards, and other considerations. If the estimated value of the going concern value is less than the total amount of allowed claims, the firm is insolvent in a bankruptcy sense.

Waterfall Distribution The reorganization value is distributed to classes following absolute priority. Strictly following absolute priority, once the value available for distribution has been exhausted, claims falling below that amount receive no distributions. For an insolvent debtor, prepetition equity holders and perhaps additional lower priority claims receive no distribution under the plan.

Figure 3.2 provides a simple illustration of the waterfall analysis based on a hypothetical situation. Suppose the company has three classes of debt, senior secured loans, senior unsecured notes, and subordinated notes, and equity interests. The revolver and term loans are *pari passu* (on equal footing). The distribution of value has high- and low-case scenarios. The table shows that in a low-case scenario, where the firm's value is worth $800, the value is split between revolver and term loans, with each recovering 80% of its claim amount, leaving no recovery to the junior claimants. In the high-case scenario, the firm's value is large enough to cover the value of claims up to subordinated notes, which receive a recovery of 50%. In both scenarios, equity holders receive zero recovery. The distribution to different classes of claims can be in the form of cash, new debt,

	Claim Amount	Distribution of Value		Recovery Rates	
		Low Case	High Case	Low Case	High Case
Distributable value	—	$800	$1,400	$800	$1,400
Secured debt					
Revolver	$200	$160	$200	80%	100%
Term loan	$800	$640	$800	80%	100%
Available to lower claims	—	$0	$400	—	—
Unsecured debt					
Senior notes	$200	$0	$200	0%	100%
Available to lower claims	—	$0	$200	—	—
Subordinated notes	$400	$0	$200	0%	50%
Available to lower claims	—	$0	$0	—	—
Equity interests	—	$0	$0	0%	0%

FIGURE 3.2   A Simple Illustration of the Waterfall Analysis

and new equity of the reorganized entity (see Chapter 5 herein for an example of Cumulus Media for how waterfall distributions work in practice).

Voting and Confirmation of the Plan   Once a Chapter 11 plan and an accompanying disclosure statement have been approved by the court and distributed, each claim holder within a class entitled to vote can cast its ballot to accept or reject the plan. A plan may specify that the rights of certain claims or interests are unaltered by the plan, for example, when secured debt is reinstated and will be paid as originally scheduled. Such classes are referred to as "unimpaired," are deemed to accept the plan, and so do not vote on the plan. Certain classes may receive no distribution under the plan, for example, for prepetition equity holders of an insolvent firm; such classes are deemed to reject the plan and again do not vote. The remaining voting classes are considered "impaired" and entitled to vote on the plan. A plan is deemed accepted by a class of creditors if at least two-thirds in amount, and more than one-half in number of voting claims within the class are cast in favor of the plan (Section 1126(c)). Shareholders (if entitled to vote) are deemed to have accepted the plan if at least two thirds in amount of the outstanding shares voted are cast for the plan (Section 1126(d)).

A hearing is held after the 60-day voting period to determine whether the plan meets various requirements such that it can be confirmed by the court. The plan must be accepted by each class of claims or interests entitled to vote or not impaired, known as a consensual plan; or, a nonconsensual plan can be confirmed over the dissent of a class of creditors through a procedure known as "cramdown" (Section 1129(b)). At least one class of impaired claims or interests must accept the plan. For example, if the only class affected by the plan comprises a mortgagee, the plan cannot be confirmed without the mortgagee's consent.

The cramdown procedures enable the court to confirm a nonconsensual plan provided it does "not discriminate unfairly" and is "fair and equitable" to each class of claims or interests impaired. In general, a plan does not discriminate unfairly if it provides a treatment to the class that is substantially equivalent to the treatment provided to other classes of equal rank. The test for what is fair and equitable differs for secured creditors (either retaining their lien or receiving a new claim with a present value equity to their original claim), unsecured creditors (either receiving payment equal to the amount of their claims or the holders of more junior claims receive nothing under the plan), or equity interests (either retaining their interests or any more junior class of interests receives nothing).

In some instances, the obstacle in reaching a consensual plan is a disagreement over the firm valuation on which the plan is based. For example, a class of junior creditors may argue that the plan value is unrealistically low, and under a more appropriate valuation methodology a higher value would justify a large share of distributions to be made under the plan (see Chapter 5 herein). In such cases, the judge may order a full valuation trial to resolve the dispute, with a finding of a sufficiently higher value meaning that the plan could not be confirmed. More typically, however, the threat of such litigation is sufficient to reach a settlement leading to a consensual plan.

A confirmed plan must also satisfy several further requirements of the Bankruptcy Code, including the "best interest" and the "feasibility" tests. The best interest test requires that the creditors must receive or retain under the plan not less than the amount they would receive or retain if the debtor were liquidated under Chapter 7. Thus, the disclosure statement typically includes a liquidation analysis based on balance sheet accounts, which in the case of a reorganization generally show a significantly lower value that would be realized under a hypothetical fire sale of assets – satisfying the best interest test and further justifying the decision to reorganize. The feasibility test requires the debtor to show that confirmation is not likely to be followed by further need for financial reorganization or liquidation not contemplated within the plan. The cash flow analysis used to determine the debtor's reorganization value can be used to demonstrate that the firm is expected to generate sufficient cash flow to meet its post-emergence obligations.

Prepackaged Chapter 11   An interesting innovation since the early 1990s is the use of a prepackaged bankruptcy, commonly known as a "prepack." Prepacks combine the time- and cost-saving attributes of an out-of-court restructuring with the voting requirements (only two thirds of a class need to approve a plan) and other advantages of an in-court Chapter 11 proceeding. Under Section 11 USC.1126(b) of the Code, debtors are permitted to negotiate with creditors prior to filing and to accept prepetition votes with proper disclosure, that is, adequate disclosure and a reasonable time for analysis, discussion, and voting. In other words, the debtor reaches agreement on a plan with most of its creditors and solicits their votes in advance of actually filing for Chapter 11. This type of bankruptcy is frequently used by large U.S. firms, accounting for as much as one third of the total Chapter 11 filings by large U.S. firms from 1990 to 2014 (Figure 3.3). The BAPCPA increased the attractiveness of prepacks by loosening restrictions on information disclosure while creditors' votes are being solicited.

A prenegotiated arrangement is similar to a prepack except that the formal solicitation of the acceptance of the plan is done after the petition is filed. Before filing, the firm asks key creditors to sign a "lock-up" agreement, typically referred to as a Restructuring Support Agreement (RSA), in which they agree to vote for the plan and other terms of the RSA once the firm is in Chapter 11.

A few dissenting creditors could effectively stall or stop a prepackaged arrangement, for example by filing an involuntary petition. However, the BAPCPA permits vote solicitation to continue after the filing of the petition as long as it began before and the process complies with applicable nonbankruptcy laws, usually the relevant securities laws. Thus, dissenting creditors now need to file their involuntary petition before solicitation begins, and the tactic of forcing a voluntary filing will no longer interfere with soliciting acceptance for a prepack.

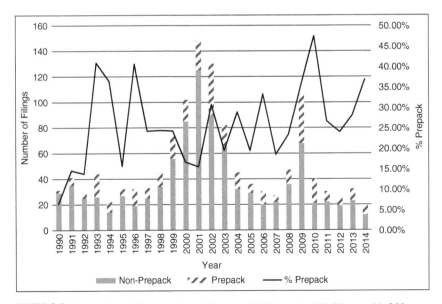

FIGURE 3.3 Prepackaged/Prenegotiated Chapter 11 Filings by U.S. Firms with $50 Million Assets

There are many advantages of a prepackaged filing when compared to the regular "free fall" Chapter 11. First, a company negotiating a prepackaged Chapter 11 has a defined exit strategy from the bankruptcy before the legal process starts, which may dramatically increase its chances of emerging as a going concern. Second, while it still may take several months to emerge, the average time in bankruptcy of these cases is far less than the almost two-year average of all Chapter 11s. Tashjian, Lease, and McConnell (1996) find evidence that the cost of a prepackaged Chapter 11 falls in between the costs of a conventional Chapter 11 and an out-of-court restructuring.

It is worth noting that an accelerated bankruptcy is not in every distressed company's best interests. For companies that suffer from more serious business problems that can be more effectively addressed through an extended stay in Chapter 11 or that have more complex debt structures that hinder soliciting votes prepetition, a prepackaged or prenegotiated Chapter 11 will be less attractive or feasible (Gilson, 2012). On the other hand, if a firm's value might significantly deteriorate upon a Chapter 11 filing, as was argued for General Motors in 2009, a prepack can enable the firm to quickly restructure and minimize its loss of value from losing customers, suppliers, and other costs.

Plan Execution Confirmation is followed by a plan "effective" date. Distributions of cash and new claims, as specified in the plan, are made to allowed claims, and prepetition claims are either cancelled or reinstated. For example, when prepetition stockholders receive no distribution under the plan, the original shares are cancelled, and newly created shares are distributed. Securities issued under a plan are exempt from registration with the SEC. Note that if a plan is approved, it is binding on all members of a given class. The plan may satisfy or modify any lien, waive any default, and merge or consolidate the debtor with one or more entities.

A plan must provide adequate means to put in place its postbankruptcy capital structure. Typically, the firm will have arranged postpetition credit known as "exit-financing," for example, a new or reinstated revolver and term loan, as well as other debt financing to fund operations post-emergence. During the 2008–2009 economic crisis, lack of availability of such financing made it difficult for some firms to reorganize.

An additional ingredient to some successful bankruptcy restructurings lies in the debtor's ability to raise new equity capital for the emerging firm. An equity infusion by existing or new investors can signal to the market that real economic value still exists in the firm's assets. Various courts have invoked a so-called "new value exception" or "new value doctrine" to the absolute priority rule, which permits existing equity holders to hold equity interest in a reorganized debtor based on additional contributions to the business even where more senior, dissenting classes of creditors are impaired. Specifically, using new value exception, the plan proponent typically includes a provision to allow equity holders to contribute to the reorganized debtor's new estate value in order to retain existing interest or to receive new equity even though one or more senior classes of claimants are not to receive full payment under the plan. Another increasingly common form of equity contribution at emergence occurs via a rights offering, whereby certain prepetition claimants are given rights to purchase additional stock in the emerging company. Rights offerings are sometimes "backstopped" by a group of creditors particularly interested in owning a majority of the postbankruptcy equity, giving them the ability to exercise rights given to other claimants less interested in contributing cash to own additional post-emergence shares.

OTHER KEY ASPECTS OF CHAPTER 11

Key Employee Retention and Incentive Plans

Key Employee Retention Plans (KERPs) and Key Employee Incentive Plans (KEIPs) are compensation schemes designed to award senior managers who are deemed critical to the restructuring and continuation of the business. A large fraction of US public firms adopted these plans in bankruptcy to retain key managers. However, there are ongoing debates on whether these plans are schemes for rewarding failed management teams at the expense of other stakeholders or optimal contracts awarded when the bankrupt firm needs to retain human capital the most. A detailed discussion of these plans is provided in Chapter 6 herein.

Asset Sales

Asset sales outside of a firm's ordinary course of business are conducted using Section 363 of the Code, which establishes a formal process for companies to sell assets on an expedited basis. Section 363 sales have several advantages compared to selling assets outside of bankruptcy or through a Chapter 11 plan. First, Section 363 sales are subject to the debtor's discretion and judge's approval, but not to a creditors' vote. In contrast, asset sales through a plan must be voted on by creditors. Second, assets are sold "free and clear of liens and encumbrances" meaning that the lender will have a lien on the proceeds of the sale only after the sale, thereby exempting the buyer from the old lender's security interest in the assets transferred. Third, because the final sale is approved by and implemented through a bankruptcy court order, the validity of the transaction is generally immune to later legal challenges.

To sell assets under Section 363, a debtor must file a sale motion describing the bidding and selling procedures. Typically, a stalking horse – an initial interested buyer with whom the debtor has entered an asset purchase agreement – is identified. Key stakeholders of the bankrupt firm, including secured creditors, unsecured creditors, and United States Trustees, among others, can file formal objections to the proposed sale. If the court approves the bidding procedures, a date is set for submitting qualified bids and for an in-court auction. A final sale hearing is held and the judge approves the sale to the winning bidder from the auction, or an auction is not held if there are no further bidders. This process can take just a few weeks to complete.

Note that judges are permitted to consider nonprice factors in choosing the winner of an auction. Judges may defer to the debtor's business judgment and the fairness of the bidding procedures to determine whether the winning sale is highest or best. For example, in Polaroid's 2009 Section 363 auction, the judge awarded the sale to the second-highest bidder, which had a better track record of buying and managing bankrupt company brands (Gilson, Hotchkiss, and Osborn, 2016). Break-up fees are awarded to stalking horse bidders if they are not the winning bidders.

Assets sold through Section 363 range from a piece of equipment (Maksimovic and Phillips, 1998) or real estate property (Bernstein, Colonnelli, and Iverson, 2018) to substantially all of the operating assets of the company (Gilson et al., 2016). Intellectual property such as patents have become a frequently sold asset through Section 363 (Ma, Tong, and Wang, 2017). The buyers can be any interested party, including industry competitors, financial institutions, and even the debtor's creditors. In fact, Section 363(k) allows secured creditors, including a DIP lender, to "credit bid" for assets using their allowed secured claims instead of cash to bid up to the value of their claim. A DIP credit agreement may contain provisions requiring the lender's approval of the bidding procedures, or of the Section 363 sale itself.

Preferences and Fraudulent Transfers

The 1978 Code and the subsequent amendments made by the BAPCPA grant the trustee or debtor the power to avoid (i.e., reserve or clawback) preferential transfers (Section 547) and fraudulent transfers (Section 548). What is the difference between these two types of transfers? We here illustrate it using a simple example. Suppose a firm has outstanding debt to two different creditors. In the first scenario, the firm pays off one creditor and then files for bankruptcy a few weeks later. In an alternative scenario, the firm decides not to pay off either creditors and instead gives the cash to a friendly equity owner. The transaction made in scenario 1 is known as a preferential transfer, which gives preferential treatment to one creditor over another. The transaction made in scenario 2 is known as a fraudulent transfer that harms both creditors. In essence, preferences refer to the redistribution of the value of the debtor between creditors. In contrast, the fraudulent transfers provision allows for bringing cash or other value back into the bankruptcy estate, which should benefit all creditors. However, these transfers can be avoided only through litigation, which is always costly.

According to the Code, preferential transfers can be avoided if such transfer was made while the debtor was insolvent and within 90 days of the bankruptcy petition date, or between 90 days and one year before the petition date if such creditor at the time of the transfer was an insider. Certain defenses are available against preference avoidance actions. For example, the transfer may not be voided if the payment was a debt incurred and paid in the ordinary course of business of both parties, or the payment was made according to ordinary business terms. As such, payments will be allowed if they meet industry standards and are in the ordinary course of that business. These provisions, as part of the amendments of BAPCPA, make it difficult for the debtor to recover preference claims against a creditor.

Transfers of assets away from the debtor can be voided if the company performed the transfer with the intent to hinder, delay, or defraud creditors, or if the company received less than a reasonably equivalent value and the company was insolvent on the date when the transfer occurred. Therefore, to determine whether a constructively fraudulent transfer occurred requires valuation analysis to determine whether the company was insolvent at the time of transfer and whether the company received reasonable equivalent value in the transaction. The statutory limitation for the fraudulent transfer look-back period is two years. However, fraudulent claims are recognized under state law. There are subtle differences in state law on the look-back period, which ranges between three and six years.

The typical fraudulent transfers involve the following transactions: asset sales before Chapter 11 filing, asset transfers within corporate entities, loan guarantee provisions between corporate entities within the parent company, and leveraged buyouts (LBOs). In fact, in nearly every LBO there is some prospect of fraudulent transfer litigation. Plaintiffs can argue that certain knowledgeable creditors, like banks, unfairly benefited from these leveraged transactions, which eventually fail and result in losses to junior creditors. The centerpiece of the claim is that banks and other "insiders" knew or should have known that the restructuring would likely result in a failure to meet the debtors' liabilities, but went along with the deal to derive its "own benefit" from upfront fees, priority payments, and so on.

Claims Trading

One of the most important advances in the bankruptcy process since the enactment of the 1978 Code is the fast development of markets for trading claims in bankruptcy. Before the act went into effect, the only claims that were routinely traded in the financial marketplace were publicly registered debt securities. The market expanded dramatically in the 1980s and continued to grow rapidly in the 1990s and 2000s. There is tremendous turnover in the ownership of debt claims once firms become distressed because of the entry of specialized distressed debt investors. Given these trends, some say claims trading has transformed U.S. corporate bankruptcy to a market-driven process.

The bankruptcy rules for the trading of claims (Rule 3001(e)) were amended in August 1991, making it easier to purchase claims. These rules apply to all claims but have a particular focus on claims that are not publicly traded. Judges will not get involved in ruling on trading of claims unless an objection is lodged. The amount paid for a claim and any other terms of the transfer does not have to be disclosed.

In general, the claims trading markets are composed of a few overlapping segments that vary in liquidity, rules, and regulation. They are public debt, bank loans, trade claims, and other claims such as tort claims, insurance claims, and derivative claims. Bonds are traded over-the-counter and are regarded the most liquid type. There is little counterparty risk involved because of the use of large financial institutions as broker-dealers. Identities of counterparties are typically not known, making comprehensive due diligence impossible.

The second type is bank loans, which are typically syndicated and traded in slices. Active trading of private claims, such as bank loans of bankrupt firms, was not evident in the 1980s but took off in the 1990s, particularly with the establishment of the Loan Syndication and Trade Association (see Chapter 2 herein). Distressed loans emerged as a new asset class. This market segment attracted enough attention that several large broker-dealers were making regular markets. Specialized distressed debt investors that intend to gain control of the bankruptcy reorganization process frequently purchase loan claims even before a bankruptcy filing. Holding prepetition loan claims grants these investors an advantageous position in supplying DIP financing.

Purchasers of trade debt have more uncertainty and challenges than the purchasers of loans and bonds. Counterparty risks and due diligence requirements are much higher for trading these claims. For example, verifying the validity of a claim may prove to be challenging, and, as a result, the purchaser has much uncertainty about how much of the claim will be allowed. This market is occupied by a group of specialized investors (e.g., Argo Partners, ASM Capital), who typically purchase claims after the schedules of claimholders or proofs of claims are filed. Trading other bilateral claims, such as tort claims and derivatives claims, possesses similar characteristics and risks to trading trade debt and are also governed by Rule 3001(c). However, the market for these claims is much smaller.

Because members of the UCC and other official committees typically have access to material nonpublic information about the debtor, members of the committee are prohibited from trading securities of the company without disclosing this fact to the counterparty. The SEC has filed several complaints against parties that traded debt claims while serving on creditors' committees. Restricted debt holders are careful to ensure the counterparty is aware that they have material nonpublic information when they trade by executing a "big-boy" letter with the counterparty, though such a letter does not fully eliminate the litigation risk. Some debt holders may choose to forgo access to material nonpublic information in favor of having freedom to trade by not joining the UCC.

Hotchkiss and Mooradian (1997) are among the earlier academic studies showing the impact of claims trading on the dynamics of the bankruptcy case.

More recently, Ivashina, Iverson, and Smith (2016) examine transfers of claims during bankruptcy using data obtained from four leading providers of claims administrative services for Chapter 11 cases. This data includes information on claims transfers from proofs of ownership transfer required to be filed (under Rule 2001(e)) for all claims that are not syndicated bank loans or public debt securities. They document that active investors, including hedge funds, are the largest net buyers of claims in bankruptcy. Trading leads to a higher concentration of ownership, particularly among claims whose holders are eligible to vote on the bankruptcy plan and among claims that represent the "fulcrum" security in the capital structure.

Accounting and Tax Issues

This section introduces two accounting and tax issues that are particularly important to Chapter 11 companies. These issues are complex and require the expertise of tax accountants. Our objective is to provide a simple introduction.

Net Operating Losses (NOLs) are negative taxable income that can be used to offset taxable income in profitable years (Section 172 of the Internal Revenue Code (IRC)) and thus are an extremely important element in any reorganization, especially if the value of the new firm is relevant, as it almost always is. In contrast, net operating loss carryforward questions are irrelevant in a straight liquidation. Theoretically, as the value of a firm is the discounted expected value of its future after-tax earnings plus the present value of tax loss carry-forwards, preserving NOLs in Chapter 11 reorganization is an important concern for a company.[12]

NOLs can be carried back up to two years and forward up to 20 years for most taxpayers. Since a bankrupt firm most likely incurs losses before filing, NOLs are typically applied to future tax years. NOLs are used on a first-in and first-out basis (i.e. oldest NOLs are applied first). The Tax Reform Act of 1986 made sweeping changes in the rules governing the use and availability of NOL carryforwards, following certain changes in the stock ownership of a loss corporation. Specifically, if the percentage of stock owned by one or more 5% shareholders increases by more than 50% within a three-year period, annual limitation on use of NOLs (equal to the product of market value of equity prior to ownership change multiplied by the long-term exempt rate set up by the IRS) is applied. That is, the Chapter 11 firm must track the stock ownership of (only) 5% percent shareholders. All shareholders with less than 5% ownership are considered a single 5% shareholder.

Generally, preserving NOLs is easier in a formal bankruptcy than in a private workout (see Chapter 4 herein) as the bankruptcy entity can benefit from two statutory carveouts (the insolvency and bankruptcy exceptions) from Section 382 limitations. The first carveout (Section382(l)(5)) allows the debtor that undergoes a change of ownership in bankruptcy to emerge without Section 382 limitations if:

- Shareholders and creditors of the company end up owning at least 50% of the reorganized debtor's (voting power and value of) stock of the newly reorganized company in exchange for discharging their interest in and claims against the debtor, and
- 50% of equity was given to "old and cold" creditors (defined as those holding claims for at least 18 months as of the filing date), ordinary business creditors, and equity holders.

This statutory relief from Section 382(l)(5) may be violated if there is active trading in the bankrupt firm's debt in Chapter 11 (or even before petition). To preserve the value of NOLs, a debtor company can request the court to issue orders that place restrictions on claim trading. Furthermore, this statutory relief is violated and thus the NOLs are lost if there is an actual ownership change within two years of the company's emergence from bankruptcy. The court may impose restrictions on stock transfer within two years after emergence. The second statutory carveout (Section 382(l)(6)) allows the debtor to use the value of equity after the change of ownership to calculate the annual deduction limit. This provision is in consideration that equity is often worthless for firms filing for bankruptcy.

Cancellation of Indebtedness (COD) Generally, cancellation of indebtedness by a creditor for less than the amount owed results in taxable COD income. However, COD income is not taxed if the debt is cancelled in bankruptcy, and taxed a certain amount when the taxpayer is in insolvency but not bankruptcy (defined by the IRC as liabilities in excess of the fair market value of assets). However, COD income reduces certain tax items, starting with NOLs. Some Chapter 11 companies have lost significant NOLs because of the deduction of COD income.

Fresh Start Accounting An accounting rule enacted in 1991, known as the "fresh start rule," enables emerging Chapter 11 companies that are insolvent before emergence and that distribute more than 50% of their stock to the old creditors, to be treated as a new business. Assets can be written-up to market value rather than carried at historical cost. The valuation process is somewhat complex, especially determining the market value upon emergence. A market value is determined based on future cash flow estimates. As much of this value as possible is assigned to tangible assets, like plant and equipment. The remainder is a new asset called "reorganization value in excess of amounts allocable to identifiable assets," such as "goodwill." The reorganization thereby creates an entity with immediate positive equity. The debt to equity ratio is vastly improved under the "*fresh start*" format.

ABI COMMISSION REPORT OF 2014

In 2014, the Commission to Study the Reform of Chapter 11, established by the American Bankruptcy Institute (ABI) released its final report and recommendation after undertaking a three-year study (http://commission.abi.org/full-report). The Commission was composed of prominent bankruptcy attorneys and financial advisors, retired judges, law professors, and US Trustee offices. The Commission's task was to look for improvements to the existing Bankruptcy Code given the 2008–2009 economic crisis and fast evolution of financial markets in the past decades. The Commission's mission statement reads:

> *In light of the expansion of the use of secured credit, the growth of distressed-debt markets and other externalities that have affected the effectiveness of the current Bankruptcy Code, the Commission will study and propose reforms to Chapter 11 and related statutory provisions that will better balance the goals of effectuating the effective reorganization of business debtors – with the attendant preservation and expansion of jobs – and the maximization and realization of asset values for all creditors and stakeholders.*

According to the report, Chapter 11 works to rehabilitate companies, preserve jobs, and provide value to creditors only if distressed companies and their stakeholders actually use the Chapter 11 process to facilitate an in-court or out-of-court resolution of the company's financial distress. Chapter 11 in turn needs to offer tools to resolve a debtor's financial distress in a cost-effective and efficient manner. The following recommended major changes are summarized in the report:

Reduce barriers to entry by providing debtors more flexibility in arranging debtor-in-possession financing, clarifying lenders' rights in the Chapter 11 case, disclosing additional information about the debtor to stakeholders, and providing a true breathing spell at the beginning of the case during which the debtor and its stakeholders can assess the situation and the restructuring alternatives

Facilitate more timely and efficient diligence, investigation, and resolution of disputed matters through an estate neutral – that is, an individual that may be appointed depending on the particular needs of the debtor or its stakeholders to assist with certain aspects of the Chapter 11 case, as specified in the appointment order

Enhance the debtor's restructuring options by eliminating the need for an accepting impaired class of claims to cram down a Chapter 11 plan and by formalizing a process to permit the sale of all or substantially all of the debtor's assets outside the plan process, while strengthening the protection of creditors' rights in such situations

Incorporate checks and balances on the rights and remedies of the debtor and of creditors, including through valuation concepts that potentially enhance a debtor's liquidity during the case, permit secured creditors to realize the reorganization value of their collateral at the end of the case, and provide value allocation to junior creditors when supported by the reorganization value; and

Create an alternative restructuring scheme for small and medium-sized enterprises that would enable such enterprises to utilize Chapter 11 and would enable the court to more efficiently oversee the enterprise through a bankruptcy process that incentivizes all parties, including enterprise founders and other equity security holders, to work collectively toward a successful restructuring

Responses to the report from industry organizations and participants were mixed, and, at the time of this text, it is unclear if any such recommendations will be adopted in legislation. Still, a valuable aspect of this review is that it highlights the provisions of Chapter 11 for which there is ongoing debate as to their role in an efficient bankruptcy process.

NOTES

1. As recently as January 2019, PG&E, a large utilities provider in California, planned a bankruptcy filing due to $30 billion in expected liability costs related to recent wildfires.
2. Gilson (2009) provides a detailed note on the major changes of BAPCPA that affect businesses operating in Chapter 11. Many law firms produced informative summaries of the new Act (e.g., "New Bankruptcy Law Amendments: A Creeping Repeal of Chapter 11" (Skadden, Arps, Slats, Meagher & Flom and affiliates, March 2005) and "Immediately Effective Bankruptcy Code Amendments" (Davis, Polk, and Wordell, April 30, 2005).
3. United States Senators John Cornyn and Elizabeth Warren introduced the Bankruptcy Venue Reform Act on January 8, 2018, to prevent the forum shopping. According to the proposed bill, corporations will not be permitted to file for bankruptcy simply on the basis of their state of incorporation (which is Delaware, for many US corporations) or in a district where a bankruptcy case is pending against one of the corporation's affiliates.
4. BAPCPA took a harsh stance toward "serial bankruptcies" or "Chapter 22s." Under this act, if a company files for Chapter 11 or 7 within a year after emerging from bankruptcy, the automatic stay is lifted in 30 days unless the court grants an extension.
5. According to the US Trustee program website (https://www.justice.gov/ust), "The United States Trustee Program is the component of the Department of Justice responsible for overseeing the administration of bankruptcy cases and private trustees under 28 USC. § 586 and 11 USC § 101, et seq. We are a national program with broad administrative, regulatory, and litigation/enforcement authorities whose mission is to promote the integrity and efficiency of the bankruptcy system for the benefit of all

stakeholders–debtors, creditors, and the public. The USTP consists of an Executive Office in Washington, DC, and 21 regions with 92 field office locations nationwide." Alabama and North Carolina are not administered by the US Trustee program but utilize the bankruptcy administrator, performing a similar function to the US Trustee.

6. See "Declaration of David A. Brandon, Chairman of the Board and Chief Executive Officer of Toys 'R' US, Inc., in Support of Chapter 11 Petitions and First Day Motions," document #20 filed on September 19, 2017 on court docket of the U.S. Bankruptcy Court for the Eastern District of Virginia.

7. See Skeel (2003a) for a comprehensive review of the origin and development of DIP financing.

8. For their sample of 658 large Chapter 11 cases between 1996 and 2013, they show that DIP loans are provided by prepetition bank lenders for 40% of cases and by alternative investors for 13% of cases. However, there is a clear upward trend in the presence of alternative investors in this space.

9. See "DIP Loans to Distressed Companies Mount with Bankruptcies," Lynn Adler, Reuters, April 28, 2016.

10. See "Apollo Gains on Wall Street with Westinghouse Bankruptcy Loan," Eliza Ronalds-Hannon and Tiffany Kary, Bloomberg, April 4, 2017.

11. The ability of lenders/lessors whose loans are secured by aircraft or vessels to repossess collateral is governed by Section 1110 of the Code. The lenders/lessors have the rights to seize aircraft or vessels after the 60-day automatic stay (grace period) after bankruptcy filing. The lease is deemed to be rejected if the debtor does not make payments after 60 days.

12. The issue of preserving NOLs in Chapter 11 is limited to C corporations as most other business entities, such as S corporations and limited liability companies (LLCs), are pass-through tax entities and, thus, there are no NOLs at the company level to preserve.

CHAPTER 4

Restructuring Out-of-Court and the Cost of Financial Distress

If restructuring in bankruptcy is very costly, firms should have strong incentives to restructure out of court. What options does a firm have for restructuring out of court, and when are they or are they not feasible? Why might some firms choose to restructure in court?

When a firm is unable to complete an out-of-court reorganization, it may have no other choice but to enter a more costly, court-supervised bankruptcy proceeding. Thus, the structure of Chapter 11 that we have reviewed in the prior chapter of this text serves as the backdrop in negotiating a potential out-of-court restructuring. Academic scholars have long argued that when the costs of bankruptcy are high, claimants of a distressed firm should be able to negotiate out of court without affecting the value of the underlying firm. Many others, however, point out that there can be significant impediments to completing an out-of-court restructuring, such as creditor holdouts or conflicts between creditor groups.

In this chapter, we first review the rigorous academic studies investigating the costs of financial distress, and in particular the costs of bankruptcy. We then introduce private workouts and out-of-court restructurings, comparing the structure and benefits of these options to an in-court restructuring. We wrap up the chapter by discussing the major challenges facing out-of-court restructuring and summarize relevant academic studies on the topic.

COSTS OF FINANCIAL DISTRESS

Understanding the costs of financial distress, and particularly the costs of legal proceedings under the Bankruptcy Code, is important for many reasons. Academics have shown that distress costs are an important determinant of the pricing of a firm's debt and of its capital structure. Under the "trade-off theory" of capital structure, if expected distress costs are significant, the optimal leverage for a company may be lower. The expected costs of distress are also significant drivers of a firm's restructuring choices.

There has, however, been some debate over time as to how significant these costs might be. In recent years, the dollar magnitude of professional fees has caught the attention of the public, particularly in many of the multibillion-dollar Chapter 11 cases. For example, fees paid to advisors in the Lehman case – the largest bankruptcy case in U.S. history – nearly $6 billion,[1] and fees in the Enron case hit $1 billion. One must wonder whether these cases are outliers and whether bankruptcy is costly for a more typical firm.

Practitioners have observed that the costs of an in-court reorganization for small firms can often exceed any remaining firm value, explaining why many smaller cases are ultimately converted to a Chapter 7 liquidation. For large and complex firms, debtors often need to be advised by a large number of professionals with specific skills (see Figure 4.1 for various roles of professionals in Chapter 11 cases). In addition, these large cases often have multiple official committees, the professional fees and expenses of which are paid from the bankruptcy estate.

The costs of financial distress and bankruptcy are typically classified as either direct or indirect. *Direct costs* include out-of-pocket expenses for lawyers, accountants, restructuring advisors, turnaround specialists, expert witnesses, and other professionals. *Indirect costs* include a wide range of unobservable opportunity costs. For example, many firms suffer from lost sales and profits caused by customers choosing not to deal with a firm in bankruptcy. The bankrupt firm may also suffer from the increased costs of doing business, such as higher debt costs and poorer terms with suppliers while in a financially vulnerable position. Indirect costs also include the loss of key employees or opportunities due to management's diversion from running the business. Another potential source of indirect costs is the fire sale discount of assets sold by firms in financial distress and bankruptcy.

While direct costs are relatively easy to identify, it has not been easy for researchers to obtain the information needed to study them in a systematic way. Indirect costs are not directly observable, but there have been several studies that provide useful estimates of their potential magnitude and determinants. We do not intend to discuss asset-pricing studies that try to gauge the expected costs of distress from asset prices, but focus on those studies that provide ex post estimates of bankruptcy costs. A full list of the studies to be discussed in this section and summary of their findings is presented in Figure 4.2.

Direct Costs

Academic studies that document direct costs focus almost entirely on costs of in-court bankruptcy proceedings. One exception is Gilson, John, and Lang (1990),

Service	Investment Bankers	Crisis Managers	Accountants
M&A	●	◕	○
Capital raising	●	◕	○
Valuation	●	◕	○
Debt capacity/capital structuring	●	◕	○
Negotiation with creditors	●	◑	○
Financial modeling	●	◕	◕
Liquidation analysis	●	◑	●
Bankruptcy Court testimony	●	◑	◕
Strategic business analysis	◑	●	◕
Analysis of financial controls	○	●	●
Day-to-day business analysis	○	●	◕
Audits	○	○	●
Pension issues	○	○	●
Financial reporting – SEC and Bankruptcy Court	○	○	●
Operate plants or business	○	◕	○
Hire, fire, or manage employees	○	●	○
Sell company's goods or services/collect receivables	○	◕	○
Collect receivables	○		○

FIGURE 4.1   Role of Professionals in Chapter 11 Cases
Source: Lazard.

who estimate direct costs in the range of 0.65% of assets for a sample of large public companies that undergo out-of-court exchange offers (as defined below). These results suggest relatively low direct costs for this type of restructuring.

The difficulty in measuring direct costs for firms restructuring in bankruptcy is that there is no centralized source listing all filing firms and information on costs (with the notable exception of the Administrative Office of the U.S. Courts and the Executive Office for U.S. Trustees, which do not make information publicly available). Researchers have only been able to compile this information by obtaining documents from individual federal bankruptcy courts. Studies of direct costs, therefore, have been based on samplings of cases, often for larger firms.

Study	Sample Used to Calculate Costs	Bankruptcy/ Workout	Time Period	Estimated Costs
Direct costs				
Warner (1977)	11 bankrupt railroads; estimated mean market value $50 million at filing	Section 77 of the Bankruptcy Act	1933–1955	Mean 4% of market value of firm one year before default
Ang, Chua, and McConnell (1982)	86 liquidations, Western District of Oklahoma; estimated mean prebankruptcy assets $615,516	Bankruptcy liquidation	1963–1979	Mean 7.5% (median 1.7%) of total liquidating value of assets
Altman (1984a)	19 Chapter X and XI cases; mean assets $110 million before filing	Chapter X and Chapter XI	1974–1978	Mean 4% (median 1.7%) of firm value just before bankruptcy for 12 retailers; 9.8% (6.4%) for 7 industrial firms
Gilson, John, and Lang (1990)	18 exchange offers (from a sample of 169 distressed firm restructurings)	Workout	1978–1987	0.65% average offer costs as a percentage of book value of assets (maximum 3.4%)
Weiss (1990)	37 Chapter 11 cases from 7 bankruptcy courts; average total assets before filing $230 million	Chapter 11	1980–1986	Mean 3.1% (median 2.6%) of firm value before filing
Lawless, Ferris, Jayaraman, and Makhija (1994)	57 small-firm Chapter 7 cases (mean asset value at filing $27,797) and Chapter 11 cases (mean asset value at filing $409,102) filed in Western District of Tennessee	Chapter 7 and Chapter 11	1981–1991	43% of firm value at filing in Chapter 7 and 22% in Chapter 11
Tashjian, Lease, & McConnell (1996)	39 prepackaged Chapter 11 cases; mean book value of assets FYE before filing $570 million	Prepackaged Chapter 11	1986–1993	Mean 1.85%, median 1.45% of book value of assets at fiscal year-end before filing

FIGURE 4.2   Studies on the Estimates of Direct and Indirect Costs of Distress

Study	Sample Used to Calculate Costs	Bankruptcy/ Workout	Time Period	Estimated Costs
Betker (1997)	75 "traditional" Chapter 11 cases; 48 prepackaged Chapter 11 cases; 29 exchange offers; mean assets FYE before restructuring $675 million	Chapter 11 and workout	1986–1993	Prepackaged bankruptcies: mean 2.85% (median 2.38%) of prebankruptcy total assets; traditional Chapter 11s: mean 3.93% (median 3.37%); exchange offers: 2.51% (1.98%)
Lawless and Ferris (1997)	98 Chapter 7 cases from 6 bankruptcy courts; median total assets $107,603	Chapter 7	1991–1995	Mean 6.1% of total assets at filing (median 1.1%)
Luben (2000)	22 Chapter 11 cases; median assets $50 million	Chapter 11	1994	Cost of professional fees in Chapter 11 averages 1.8% (median 0.9%) of total assets at beginning of case; 2.5% excluding prepacks
LoPucki and Doherty (2004)	48 Delaware & S.D.N.Y Chapter 11 cases; mean assets at filing $480 million	Chapter 11	1998–2002	Mean professional fees equal 1.4% of assets at beginning of case
Bris, Welch, and Zhu (2006)	225 Chapter 11 cases (mean prebankruptcy assets $19.8 million) and 61 Chapter 7 cases (mean prebankruptcy assets $501,866) filed in Arizona and S.D.N.Y.	Chapter 7 and Chapter 11	1995–2001	Chapter 7: mean 8.1%, median 2.5% of prebankruptcy assets; Chapter 11: mean 16.9%, median 1.9% of prebankruptcy assets
LoPucki and Doherty (2008)	74 large Chapter 11 filings in various courts (mean assets $510 million)	Chapter 11	1998–2003	Mean 1.1% of prebankruptcy assets
Indirect costs				
Altman (1984a)	19 Chapter X and Chapter XI cases	Chapter X and Chapter XI	1974–1978	Difference between estimated and actual profits; mean 8.2% (median 5.7%) of firm value measured just before bankruptcy for 12 retailers, 13.9% (9.7%) for industrial firms

FIGURE 4.2   *(Continued)*

Study	Sample Used to Calculate Costs	Bankruptcy/ Workout	Time Period	Estimated Costs
Opler and Titman (1994)	Highly leveraged firms in distressed industries	Distressed firms	1974–1990	Highly leveraged firms in distressed industries experience 26% greater sales decline than firms with low leverage; financial distress costs are positive and significant
Borenstein and Rose (1995)	4 US airlines in Chapter 11	Chapter 11	1989–1992	Lower pricing only in financial distress before Chapter 11 filing (but not in Chapter 11)
Andrade and Kaplan (1998)	31 highly leveraged transactions that subsequently become distressed	LBOs	1987–1992	10%–20% of firm value
Maksimovic and Phillips (1998)	302 Chapter 11 cases (owning 1,195 plants)	Chapter 11	1978–1989	No real economic costs attributable to Chapter 11 itself
Pulvino (1999)	27 US airlines, 8 of which are in Chapter 7 or 11	Chapter 7 and Chapter 11	1978–1992	Prices bankrupt airlines receive for sales of used aircraft are lower than prices received by distressed but nonbankrupt firms
Ciliberto and Schenone (2012)	29 airlines in Chapter 11 and 2 in Chapter 7	Chapter 7 and Chapter 11	1992–2007	Bankrupt carriers drop prebankruptcy routes by 25% and drop prices by 3.1% in bankruptcy reorganization
Hortaçsu, Matvos, Syverson, and Venkataraman (2013)	Distressed auto manufacturers	Distressed firms	2006–2008	A 30% increase in GM's CDS spread would cost GM North America 10% of its value
Phillips and Sertsios (2013)	21 airlines with 13 never in bankruptcy, 7 in bankruptcy once, and 1 in bankruptcy twice	Chapter 11	1997–2008	Bankrupt carriers drop service quality (in mishandled bags) in financial distress and reduce prices

FIGURE 4.2 (*Continued*)

Study	Sample Used to Calculate Costs	Bankruptcy/ Workout	Time Period	Estimated Costs
Taillard (2013)	15 large firms in financial distress due to asbestos litigation (12 in Chapter 11)	Distressed firms and Chapter 11 firms	2000–2002	A net gain of 5%–35% of value pre-distress
Brown and Matsa (2016)	145 firms during the Great Recession (2008–2010)	Distressed firms	2008–2010	Financial distress results in fewer and lower-quality applicants for jobs at distressed firms
Graham, Kim, Li, and Qiu (2016)	190 Chapter 7 and Chapter 11 cases in UCLA – LoPucki's BRD	Chapter 7 and Chapter 11	1992–2005	10% reduction in employee earnings

FIGURE 4.2 *(Continued)*

One of the earliest attempts to measure direct costs is by Warner (1977). He examines payments for legal fees, professional services, trustees' fees, and filing fees for 11 bankrupt railroads filing under Section 77 of the Bankruptcy Act between 1933 and 1955. These cases took, on average, 13 years to settle, and the direct costs are estimated to average 4% of the market value of the firm one year before default.

Weiss (1990) obtained documents from seven bankruptcy courts, including the Southern District of New York. Based on his examination of 37 cases between 1980 and 1986, all of which were NYSE or AMEX firms, he estimates that direct costs of bankruptcy average 3.1% of the book value of debt plus the market value of equity at the fiscal year end before the bankruptcy filing, with a range from 1 to 6.6%. Weiss interprets these figures as relatively low direct costs, which would be expected to have little or no impact on the pricing of claims before bankruptcy. Several other studies report average direct costs in the range of Weiss's study. These include Ang, Chua, and McConnell (1982) (7.5% of the total liquidating value of assets), Altman (1984a) (6.1% of firm value for his full sample), Betker (1997) (3.9% of prebankruptcy total assets for nonprepackaged bankruptcies), and Lawless and Ferris (1997) (6.1% of assets at filing).

Three legal studies report professional fees for relatively large public companies in Chapter 11. Lubben (2000) reports, for 22 firms filing in 1994, that the cost of professional fees in Chapter 11 is 1.8% of a distressed firm's total assets, with some cases reaching 5%. LoPucki and Doherty (2004) find professional fees equal to 1.4% of a debtor's total assets at the beginning of the bankruptcy case for 48 large cases filed in Delaware and New York Southern District. In addition, LoPucki and Doherty (2008) examine the legal fees of 74 large Chapter 11 filings in various courts and find that they account for 1.1% of prebankruptcy assets, largely consistent with their earlier estimates.

There are a number of studies that pay particular attention to estimating the direct costs of smaller nonpublic firms and examine both Chapter 7 and Chapter 11 cases. Lawless, Ferris, Jayaraman, and Makhija (1994), studying 57 small Chapter 7 and Chapter 11 cases filed between 1981 and 1991, find that direct costs account for 43% and 22% of firm value at filing in Chapter 7 and Chapter 11, respectively. These numbers are unusually large compared to those for large firms. Bris, Welch, and Zhu (2006) examine over 300 cases from two bankruptcy courts, Arizona and Southern District of New York. They find that, for Chapter 7 cases, direct bankruptcy expenses are estimated to have a mean of 8.1% of prebankruptcy assets (median of 2.5%). However, bankruptcy professionals (attorneys, accountants, and trustees) regularly end up with most of the postbankruptcy firm value in Chapter 7. Based on their estimates of postbankruptcy remaining value, in 68% of the Chapter 7 cases the bankruptcy fees "ate" the entire estate. For Chapter 11 cases, Bris et al. find that direct costs have a mean of 16.9% of prebankruptcy assets (median 2%). The only other published study to consider smaller Chapter 7 and Chapter 11 cases is Lawless et al. (1994); in comparison with the earlier studies, their findings suggest that there may be substantial fixed costs associated with the bankruptcy process.

Overall, several important facts emerge from these studies. First, there is likely to be an important scale effect. While much of the research on this question focuses on larger public companies, smaller firms may be unable to survive a reorganization process given the magnitude of fees relative to their assets. Second, the dollar amount of fees for large public companies can be tremendous, even though, as a percentage of assets, these fees are not large. Still, even when the percentage of direct costs is low, the indirect costs of financial distress may be significant. Third, as more data becomes publicly available in electronic form, our ability to measure and monitor these costs will improve over time.

Indirect Costs

In contrast to direct costs, indirect costs are not easily observable and are therefore difficult to specify and empirically measure. However, researchers have developed several approaches to infer the likely magnitude of these costs. A key measurement problem is that we cannot distinguish whether the poor performance of a firm is caused by the financial distress itself (and therefore is an indirect cost) or whether it is caused by the same economic factors that pushed the firm into financial distress in the first place. These studies therefore attempt to identify whether firm performance reflects the costs of financial distress, the costs of economic distress, or an interaction of the two. Moreover, these studies employ unique samples and experiments to distinguish the costs of financial distress from the costs of bankruptcy.

Altman (1984a) is the first to provide a proxy methodology for measuring the indirect costs of bankruptcy. He defines indirect costs as those lost sales and profits caused by customers choosing not to deal with a firm that has high likelihood of bankruptcy as well as the increased costs of doing business (e.g., higher debt costs and poorer terms with suppliers) while in a financially vulnerable condition. The firms examined include samples of 19 industrial and retail firms that went bankrupt in the 1970–1978 period and a second group of seven large, industrial bankruptcies from the early 1980s. He finds that indirect costs average 10.5% of firm value measured just before bankruptcy. The combined direct and indirect costs average 16.7% of firm value, indicating that total bankruptcy costs are not trivial.

Subsequent to Altman's initial work, several other studies have attempted to isolate indirect distress costs, each using quite different methodologies and data sets. Their key insight is to recognize that it is important to separate the effects of financial from economic distress. For example, while Altman documents large declines in profitability, he cannot distinguish them from negative operating shocks.

Andrade and Kaplan (1998), using methodology similar to that of Kaplan (1989, 1994), examine 31 firms that become distressed subsequent to a management buyout or leveraged recapitalization between 1980 and 1989. At the onset of distress, having recently completed a highly leveraged transaction, the firms in their sample are largely financially distressed but not economically distressed. Thus, their research design provides an opportunity to isolate the costs of pure financial distress.[2] Based on changes in firm value over time, they estimate the net costs of financial distress to be 10 to 20% of firm value; for firms that do not also experience an adverse economic shock, the costs of financial distress are negligible. In addition, they find that distress costs are concentrated in the period after a firm becomes distressed, but before it enters Chapter 11, suggesting that it is not Chapter 11 itself that contributes to indirect costs.

Andrade and Kaplan also examine qualitative aspects of the behavior of distressed firms. A number of firms are forced to cut capital expenditures substantially, sell assets at depressed prices, or delay restructuring or filing for Chapter 11 in a way that appears to be costly. Taillard (2013) follows Andrade and Kaplan's methodology to infer the indirect costs of financial distress using 15 large, financially distressed firms that are exposed to asbestos litigation around 2000. Surprisingly, he finds that distress creates a net gain of 5 to 35% of firm value. However, the specific sample employed by the study limits the scope of applications.

Opler and Titman (1994) recognize the reverse causality problem in studies that attempt to relate performance declines and financial distress. Their approach is to identify depressed industries that have experienced economic distress, based on negative industry sales growth and median stock returns below −30%. Within those industries, they investigate whether firms that are highly leveraged before the onset of the distressed period fare differently than their more conservatively financed counterparts. Their hypothesis is that if financial distress is costly, more highly leveraged firms will have the greatest operating difficulties in a downturn. They find that highly leveraged firms lose market share and experience lower operating profits than their less leveraged competitors. Although they do not provide specific estimates of the level of indirect costs, their tests minimize the reverse causality problem, making it difficult to interpret some of the previous work. They interpret their findings as being consistent with the view that the indirect costs of financial distress are significant and positive.

The airline industry has been a fertile ground for studying the effect of financial distress and bankruptcy on product market outcomes such as pricing, market share, and product quality (such as flight delays, cancellations, and mishandling of bags) as well as the transaction prices of aircraft, largely due to the fact that the industry is closely regulated. Borenstein and Rose (1995) investigate the pricing behavior of four major airlines in bankruptcy (Eastern, Continental, America West, and TWA) from 1989 to 1992 and find that bankruptcy filing hardly affects airline pricing, although financial distress preceding a Chapter 11 filing appears to be associated with lower prices. In contrast, rivals increased prices in the quarter when these airlines declared bankruptcy. Ciliberto and Schenone (2012) engage a much larger sample of 31 airlines that filed for Chapter 11 or Chapter 7 between 1992 and 2007 to study a variety of product market outcomes. They find that bankruptcy carriers permanently drop approximately 25% of their prebankruptcy routes and lower the average frequency of flights within a route by 21%. In addition, they find that airlines in bankruptcy drop prices by 3.1% but immediately increase prices by 5% after emerging. Using quarterly data of 21 airlines from 1997 to 2008, Phillips and Sertsios (2013) find that prices decline in financial distress and further show that service quality, measured by the number of mishandled bags and flight delays, declines in financial distress but increases in bankruptcy, suggesting that airlines invest in customer retention and reputation in bankruptcy.

Pulvino (1998, 1999), on the other hand, studies sales of aircraft by distressed versus nondistressed airlines. He finds that financially constrained airlines receive lower prices than their unconstrained rivals when selling used aircraft. In contrast to Andrade and Kaplan (1998), however, he further finds that, for airlines in either Chapter 7 or Chapter 11, the prices that the bankrupt airlines receive for their used aircraft are generally lower than the prices received by distressed but non-bankrupt firms (14–46% discount). In a similar vein, using prices of durable goods transactions, Hortaçsu, Matvos, Syverson, and Venkataraman (2013) examine the changes in used car prices in response to spreads in credit default swaps (CDS). Their argument is that financial distress disrupts a durable goods producer's provision of complementary goods and services such as warranties, parts, and upgrades. Financial distress threatens the loss of these valued amenities and thus reduces customers' willingness to pay for a firm's core products. They find that used car prices are lower when a car manufacturer's CDS spread is higher. The effect is more pronounced for cars with longer expected service lives.

While these studies indicate that financial distress is costly, they differ in their conclusions as to how bankruptcy status, itself, influences these costs. The idea that financial distress, and not Chapter 11 per se, leads to a loss in value is further supported by Maksimovic and Phillips (1998). These authors use plant-level data, obtained from the U.S. Census Bureau, to examine the productivity and plant-closure decisions of bankrupt firms. They find that Chapter 11 status is much less important than industry conditions in explaining the productivity, asset sales, and closure conditions of Chapter 11 bankrupt firms. In declining industries, the productivity of plants in Chapter 11 bankruptcy and subsequent to emerging does not significantly differ from that of their industry counterparts, nor does it decline during Chapter 11. This suggests that few real economic costs are attributable to Chapter 11 itself and that bankruptcy status is not important to indirect costs. This further suggests that firms that complete an out-of-court restructuring may still incur significant indirect costs of financial distress.

A few recent studies examine the indirect costs of bankruptcy that are related to the loss of human capital and labor. For example, Graham, Kim, Li, and Qiu (2016) show that employees experience a 10% reduction in earnings after a firm files for bankruptcy. Affected employees are likely to work fewer hours and even leave the local labor market. Relatedly, Brown and Matsa (2016) document a strong, positive relation between corporate distress (measured by CDS spreads) and the number and quality of applicants for a given job position at the distressed firm, reflecting the difficulty of a distressed firms to attract new skilled labor. These and other related studies on labor issues are discussed in detail in Chapter 6.

Summary and Recent Observations

In sum, this work suggests that while the direct costs of bankruptcy are only a small percentage of firm value for large, public companies, for smaller firms the costs may be prohibitive and lead to liquidation. Further, research shows that, even for larger firms, the indirect costs of distress can be significant. It is also important to consider that indirect costs are not limited to firms that actually default or enter Chapter 11. In fact, bankruptcies often create negative externalities to other firms such as suppliers, firms in the same industry, and even those that operate in the local market.

Difficulties in restructuring complicated capital structures have also been evident from many recent Chapter 11 cases and have contributed to the rising costs. For example, a number of large cases have involved firms with multiple bank groups and bond issues at different levels in corporate structure and with lenders having different levels of security (e.g., first lien, second lien), a situation that increases the likelihood of conflicts between claimants. The size of creditor groups may make reaching a consensus difficult. Conflicts also occur between prebankruptcy and more recent entrants into a case, such as vulture investors, who may have purchased claims at prices substantially below par. When conflicts between claimants lead to increased difficulty in negotiating a restructuring, both the direct and the indirect costs of distress are likely to increase.

RESTRUCTURING OUT OF COURT

In this section, we discuss the options for restructuring debt out of court, the advantages and disadvantages of out-of-court restructuring over bankruptcy restructuring, the major challenges faced by companies of successfully restructuring, and how to structure an exchange offer. We reference the most related academic studies on these topics in our discussions.

Restructuring Options

A company that intends to restructure its debt out-of-court can do so through recapitalization via a capital injection, debt repurchase through open market purchase or a tender offer, and structuring an exchange offer. For some companies, all three options are available, while for others that are severely distressed, the exchange offer may be the only available option.

Capital Injection The option that a company often considers first to resolve temporary financial distress is to seek current security holders or outsiders for capital infusion. Even though a capital injection results in large dilution to existing equity ownership, it may provide the necessary financial relief for the company and can allow the company to grow out of its distress. Therefore, capital injections are expected to have a positive effect on the value of existing equity.

For more severely distressed or insolvent firms, absent a balance sheet restructuring to reduce the level of debt, investors may be unlikely to infuse equity capital. In such a case, some portion of the cash invested goes to reducing the impairment of the debt holders, and only the portion of the cash invested that exceeds the debt holders' impairment contributes to the value of the equity investor's stake. Thus, investing in equity of an overlevered company simply transfers wealth from the equity investor to the debt holders, and creates an immediate loss of value for the investor. In economics, this phenomenon is termed "debt overhang" and makes it impossible for such a firm to issue additional equity (Myers, 1977).

Debt Repurchase A financially distressed company can acquire its outstanding debt through open market purchases or in privately negotiated transactions (*open market purchase*), or it can make a tender offer to purchase some or all of its outstanding debt (*tender offer*). Either option allows the company to retire some of its debt at a significant discount to the face value of the debt. A cash tender offer requires the company's compliance with the tender offer rules of the Securities Exchange Act of 1934. To avoid the repurchase being classified as a tender offer and the required compliance, a company can make the purchase in an open market transaction. The company must carefully structure its repurchase program, such as approaching a limited number of holders and repurchasing on separate terms and prices from different holders. A repurchase that is later found to be noncompliant could expose the company to sanctions.

Distressed Exchange   When a company cannot solicit investors for a capital infusion or does not have enough cash to restructure debt through capital infusion or debt repurchase, making a distressed debt-equity exchange becomes an attractive option. With this option, the company makes an offer to all (or some) classes of debt to exchange newly issued debt and newly issued or existing equity securities for the outstanding debt securities. The fair value of the new securities given to old debtholders is most likely at a discount to the face value of the old debt securities. The issuance of new securities in an exchange offer is considered an offering of new securities under the Securities Exchange Act of 1934 and must be registered unless an exemption is given. The distressed exchange may be coupled with a prepackaged filing. If a company fails to get a sufficient amount of debt exchanged, it may choose to file for a prepackaged or a prenegotiated Chapter 11 (see Chapter 3 herein).

The Advantage and Disadvantages of Out-of-Court Restructuring versus Bankruptcy

As discussed in Chapter 3, there are many benefits of an in-court bankruptcy restructuring to the debtor. For example, the automatic stay stops secured lenders' foreclosure and suspends all interest and principal payments on prepetition debt until a reorganization or liquidation plan is reached. Through the assumption, rejection, and reassignment of leases, a company is able to close underperforming plans and stop undesirable leases. Compared to an in-court restructuring, however, an out-of-court workout has many desirable features. The most important feature is the savings on professional fees and expenses, which may amount to millions of dollars for large companies.

The estimates of direct costs for exchange offers documented by Gilson, John, and Lang (1990) are quite low compared to those for companies going through bankruptcy. Importantly, the development of prepackaged bankruptcies, which are viewed as a hybrid of a more traditional Chapter 11 and an out-of-court restructuring (see Chapter 3 herein), has given firms the ability to negotiate a Chapter 11 plan before filing, allowing them to exit bankruptcy within months rather than years, thereby resulting in significant savings. In studying prepackaged cases filed in the 1980s and early 1990s, Tashjian, Lease, and McConnell (1996) and Betker (1997) suggest that prepackaged bankruptcy costs are lower than in traditional Chapter 11 and are comparable to those in out-of-court restructuring.

Compared to traditional bankruptcy, out-of-court restructuring creates less damage to a company's reputation; customers, suppliers, and employees are more likely to view the viability of the business more favorably in an out-of-court restructuring versus an in-court transaction. Further, management does not spend time dealing with court hearings or gathering information for court-appointed committees. A less talked-about, undesirable feature of Chapter 11 filing, compared to out-of-court restructuring, is that a formal bankruptcy filing could lead to a substantial increase in the amount of claims because claims that otherwise would not be immediately made, such as unfunded pensions, employee injuries, and disputed payments, would be filed. One caveat about the in-court restructuring is that the discretion and biases of judges may create an uncertain outcome. Given the advantages of the out-of-court restructuring and the disadvantages of Chapter 11, many companies opt to seek a private workout with lenders. However, they face significant hurdles and challenges in structuring a successful out-of-court restructuring.

Aside from bankruptcy cost concerns, tax implications need to be carefully considered. Debt repurchases and exchange offers may result in the recognition of cancellation of debt (COD) income, which is not charged in a bankruptcy restructuring unless the company applies net operating losses (NOLs) for tax purposes. However, the actual tax to be charged on COD income depends on whether the company was "insolvent" and the excess of the amount owed on the outstanding debt over the fair market value of newly issued debt and equity.

Challenges of Out-of-Court Restructuring

Out-of-court restructuring has a number of challenges, such as creditor coordination, the holdout problem, empty creditors, information problems, and conflicts of interests. These challenges make it difficult for companies to successfully restructure debt claims out of court. However, studies have found that there are potential solutions to mitigate these challenges to facilitate a cost-effective workout.

Creditor Coordination Academic studies point out that coordination problems among public bondholders introduce investment inefficiencies and affect the likelihood of workout (Gertner and Scharfstein, 1991). They find that the fraction of bank debt in the capital structure and the complexity of debt structure affect the ways in which distressed companies restructure (Gilson, John, and Lang, 1990; Asquith, Gertner, and Scharfstein, 1994; Bolton and Scharfstein, 1996). Particularly, companies with both secured private debt and unsecured public bonds find it more challenging to achieve out-of-court restructuring due to the creditor coordination problem.

Holdout Problem It is important to note that a distressed exchange offer is binding only on those creditors who agree to participate. Creditors that hold blocks of a claim may have the intention of not participating in the out-of-court restructuring and subsequently benefiting from the full contractual payment for the securities that are not included in a successful out-of-court restructuring. This situation is known as the *holdout problem*. A holdout can complicate a company's efforts to obtain the minimum acceptance rate for its exchange offer to be effective.

However, these holdout creditors will gain little and may lose their investment if the restructuring is unsuccessful and the firm is forced into bankruptcy or liquidation. Furthermore, a natural result of a successful exchange offer is a huge reduction in the liquidity of the old securities that remain outstanding, and this may have an adverse impact on trading ability. This potential outcome would encourage holders to exchange their debt.

In addition to these natural concerns for debtholders, companies can adopt other strategies to mitigate the holdout problem. First, the Trust Indenture Act of 1939 requires unanimous debtholder consent before a firm can make changes to terms such as the principal, interest, or maturity of the public debt. However, for nonpricing-related terms, majority or super-majority is typically required, depending on the specific contract. Companies often solicit "exit consents" simultaneously with an exchange offer. For example, companies may propose amendments to an old indenture agreement to subordinate the old securities. If the consent solicitation were successful, debtholders who refuse to accept the offer would continue to hold their claims with the payment terms intact but are likely to lose seniority or other provisions that are important to protect their claims.

A second strategy is that a company can create incentives for those holders who tender early. For example, the company can establish an early tender payment or early consent payment. These are important means of encouraging holders to tender their debt securities early.

Third, a more coercive strategy is to couple an exchange offer with a disclosure statement and solicitation of votes for a prepackaged Chapter 11. Since approval of a plan requires only half in number and two thirds in value of each class of claim and it is binding for all creditors, this strategy allows a company to create a credible threat to holdout creditors. They may be convinced that they will be forced to accept the exchange in a bankruptcy.

Credit Default Swap (CDS) and Empty Creditors CDS allows investors to insure against potential losses in their holdings of (unsecured or secured) claims on borrower default. The occurrence of a credit event specified in a contract, typically the filing of bankruptcy of the reference entity, triggers contractual payment from the CDS seller to the buyer to compensate the insurance buyer investors for the loss on their bondholding. The vast majority of the CDS market involves senior unsecured obligations. Recently, legal scholars and economists (Hu and Black, 2008a, 2008b; Bolton and Oehmke, 2011) noted that the existence of "empty creditors," those that hold credit default swaps and have strong incentive not to negotiate with a company out of court, can significantly impede a company's ability to succeed in private workouts.

An *empty creditor* is an individual (or institution) who may have the contractual control but, by simultaneously holding CDS, faces little or no economic exposure if the debt goes bad. Since, according to most CDS contracts, only a bankruptcy event (not an out-of-court restructuring)[3] would trigger the contract payment from the CDS seller to the buyer, bankruptcy filing is more attractive than an out-of-court restructuring for creditors holding CDS as insurance. These parties have incentive to push the company into bankruptcy to receive full payment. At times, the amount of CDS contracts is so large relative to the amount of underlying bonds outstanding that it is hard to believe that the holders bought these contracts for insurance purposes. For example, in Delphi's 2005 bankruptcy filing, the company had approximately $2 billion of senior debt outstanding. However, the total CDS outstanding was about 14 times larger (about $28 billion).

In a number of high-profile bankruptcy cases (e.g., Six Flags' 2009 bankruptcy and Caesars' 2014 bankruptcy), the empty creditor problem was blamed for causing the out-of-court restructuring attempt to fail. However, academic studies seem to find mixed evidence of the presence of empty creditor problems. On the one hand, Subrahmanyam, Tang, and Wang (2014) show that firms with CDS outstanding have a higher probability of filing for bankruptcy. Danis (2016) shows that the participation rate among public bondholders in distressed exchange offers is significantly lower if a firm has CDS being traded on its debt. On the other hand, Bedendo, Cathcart, and El-Jahel (2016) show that a firm's choice between out-of-court restructuring and bankruptcy is not related to whether there are CDS contracts on the issuer's debt.

Other Challenges Impeding the Success of Out-of-Court Restructuring Information asymmetry exists between company management and debtholders, and sometimes bankruptcy cannot be avoided, even if it imposes larger deadweight costs (Giammarino, 1989). Private information can interfere with the efficiency of recontracting with debtholders as debtholders would rather appeal to a costly arbitrator than put their trust in management and shareholders. Brown, James, and Mooradian (1993) suggest that offering equity to bondholders in distressed exchange offers conveys private information about a company's prospects; however, offering equity to private lenders who are better informed conveys favorable information to outsiders. Their study of market reactions confirms the signaling effect of offering equity to private versus public lenders.

Another challenge that companies face is how to align the interests of debtholders and equity holders. These holders have different preferences about a firm's risk-taking for investment decisions in distress. Debtholders may have a stronger incentive to take over the restructuring process to enforce governance changes, thereby realigning the interests of management with creditors (not the old shareholders), even though it means a costly in-court restructuring. A study by Chu et al. (2018) finds that the incentive alignment of creditors and shareholders, measured by financial institutions' simultaneous holding of debt and equity, is strongly associated with the likelihood of restructuring out of court.

Structuring an Exchange Offer

To structure an exchange offer, a distressed company first needs to carefully examine its capital structure, and study the incentives and motives of various classes of claim holders. It is important to identify the classes of debt which are indifferent to in-court versus out-of-court restructuring and those who would be most affected by a bankruptcy proceeding. Second, the company needs to investigate investors' views on the company's prospects through their pricing of debt and equity securities. For example, perhaps the investors are too pessimistic and thus underprice the debt and equity. Security pricing may create both opportunities and challenges for a successful distressed exchange. More important, the company needs to carefully review the terms and conditions of its outstanding debt to determine whether covenants in its credit agreements or bond indentures restrict its ability to redeem securities at a discount.

Analyzing Security Holders' Incentives Typically, an out-of-court restructuring for a small to medium-sized company means negotiating with loan lenders to obtain their consent for maturity extensions or refinancing of the loans as the company's liabilities are mostly composed of bank loans and trade debt. However, it is more challenging for a large firm to strike a deal with the holders of many classes of debt due to the many challenges we discussed above. It must negotiate with not only bank lenders but also institutional lenders and bondholders, who often have a tight timeline for completing the negotiation process and striking a deal.

In an out-of-court restructuring, the incentives of all creditors should generally be aligned. That requires the enterprise value of the restructured company (i.e., the "size of the pie") to be greater than it would be if restructuring in a formal bankruptcy, and as a result, all claimants will likely benefit by avoiding bankruptcy filing. If this condition does not hold, at least one class of claimants will likely lose out and oppose an out-of-court restructuring. Here we separately analyze the incentives of secured lenders, unsecured bondholders (e.g., senior unsecured bondholders), trade credit holders, equity holders, and management.

The incentive of secured lenders depends on whether they are sufficiently protected. If their claims are adequately protected (i.e., overcollateralized), they can recover their claims regardless of the restructuring outcome (out-of-court restructuring or Chapter 11). In addition, secured claims accrue interest in bankruptcy, and prepetition lenders are in an advantageous position for providing post-petition financing. Secured loan lenders may even have a slight preference for a firm to restructure in court. They normally have no incentive to make concessions to enable an out-of-court deal to work. Warrants and other option-like securities are awarded at times to lenders as a sweetener so that they will agree to a workout deal. In contrast, if the claims of secured lenders are not sufficiently protected (i.e., undercollateralized) – that is, when junior creditors are out-of-money and secured lenders are expected to be most affected by bankruptcy costs – they have a stronger preference to strike an out-of-court deal.

James (1995) examines cases where banks take concessions by taking equity in debt restructurings. He finds that, for firms with public debt outstanding, banks never make concessions unless public debtholders make concessions. However, when banks take equity in restructuring, they tend to obtain a large proportion of the firms' stock. Subsequent work (James, 1996; Demiroglu and James, 2015) finds that banks' participation and forgiveness have a significant impact on the likelihood of success of exchange offers. The effect of bank participation on restructuring outcomes depends on whether loans are diffusely held and securitized. For public unsecured bondholders, two conditions need to be satisfied for their incentives to be aligned with an out-of-court restructuring: (1) their claims are out-of-money, and (2) restructuring in Chapter 11 must be more costly. When both of these conditions hold true, unsecured creditors bear the costs of a Chapter 11 process. Note that unsecured creditors do not accrue interest in Chapter 11. This provides even stronger incentives for them to agree to an out-of-court restructuring. For trade creditors, their preference is to have a successful workout, even though they play a minimal role in the out-of-court restructuring process.

Although filing for Chapter 11 is not often preferred by shareholders, evidence from Chapter 11 bankruptcies suggests that shareholders can sometimes extract payments before other creditors are paid in full, resulting in deviations from the absolute priority rule of the U.S. Bankruptcy Code (see Chapter 3). At times, the court may grant approval of official equity committees to represent shareholders' interests in bankruptcy proceedings. This practice provides equity holders with some bargaining power in a private workout.

Management typically prefers an out-of-court restructuring to Chapter 11; it is seldom in management's interests to file for bankruptcy immediately if an out-of-court restructuring can be achieved quickly and economically. In Chapter 11, management is restricted from taking actions which are outside the firm's ordinary course of business without the court's approval. However, management may threaten to file for bankruptcy if the out-of-court restructuring fails because Chapter 11 affords certain benefits, including the selection of a debtor-friendly bankruptcy court, the ability to continue running the business, and the exclusivity period to file a plan, putting management in a powerful bargaining position. Recent evidence on the turnover and compensation of chief executive officers (CEOs) around bankruptcy filing (Eckbo, Thorburn, and Wang, 2016) shows that more than 80% of incumbent CEOs (i.e., those in position at the beginning of three years before bankruptcy filing) are replaced from two years before bankruptcy to plan confirmation. The evidence suggests that a CEO's bargaining and control power has diminished significantly in recent decades, partly due to enhanced creditor control of the bankruptcy process.

Other Factors to Consider The key to a successful exchange offer is to make the new securities more attractive than the old securities. For example, the new debt securities may carry higher coupon rates or higher seniority. Furthermore, a company needs to identify the class of debt for which the holdout problem may be the most severe. The company also needs to identify which class of claims is likely to be the fulcrum security, the security that is most likely to receive major equity ownership after restructuring, in the capital structure.

When offering new debt at a discount, the new debt should be senior to the existing debt with respect to security or maturity and carry higher coupons than the existing debt. Exchanging debt securities with debt securities is often not sufficient for a company to significantly reduce its leverage ratio, and, therefore, equity is used in combination with debt securities. However, equity can be a challenging currency for debtholders to accept in an exchange offer because, by exchanging their claims, they are effectively subordinated to other debtholders that do not participate or do participate and receive new debt claims. Therefore, before the exchange offer, it is essential to design and award new securities to maintain the seniority structure of the debt claims. For example, the company may issue new senior debt to old senior debtholders, while awarding new junior debt and a significant portion of equity to junior debtholders.

In this situation, it is important to identify the institutional debtholders (e.g., banks, insurance companies, hedge funds). Are they passive investors? Are they par holders? Did they purchase debt at below face value in the second market? Are they "loan-to-own" lenders? Different types of investors have divergent interests in the restructuring outcome. They often believe that they should be paid in full before any junior holders receive any value. At times, debtholders may reject proposals that they deem unfair even though accepting such proposals is in their best economic interests.

The ultimate success of an exchange offer rests on management's careful consideration and assessment of the debt structure, the going-concern value, the ongoing financing requirements, the heterogeneous incentives of different classes of creditors, the preferences of managers and shareholders, the bargaining power of the different parties, and the quality and credibility of management. All these factors interact to shape the final restructuring outcome. Financial modeling and tools are only some of the elements involved in out-of-court restructuring. The process requires significant bargaining and negotiation skills.

There are a number of good examples of large firms successfully accomplishing an out-of-court restructuring to avoid a bankruptcy filing. One notable case is the Ford Motor Company's restructuring in March 2009. The company used a combination of a debt repurchase program and exchange offers to successfully reduce its indebtedness by about $10 billion and cut its annual interest expenses by more than $500 million. The outcome was that Ford was the only one of the Big Three US automakers to avoid bankruptcy and gained a fast recovery after the recession of 2008–2009. Its share price increased sixfold within a year of the successful restructuring, while its two competitors, General Motors and Chrysler, entered into bankruptcy.

NOTES

1. See "Fed Says Lehman Brothers Chapter 11 Case is Costliest in History", by Katy Stech Ferek, January 16, 2019
2. Cutler and Summers (1988) attempt to isolate the effects of financial conflict from economic distress. They document the significant wealth losses (over $3 billion) associated with the Texaco-Pennzoil litigation between 1985 and 1987. Although many of the costs they estimate are not directly related to Texaco's bankruptcy case, they do show that financial conflict can have substantial effects on productivity.
3. The CDS contract defines the type of events triggering payment on the CDS. Because it is difficult to define what types of out-of-court restructurings and debt contract amendments constitute a triggering event, most recent contracts exclude these events.

CHAPTER 5

Valuation of Distressed Firms

Enterprise valuation plays a central role in negotiations over how and when to restructure a distressed firm, and significantly within Chapter 11 itself. Enabling a firm to reorganize presumes that the firm's going concern value is greater than the value that would be realized in a liquidation. The firm's estimated value determines the size of the pie to be divided among existing claimants, and drives projected payouts and recoveries. It is also critical in approving debtor-in-possession and exit financing, determining the feasibility of a plan, and in determining an appropriate capital structure for the reorganized firm. Finally, enterprise valuation is used in litigation to avoid (undo) preferential payments made by the debtor in the period prior to filing, as it is needed to demonstrate the debtor's insolvency at the time those payments were made.

Even under ideal market conditions and the sound application of valuation methodologies, different stakeholders frequently estimate widely divergent values for the firm. These estimates are based on forecasts of the future cash flow generating ability of the reorganized firm, and on the observed values of similar companies and transactions. While we are guided by well-developed financial models, the forecasts themselves are fraught with uncertainty. Adding to this the likelihood that parties to a negotiation have conflicting interests for estimating low versus high valuations, and that experts from time to time adopt ad-hoc methods to support a valuation, one can feel that "at worst [enterprise valuation] is not much more than crystal ball gazing."[1]

The magnitude of valuation disputes is perhaps most visible when the value used as the basis of a Chapter 11 plan is contested. Figure 5.1 reports the low and high ends of the range of enterprise valuations submitted by various parties to litigation in just a few large Chapter 11 cases. The higher valuations submitted to the court range from 26% to 111% over the low-end values. The enterprise value ultimately used as the basis for the reorganization can end up at the low end (as in Cumulus) or nearer to the high end (as in Exide). These cases also illustrate the tendency for senior claimants to submit lower valuations than junior claimants, and that the debtor's valuation can land on either side of the range of estimates.

This chapter provides some explanations as to why such large differences in valuation arise. Specifically, we address the following questions:

1. Are there factors specific to the restructuring process that make application of commonly accepted valuation models difficult?
2. How can commonly accepted valuation models be applied to distressed firms? What is the role of the court in guiding and/or directing the use of valuation methodologies?
3. How well do these models work in bankruptcy?
4. How do potential biases among different claimants influence the process?

In answering these questions, we summarize a set of best practices in using valuation models, and caution against methods that have little scientific standing. For simplicity we refer in our discussion to valuations in bankruptcy. However, the methods we describe and the typical conflicts between various parties apply as well to distressed firms negotiating an out of court restructuring, with the backdrop of a potential Chapter 11 filing should the out of court restructuring fail.

Why is valuation in a bankruptcy context difficult? Even within the framework of commonly accepted valuation methodologies, valuations are inherently uncertain. The basic methods for assessing firm value – via the sum of discounted future cash flows (DCF) or the use of valuation multiples from comparable companies and transactions – rely on strong model assumptions and estimates of future performance. Applying these methods to distressed firms is particularly challenging for a number of reasons.

1. *Historical performance is less useful in developing cash flow forecasts.*

For a nondistressed firm, a valuation expert can often rely on a company's recent performance to guide forecasts of future performance. Cash flows used in a DCF analysis are commonly modeled as growing at some "normal" rate above recently realized cash flows. Valuation multiples are typically applied to a cash flow benchmark such as the last 12 months or the next 12 months projected earnings before interest, taxes, depreciation, and amortization (EBITDA). Typically, a distressed firm has recently experienced a significant decline in cash flows. The firm has also often undergone substantial restructuring of its business, such as rejecting contracts, selling or discontinuing certain operations, or making other changes needed to restore profitability. Moreover, firms are frequently projected to continue their recovery for several years postbankruptcy before reaching a more normalized state. As a result, operations of the emerging firm can be substantially different from those of the prebankruptcy company.

Arguably, this can lead to more uncertainty for the projected cash flows of formerly distressed or bankrupt firms. For discounted cash flow analysis, the horizon of cash flow projections must be long enough to account for the effects of the restructuring, and valuations based on multiples should use forward projected cash flows and multiples. As argued below, the lack of relevant historical data for the firm also raises issues in estimating an appropriate discount rate for a discounted cash flow analysis.

2. *Useful market prices for the firm's claims are often unavailable.*

The total market capitalization of a nonbankrupt publicly traded firm is calculated as the sum of the market value of its equity, using its observed trading price and the number of shares outstanding, and an estimate of the net value of its debt.[2] The debt of healthy firms often trades close to par, but in distress can trade substantially lower. Prices for defaulted publicly registered debt have been publicly observable via FINRA's TRACE system only since 2005. Loans and trade debt, while accounting for a significant proportion of the capital structure, arguably are not "securities" under U.S. law, and have no established central market mechanism for reporting prices. Further, even when prices for the firm's debt or equity are observable, they may not provide useful indications of value when they are influenced by the negotiations themselves (see discussion below, as well as Chapter 3 herein).

3. *Firms in the same industry often have recently restructured or are in distress themselves.*

Industrywide distress is a common phenomenon, limiting the usefulness of some comparable companies or comparable transactions in many cases. The most recent wave of oil and gas company restructurings in the United States demonstrates the potential impact of an economic shock to an industry, but numerous other such waves have included retailers, suppliers to the automobile industry, steel manufacturers, textile manufacturers, telecommunications firms, and others. When cash flows of comparable firms are depressed (and often negative), trading market valuation multiples based on similar firms may not be feasible. Comparable transactions may be of limited use, as research has shown that purchases of distressed assets typically occur at a discount to similar nonbankrupt transactions.[3] Continued industry distress may also seriously challenge the restructuring firm, even once it emerges with lower leverage.

4. *Acceptance of potentially useful valuation methods, shown to be successful in academic studies, has been slow.*

Studies have demonstrated that the Fama-French "three-factor" asset pricing model outperforms the capital asset pricing model (CAPM) in empirical tests of the cost of capital.[4] Many practitioners incorporate a "size premium" in discount rate calculations, but the broader use of the Fama-French factors is relatively rare among bankruptcy practitioners.[5] Likewise, the adjusted present value (APV) method for estimating DCF enterprise value is well suited for bankruptcy valuations because it sidesteps estimating long-term capital structure weights required to implement a weighted average cost of capital (WACC), and can flexibly incorporate more complex tax attributes of the emerging firm. Although explained in most corporate finance textbooks, the APV model is less commonly used than more familiar DCF models (which discount free cash flows at an after tax weighted average cost of capital).

In the following section of this chapter, we review models commonly used to value distressed firms, based on accepted practices. At the same time, courts and many practitioners have accepted valuations based on unnecessarily ad hoc or even arguably invalid methods. For example, practitioners commonly incorporate company-specific risks into cost of capital estimates, even though the models to which a premium for "firm specific risk" is added are based on the idea that company-specific risk is not priced into the cost of capital. We discuss the problems with incorporating company-specific risk and other arguably ad-hoc methods for valuation.

5. *Valuations can be strategically influenced by the bargaining power of parties to the negotiations and/or judicial discretion in bankruptcy.*

Claimholders with advantages in bargaining power can successfully influence the value on which a plan is ultimately based. Junior claimants have an incentive to argue for higher estimated enterprise values, while senior claimants often prefer a lower valuation (see the examples in Figure 5.1). Valuation outcomes can then be biased toward the interests of claimholders with greater influence or alignment with debtor management. We return to this important issue below.

Even when implementing well-accepted models, there can be diverse opinions on how to treat the underlying assumptions.[6] Further, conclusions regarding value are dependent on the weighting of the discounted cash flow and other methodologies, and the sometimes necessarily subjective arguments for relying more or less heavily on a particular approach. Combined, these factors suggest that valuations negotiated in the context of a distressed restructuring may be less precise, if not more likely biased. We review evidence regarding the "accuracy" of valuations below. Valuation disputes can impede the ability to complete an out-of-court restructuring or to confirm a plan of reorganization, and avoid the costs of a prolonged stay in bankruptcy. In cases where a settlement is not negotiated and a valuation trial is held in bankruptcy court, a judicial estimate of value replaces a market-determined one.[7] When valuations are determined by bankruptcy judge – not a market – the judge can be influenced by the larger case dynamics, objectives, and desires to move the case to near-term conclusion. The judge's familiarity with valuation theory can also have an important effect on how the judge will view the evidence.

Company (Bankruptcy Year)	Low End of Value Range, ($ Millions)	High End of Value Range, ($ Millions)	Percentage Range, [(High-Low)/ Low] × 100	Midpoint Value at Confirmation ($ Millions)
Cumulus Media (2017)	Debtor: $1,500	Unsecured: $2,400	60.0%	$1,675
Breitburn Energy Partners (2016)	Debtor: $1,600	Equity: $3,800	137.5%	$1,600
Chemtura (2009)	Debtor: $1,900	Equity: $2,600	36.8%	$1,900
Calpine (2008)	Creditors: $16,300	Equity: $24,400	49.7%	$21,700
Nellson Nutraceutical (2007)	1st lien: $322	Debtor/equity: $404.5	25.6%	$320
Mirant (2005)	Debtor: $8,266	Equity: $14,549	76.0%	~9,000*
Bush Industries (2004)	Debtor: $95	Equity: $200	110.5%	$95
Exide Technologies (2003)	Debtor/seniors: $950	Creditors: $1,700	78.9%	$1,500

*Approximation. Confirmed plan excluded explicit valuation.

FIGURE 5.1 Valuation Ranges from Large Chapter 11 Bankruptcies

VALUATION METHODOLOGIES

Since the intrinsic or "true" value of the firm is unobservable, we must rely on various methodologies accepted as useful approaches to estimating value. The approaches described here are used in M&A and other corporate restructuring practices, but we emphasize concerns particular to distressed companies.

The two most widely used approaches to valuation are "relative valuation" models (comparable company and comparable transactions values), where value is derived from the pricing of comparable assets, and discounted cash flow models.[8] For clarity in our discussion, we refer to the firm being valued as the "target" firm.

Relative Valuation Models: Comparable Company and Comparable Transactions

The "comparable company" approach, sometimes also referred to as a "trading multiples" valuation, estimates the value of the target firm by applying valuation multiples of peer firms to the target. The three steps involved are to (1) identify peer or "comparable" publicly traded firms, (2) observe how comparable firms are valued by the market, and (3) apply that valuation to the target firm.

The most critical aspect of this analysis is the selection of a set of comparable companies, which requires some judgment by the analyst. Fundamentally, comparable firms should match the target in terms of risk and growth prospects.

A "pure play" peer firm would be an ideal comparable, but in most cases does not exist. Typically, an industry screen, based for example on SIC codes, produces an initial set of possible comparable companies. From this set, comparability can be determined by comparing characteristics such as size, growth, mix of businesses, bankruptcy status, profitability, leverage, cost structure, and so on. When the set of comparables has been defined, it is then useful to compare various financial performance measures for the comparable firms and target to understand how well the target firm fits within this group.

Each comparable firm must be publicly traded so that we can observe their total enterprise value (TEV) as the current market value of equity and an estimate of the value of net debt from financial statements. The ratio of TEV to a particular cash flow or balance sheet measure for the firm yields a valuation "multiple." The most commonly used metric for this analysis is some variant of the ratio of TEV to EBITDA, since EBITDA is likely to be highly correlated with firm cash flows and therefore value. Variations on the metric are often industry specific, for example EBITDAR (EBITDA minus restructuring or rental expenses) or EBITDA minus capital expenditures, to better normalize estimated cash flows across peer firms.

Valuation multiples obtained from the comparables are applied to produce an estimated range of values for the target firm. Application of the multiples is relatively straightforward if the target firm appears similar to the average comparable firm, or at least fits within the range of comparables. Additional subjective adjustments based on a belief that the comparables are fundamentally different from the target firm, however, can be ad hoc and difficult to support. As described by one of the leading valuation textbooks, "Even when a legitimate group of comparable firms can be constructed, differences will continue to persist in fundamentals between the firm being valued and this group. Adjusting for differences subjectively does not provide a satisfactory solution to this problem."[9] Distressed companies in particular may not fit well within a larger competitive norm, which can lead to skepticism in values estimated from this approach. Often, proponents of a higher end valuation will favor the use of a given set of comparables, while proponents of a lower-end valuation will point to the frailties of the target business relative to the comparables.

Beyond potential difficulties in agreeing on a set of comparable firms, one needs to be careful in how multiples are applied to firms undergoing a reorganization such as in Chapter 11. Since historical data for the target firm can be of limited relevance, it is typically more appropriate to apply a "forward" multiple to forecasted performance for the reorganized firm. For example, near term post-emergence EBITDA may be low when restructuring of the business is ongoing; applying a valuation multiple to a depressed number would understate the firm's long-run growth prospects. In this case, the multiple should be applied to the first projected year that represents normalized operations (reflecting the benefits of rejected leases, or improved margins after discontinuing unprofitable business lines, as examples).

Comparable Company	Market Value of Equity[1]	TEV	EBITDA 2017[2]	EBITDA 2018P	TEV/EBITDA Multiple	
					CY17	CY18P
Entercom Communications	$1,421	$3,075	$375.0	$389.2	8.2 x	7.9 x
Beasley Broadcast Group	$306	$518	$50.3	$50.3	10.3 x	10.3 x
Salem Media Group	$106	$367	$48.3	$51.0	7.6 x	7.2 x
Saga Communications	$229	$201	$26.1	$26.1	7.7 x	7.7 x
Multiple: average					8.5 x	8.3 x
						Implied value (avg.)
Cumulus				$236		$1,953

[1] As of March 2018.
[2] EBITDA is estimated from analyst reports, CIQ, and public filings, as reported by the financial advisor to the unsecured creditors committee (p. 566 of docket 657).

FIGURE 5.2 Cumulus Media Inc.: Comparable Company Analysis Based on EBITDA Multiple (Dollars in Millions)

An example of this analysis is given in Figure 5.2 for Cumulus Media, Inc., which filed for Chapter 11 in November 2017.[10] Our calculations use only publicly disclosed financial projections and other information provided in the debtor's disclosure statement and other documents filed with the bankruptcy court.[11] Based on available information, the comparable publicly traded firms we use for this example are selected from those used by advisors to the debtor and to the unsecured creditors committee; one source of disagreement between these groups was the extent to which certain comparables were not "pure plays" (firms having business operations only similar to those of Cumulus) or were dissimilar to Cumulus in other important dimensions. The table shows multiples based on the prior calendar year as well as projected EBITDA. Applying these multiples to the projected EBITDA of Cumulus produces a range of estimated total enterprise value for Cumulus from $1.8 to $2.4 billion, with an average of over $1.9 billion.

The comparable companies approach is widely used and relatively easy to implement. It is most useful when a large number of comparable firms are publicly traded, which is also typically associated with greater analyst coverage of the industry. Any relative valuation approach, however, will build in errors (overvaluation or undervaluation) that the market itself makes in valuing these firms.

The second widely accepted relative valuation model is the "comparable transactions" approach, also referred to as a comparable M&A analysis. The approach and implementation are very similar to the comparable company approach, except that prices paid in recent acquisitions of companies comparable to the target are used to determine the valuation multiple. There are two important limitations in applying this approach for distressed firms. The first is that only recent acquisitions of comparable firms under similar market conditions should be included, possibly limiting the number of relevant transactions. For example, a number of retailing companies were purchased in leveraged buyouts in the mid 2000s, followed just a few years later by a sharp industry decline and a number of distressed restructurings; valuation multiples from the prior buyouts would not be useful following the decline in the market/economic environment.[12] The second is that acquisition prices typically reflect a "control premium," sometimes based on the value of expected synergies or other benefits to the acquirer, leading to somewhat higher estimates of value. However, when financial distress amongst similar firms is clustered in time, comparable transactions may largely consist of other distressed firms or even sales in bankruptcy; if so, this approach can produce a lower estimate of value in comparison to a comparable company or DCF valuation. As show in Figure 5.3, the comparable transactions method produces a midpoint total enterprise value of approximately $2.2 billion for Cumulus Media.

Discounted Cash Flow

Discounted cash flow (DCF) is a forward-looking approach that estimates firm value as the discounted value of expected future cash flows. As such, it is sensitive to a number of assumptions used to derive the cash flows or discount rate. In contrast to the relative valuation models, however, this approach requires that the analyst be explicit about these important assumptions. DCF methods are considered by some experts to be the most useful measure of intrinsic value.[13] It is also arguably the most consistent with bankruptcy law's "fresh start" evaluation of the business undergoing transformation (see Chapter 3 herein and discussion of fresh start values below).

Announcement Date	Target	Acquirer	Transaction Value	LTM EBITDA	LTM EBITDA Multiple
February 2017	CBS Radio	Entercom	$2,850	$351.9	8.1 x
July 2016	Greater Media	Beasley	$240	$22.0	10.9 x
		Cumulus 2018P EBITDA	Average multiple	Implied value	Value range
		$236	9.5	$2,242	$1,912–$2,572

FIGURE 5.3 Cumulus Media Inc.: Comparable Transactions Analysis Based on EBITDA Multiple (Dollars in Millions)
Source: Capital IQ and docket #657.

The total enterprise value from a DCF model is estimated as:

DCF value = Present value (cash flows during projection period)
+ Present value (terminal value)

This method requires detailed projections of post-reorganization cash flows for the reorganized company. In many bankruptcy cases, detailed projections used as the basis for a plan are provided along with the disclosure statement provided to voting classes. The terminal value captures the value of all cash flows that would occur subsequent to the projection period.

The two important components of the DCF model are therefore the projected cash flows and the discount rate applied to those cash flows. The most commonly used approach is to discount "unlevered free cash flows" at an after tax weighted average cost of capital (WACC). The free cash flows are the total after-tax cash flows generated by the firm that are available to all providers of the company's capital, both creditors and shareholders. The WACC reflects all investors' (both debt and equity) opportunity cost for investing in assets of comparable risk. However, there are several issues specific to valuation of firms in financial distress or bankruptcy that lead us to suggest a second DCF method, known as the "adjusted present value" (APV) method, or the closely related "capital cash flow" method.[14] The APV method is often implemented more easily and accurately than the free cash flow approach when the firm's capital structure changes significantly during the forecast period. Further, it can be better suited for the complicated tax situations of firms in a distressed restructuring.

DCF Method #1: Discounting Free Cash Flows at the Weighted Average Cost of Capital

The first key input to this DCF model is the projection of free cash flows. Free cash flow (FCF) is defined as the sum of the cash flows generated by firm which are available to all providers of capital, including debt, preferred stock, and equity, and is calculated as:

FCF = EBIT(1 − tax rate) + Depreciation and other noncash charges −
Capital expenditures − Net working capital spending

FCF excludes cash flow from nonoperating assets, which are generally valued separately. Free cash flow is before financing and is unaffected by the company's projected capital structure, and so is also often referred to as "unlevered free cash flow." It can be thought of as the after-tax cash flow that would be available to the firm's shareholders if the firm had no debt. In other words, the tax benefits from the deductibility of interest on debt and other tax shields are not included in the cash flows themselves, and tax payments are estimated as projected EBIT times the firm's effective tax rate. While the tax benefits of the debt structure do not enter the free cash flow calculation, they do affect the discount rate and therefore the estimated enterprise value.

Since free cash flows are flows to all providers of capital, the appropriate discount rate is a blend of the required rates of return on debt and equity, weighted by the contribution of those sources of capital to the firm's total market value. The resulting weighted average cost of capital (WACC) is therefore:

$$\text{WACC} = r_d(1 - t) * \frac{D}{V} + r_e * \frac{E}{V}$$

where

r_d = expected yield on the firm's debt postrestructuring

r_e = cost of levered equity capital

D/V = proportion of market value of the reorganized firm financed with debt

E/V = proportion of market value of the reorganized firm financed with equity

t = effective tax rate of the reorganized firm

A common approach to estimate the cost of equity capital, r_e, is to use the Capital Asset Pricing Model (CAPM), $r_e = r_f + \beta_e * (r_m - r_f)$, where r_f is the risk free rate, β_e is the firm's equity beta, and $(r_m - r_f)$ is the market risk premium. Estimates of market risk premia are available from published sources, based on historical returns of equity over treasury returns, survey evidence, or implied by current market valuations.[15] Current treasury bond yields are used for the risk free rate. As discussed above, inclusion of an additional company specific risk premium is inconsistent with the assumptions and application of the CAPM model; an alternative and less ad hoc approach is to demonstrate the impact of uncertainty directly in scenarios of projected cash flows.

For a distressed firm, measuring an equity beta (β_e) using the firm's historical stock returns is problematic. Historical stock returns are generally negative as the debtor heads into financial distress, and bear little resemblance to equity returns expected following a successful restructuring. Substantial asset restructuring also makes historical performance less relevant. In many other cases, distressed firms simply have no traded stock, either because it has ceased to publicly trade or because the firm's stock was never listed. The best alternative is to estimate betas from comparables, using their historical data or estimates from public data providers. The equity beta obtained from comparables must be adjusted for differences in financial leverage between the comparables and postreorganization firm.[16]

An example of the WACC calculation is given in Figure 5.4 for Cumulus Media. The beta is estimated from comparables, and adjusted for the leverage of

Inputs:	Symbol	Value
Risk-free interest rate (20-year government bond yield)	r_f	3.0%
Market risk premium[1]	$(r_m - r_f)$	6.4%
Mean unlevered beta of comparables[2]	$\beta_{unlevered}$	0.86
Debt beta	β_{debt}	0.40
Tax rate	t	21%
Projected equity-to-value ratio	E/V	23%
Projected debt-to-value ratio	D/V	77%
Levered equity beta for reorganized Cumulus	βe	2.48
Pretax cost of debt	r_d	7.5%
Cost of equity $= r_f + \beta_e * (r_m - r_f)$	r_e	18.9%
WACC $= r_e * E/V + r_d(1 - t) * D/V$		8.96%

[1]Damodaran, 2018, http://pages.stern.nyu.edu/~adamodar/.
[2]Levered/unlevered betas calculated as: $\beta_L = (\beta_U - \frac{D}{V}\beta_D)/\frac{E}{V}$

FIGURE 5.4 Cumulus Media Inc.: Example of Discount Rate (WACC) Derivation

the reorganized firm. This analysis produces a WACC of approximately 9% for Cumulus.

Lastly, the DCF model requires calculation of a terminal value (TV), representing the value of all free cash flows expected to be produced after the final projection year. Two approaches to calculating the TV are commonly used. The first uses a comparable company approach, for example applying a multiple of EBITDA to the projected free cash flow for the first year following the projection period. The cash flow used in the terminal value should represent normalized operations that are reasonably sustained indefinitely. The second common approach is to use a "growing perpetuity" model:

$$Terminal\ value_T = \frac{FCF_{T+1}}{WACC - g}$$

Again, the free cash flow in year T + 1 (of a T-year projection period) should reflect normalized long-term operating performance (what is "normalized" outside of the projection period is itself sometimes a subject of debate). The key input to this model is g, the assumed long-term growth rate for cash flows. Detailed discussion of the assumptions behind the growing perpetuity approach is given by Koller, Goedhart, and Wessels (2010) (p. 253). These assumptions are particularly important because the terminal value typically accounts for an even greater proportion of total enterprise value when cash flows are projected to significantly increase over time. Gilson, Hotchkiss, and Ruback (2000) value 63 companies in Chapter 11 and find for the median firm in their sample that the terminal value accounts for 70.5% of total value. A small change in the assumed terminal value growth rate can have a significant impact on estimated value.

The free cash flow DCF method is used in Figure 5.5 to value Cumulus Media. Using a range of discount rates, and applying a range of terminal value growth rates from −1 to +1%, the estimated values range from $1.75 to $2.6 billion. Interestingly, the financial advisors to the debtor applied long-term growth rates as low as −3% in calculating the terminal value, based on the declining broadcasting

	FY2018E[1]	FY2019E	FY2020E
Net revenue	$1,171,684	$1,206,296	$1,261,215
Less: Content costs and SG&A	896,254	919,947	951,017
Less: Other corporate expenses	39,596	40,040	40,493
EBITDA	$235,834	$246,309	$269,705
Less: Depreciation and amortization	65,555	64,769	64,872
EBIT	$170,279	$181,540	$204,833
Less: Cash taxes (= EBIT*T)[2]	35,759	38,123	43,015
Plus: Depreciation, amortization, and stock-based compensation	67,838	67,813	67,916
Less: Capital expenditures	25,000	20,000	20,000
Less: Change in working capital	2,808	10,229	12,779
Less: Other (LMA, restructuring fees, and other)	6,046	10,245	3,646
Unlevered free cash flow (FCF)	**112,336**	**170,756**	**193,309**

Calculation of Enterprise Value

Discount rate (from Figure 5.4)	8.96%	
Total present value of FCF	**407,886**	
Terminal value growth rate	1%	
Terminal value		2,453,909
Present value of terminal value	**1,952,172**	
Total enterprise value	**2,360,058**	

Sensitivity to WACC and terminal growth rate

	WACC: 8%	9%	10%
Terminal value growth rate: −1%	$2,146,229	$1,928,434	$1,750,338
1%	2,686,032	2,347,049	2,083,503

[1]FY2018 as shown includes only cash flows after assumed 4/20/2018 effective date.
[2]Assumes a 21% effective tax rate.

FIGURE 5.5 Cumulus Media Inc.: Discounted Cash Flow Valuation (Dollars in Thousands)
Source: Authors calculations based on information provided in the Disclosure Statement for the First Amended Joint Plan of Reorganization of Cumulus Media Inc. and Its Debtor Affiliates Pursuant to Chapter 11 of the Bankruptcy Code, filed February 12, 2018.

industry. Even under the most pessimistic outlook for the industry, the existing company would continue to generate positive cash flow for decades.

DCF Method #2: Adjusted Present Value (APV)

The adjusted present value (APV) approach follows directly from the work of Modigliani and Miller (Berk and DeMarzo, p. 648). The total enterprise value equals the sum of the values of the operating assets plus the present value of tax shields. Relating this to our discounted cash flow valuation model:

$$TEV_0 = \sum_{t=1}^{T} \frac{Free\ cash\ flow + Terminal\ value}{1 + WACC_{pre-tax}} + \sum_{t=1}^{T} \frac{Tax\ shields}{1 + r_{tax\ shields}}$$

The cash flow forecasts directly incorporate projected tax shields, which are generally tax savings from the deduction of interest payments and use of net operating losses. Unlevered cash flows and terminal value are calculated identically to those described above, but the discount rate for these cash flows is the unlevered WACC (using the pretax cost of debt). This rate can be calculated simply by using the CAPM for the unlevered firm, and is equivalent to a pretax WACC:[17]

$$WACC_{pretax} = r_f + \beta_{unlevered} \times (r_m - r_f)$$

The discount rate for tax shields depends on the riskiness of realizing those tax shields. For interest tax shields, when the firm is projected to have a target debt ratio – meaning a constant projected D/V ratio – tax shields are discounted at the same unlevered cost of capital as applied to the unlevered free cash flows; otherwise, tax shields are discounted at the pretax cost of debt. Since the firm's ability to use NOLs to offset taxable income is limited to its net income, projected savings from the use of NOLs are correlated with net income; this suggests that the levered cost of equity be used to discount noninterest tax shields such as NOLs. For firms emerging from Chapter 11, the amount of NOLs that can be used to offset taxable income each year is often capped at a dollar amount (see Chapter 3 herein).

DCF Method #3: Capital Cash Flows (CCF)

A variant on the APV approach, known as the "capital cash flow" approach, uses a single discount rate for both the unlevered firm cash flows and tax shields (including tax savings from the use of NOLs – see Ruback 2002):

$$Value_{Enterprise} = \sum \frac{Capital\ cash\ flows}{1 + WACC_{unlevered}}$$

This approach assumes that debt is maintained as a fixed proportion of value, so that interest and other tax shields have the same risk as the firm.[18] During the projection period, capital cash flows are calculated using the formula:

Net income

+ Cash flow adjustments
+ Cash and noncash interest
= Capital cash flows

Cash flow adjustments include adding back depreciation, amortization, deferred taxes and after-tax proceeds from asset sales, and subtracting working capital investment and capital expenditures. A benefit of this approach is that capital cash flows can utilize the company's own projections of net income, which incorporates expected tax payments. For firms with complex tax shields, this can be a strong advantage over approaches where tax shields must be estimated by the analyst.

Figure 5.6 shows a CCF valuation for Cumulus, which in this case produces a range of value close to that of the prior DCF valuation, the differences coming from the estimated tax shields and the discount rate applied to those tax shields.

Fresh Start Accounting Estimates of Value

Statement of Opinion (SOP) 90-7, Financial Reporting by Entities in Reorganization under the Bankruptcy Code, requires "fresh start accounting" for all firms that filed for Chapter 11 on or after January 1, 1991 or that had a plan of reorganization confirmed on or after July 1, 1991.[19] This directive requires some firms to restate their assets and liabilities at their going-concern values. Fresh-start accounting must be adopted when (1) the going concern value of the debtor's assets at reorganization is less than the value of all allowed prepetition liabilities and postpetition claims, and (2) prepetition stockholders retain less than 50% of the reorganized firm's voting common shares. Fresh-start values are generally estimated using DCF and comparable company methods.

Fresh-start values are contemporary estimates of values, produced by the firm's managers and accountants, which emerge from the administrative bankruptcy process. Disclosure statements rarely describe the assumptions used to generate fresh start estimates of value. In some cases they not coincide exactly with assumptions given with management's cash flow projections in the same document.

Summarizing Value from Various Estimates

Each method produces a different range of value for the target firm – an overall value must weight these various estimates. For each method, the valuation is

	FY2018E	FY2019E	FY2020E
Net income[1]	−29,230	86,661	99,024
Plus: Depreciation, amortization, and stock-based compensation	67,838	67,813	67,916
Less: Change in working capital	2,808	10,229	12,779
Less: Capital expenditures	25,000	20,000	20,000
Plus: Interest expense	96,336	75,919	72,046
Capital cash flows	107,136	200,164	206,207

Risk-free rate (20-year government bond yield)	r_f	3.00%
Market risk premium[2]	$(r_m - r_f)$	6.40%
Tax rate	t	21%
Pre-tax (unlevered) cost of capital	r_u	10.0%
Total present value of FCF		**438,137**

Terminal value growth rate	1%		
Terminal value			2,314,101
Present value of terminal value	**1,823,478**		
Total enterprise value	**2,261,615**		

Sensitivity to WACC and terminal growth rate

	WACC: 8%	9%	10%
Terminal value growth rate: −1%	$2,322,819	$2,090,533	$1,900,531
1%	2,906,072	2,543,538	2,261,615

[1] Assumes a 21% effective tax rate; excludes gain on early retirement of debt at reorganization.
[2] Damodaran, 2018, http://pages.stern.nyu.edu/~adamodar/.

FIGURE 5.6 Cumulus Media Inc.: Capital Cash Flow Valuation (Dollars in Thousands)
Source: Authors' calculations based on information from the Disclosure Statement for the First Amended Joint Plan Of Reorganization of Cumulus Media Inc. and Its Debtor Affiliates Pursuant to Chapter 11 of the Bankruptcy Code, filed February 12, 2018.

only as good as the cash flow forecasts on which it is based. Sensitivity analysis and alternative projections based on different assumptions are generally more useful than ad hoc adjustments, as they force the analyst to be explicit about perceived risks.

Experts can have differing views as to which method is more or less relevant to a particular case, and there is no theoretical guidance to help resolve differences of opinion in weighting these methods. When agreement regarding conclusions about value is not reached, the bankruptcy judge has discretion to come up with his or her own valuation based on information provided by the experts. It is not unusual for a judge to pick and choose aspects of different analyses, or to take one side's expert at its word. Thus, while the 1978 bankruptcy reforms largely removed the SEC and the judge from a formal valuation, judges still need sufficient expertise to make rulings based on the analyses produced in many cases. Interestingly, in the Exide case (2002, SDNY), the judge appointed his own advisor to assist in his decision regarding valuation, though the parties involved reached a settlement prior to the scheduled valuation trial.

Liquidation Values

To satisfy the best interests test required to confirm a bankruptcy plan (§1129(a) of the Bankruptcy Code – see Chapter 3 herein) the debtor must show that its reorganization plan is in the best interest of all claimants. That is, each creditor class must get at least as much as they would under the absolute priority rules in a Chapter 7 liquidation. Thus, firms must provide estimates of the value that would be realized in liquidation. Liquidation value will be low if asset specificity is high – that is, there are few alternative uses for the assets – or if the secondary market for assets is thin (as is often the case).

For this analysis, each asset on the balance sheet is assigned an estimate of the proceeds that would be received in a hypothetical conversion to Chapter 7. The amounts projected to be recovered, net of fees and expenses, are available for distribution in priority order to the firm's claimants. Liquidation also results in additional costs such as the compensation of a bankruptcy trustee to oversee the process, legal and other professional fees, asset disposition expenses, litigation costs, and other claims that would arise should the business be discontinued.

If a reorganization is the hoped-for outcome, there is clearly concern that liquidation values presented along with a reorganization plan will be understated. Aldersen and Betker (1995) examine projected liquidation values for 88 firms that completed reorganizations under Chapter 11, and find they are on average about one third of estimated going concern value. To some extent, these estimates may provide an upper bound on expected liquidation costs, as few parties benefit from objecting to an understated liquidation value. Aldersen and Betker do find, however, that firms with high estimates of liquidation costs have lower debt levels at emergence, making future financial distress less likely.

Figure 5.7 shows the liquidation analysis for Cumulus Media, as described in the firm's disclosure statement. Under the hypothetical liquidation, secured creditors are projected to receive a partial recovery for their claims, and no proceeds are available to make distributions to unsecured claims and prepetition equity interests. The analysis is used to demonstrate that the reorganization plan, which produces higher expected recoveries, satisfies the best interests test needed to confirm the reorganization plan.

	Estimated Book Value	Estimated Recovery Rate:		Midpoint Estimated Liquidation Value
		Low	High	
Unrestricted cash	65,200	100.0%	100.0%	65,200
Accounts receivable, net	221,634	70.0%	85.0%	171,767
Prepaid expenses and other current assets	10,995	35.0%	50.0%	4,673
Current assets	297,830	75.3%	87.0%	241,640
Cumulus Radio PP&E, FCC licenses, intangibles	1,301,863	48.2%	54.3%	666,808
Westwood PP&E, FCC licenses, intangibles	298,137	12.0%	13.4%	37,860
Corporate PP&E, intangibles, and other long-term assets	62,391	40.0%	60.0%	31,195
Long-term assets	1,662,391	41.4%	47.1%	735,863
Total proceeds from liquidation				977,503
Plus: Postconversion cash flow				107,284
Less: wind down costs and fees				91,050
Less: Payments to superpriority and administrative claims				53,995
Net proceeds available for distribution				939,742
Credit agreement claims	1,728,614	54.4%		
Unsecured claims (senior notes, credit agreement deficiency claims, general unsecured, and other)	1,510,700	0.0%		
Prepetition equity holders		0.0%		

FIGURE 5.7 Cumulus Media Inc.: Liquidation Analysis (Dollars in Thousands)

USING ESTIMATED VALUES IN BANKRUPTCY

Determining Payoffs and Recovery Rates: Waterfall Analysis

As shown in the hypothetical example in Figure 3.2 (Chapter 3 herein), new claims (debt and equity) on the reorganized firm are distributed to the existing claims, which are extinguished at emergence. The estimated value at emergence therefore determines the value of new claims that can be distributed. Secured claims, to the extent of the value of their collateral, are paid first generally in cash, new debt, or sometimes stock. Based on any remaining distributable value, payments are made to unsecured claims and prepetition equity interests in order of their priority, where higher priority claims must be paid in full or must consent to distributions that would violate absolute priority. The most junior class receiving a distribution typically receives equity in the reorganized firm.

An example of this "waterfall" structure is provided for Cumulus Media in Figure 5.8. The debtor estimated its midpoint reorganization value at $1.675 billion, meaning that claims to this amount could be distributed under its reorganization plan (after paying administrative and priority claims). The claims available for distribution consisted of a new first-lien loan (the "exit financing") for $1.3 billion, as well as the equity in the reorganized firm. Had the new loan as well as 100% of the reorganized firm's equity been distributed to the term loan lenders, the lenders would have had a projected recovery of 96.3% of their original $1.74 billion claim amount.

Under the plan in fact confirmed by the court, the term loan lenders consented to give unsecured creditors 16.5% of the reorganized firm's equity, yielding them a recovery of 91% rather than 0% in the first scenario (based on the assumed $1.675 billion reorganized TEV). This case provides an example of a contested valuation: the unsecured creditors committee, representing the more junior creditors, argued for a higher valuation of $2.2 billion. Figure 5.7 shows that under the higher firm value, the recovery rate to the term loan lenders would exceed 100%. Thus, the secured lenders would be overpaid and the plan would not have been confirmable had the judge agreed with the higher valuation. The plan was in fact confirmed in May 2018, based on the judge's ruling, which agreed with the debtor's valuation.

Other Uses of Valuation in Chapter 11

Valuation is central to other aspects of the restructuring/Chapter 11 process beyond those related to confirmation of a reorganization plan. In bankruptcy, these include:

1. For the purposes of a plan, a secured creditor has a secured claim up to the value of the collateral, and an unsecured "deficiency" claim for the portion of the debt exceeding the collateral value. Many companies, particularly those which have borrowed in the leveraged loan market, have pledged essentially all of their assets to secure their debt. Thus, valuations involving secured creditors essentially require valuation of the total enterprise.

2. Assessing adequate protection (section 363): The value of the secured creditor's collateral interest is determined at the time of the bankruptcy filing, but may diminish in value as the debtor continues to operate. That is especially true if the debtor intends to prime the creditor's liens to access DIP financing. The debtor cannot make use of collateral pledged to the creditor without providing "adequate protection," which is essentially surety that degradation in collateral positioning will not equate to a degradation in secured status. So, for example, a DIP loan may prime a prepetition lien if the collateral securing

Claim Amount ($ Millions)[1]	Waterfall Assuming Reorganized TEV = 1,675 Distributed Only to Term Loan Lenders				Distributions under Confirmed Plan and Assuming Reorganized TEV = 1,675				Distributions under Confirmed Plan, Assuming Reorganized TEV = 2,200			
	1st lien exit term loan	Equity in reorganized firm	Value of distributions	Recovery rate	1st lien exit term loan	Equity in reorganized firm	Value of distributions	Recovery rate	1st lien exit term loan	Equity in reorganized firm	Value of distributions	Recovery rate
Term loan lenders 1,740.0	1,300.0	100.0%	1,675.0	96.3%	1,300.0	83.5%	1,613.1	92.7%	1,300.0	83.5%	2,051.5	117.9%
Senior notes and general unsecured 681.8		0.0%	0.0	0.0%		16.5%	61.9	9.1%		16.5%	148.5	21.8%
Prepetition equity	0		0.0	0.0%	0		0.0	0.0%	0		0.0	0.0%
Total 2,421.8			1,675.0				1,675.0				2,200.0	

Debtor/judge valuation:

TEV of reorganized firm	1,675
− Debt at emergence	1,300
Equity value at emergence	375

Unsecured creditor's committee valuation:

	2,200
	1,300
	900

[1] Administrative and other priority claims are estimated and assumed for the purpose of this example to be paid out in cash at the effective date of the plan.

FIGURE 5.8 Cumulus Media Inc.: Distributions under Plan of Reorganization and Waterfall Analysis

both financings is of sufficient worth that the "equity cushion" can itself serve as adequate protection. Disputes over adequate protection are relatively rare.

3. Postpetition interest (506(b)): A creditor which is oversecured is entitled to be paid interest and fees provided for in its contract.

4. Fraudulent transfer and preference litigation (see Chapter 3 herein): One of the key issues in these cases is to determine whether the firm was insolvent when it transferred assets prior to filing for bankruptcy.

5. Appointment of an official committee of equity security holders. Stock-holders do not have an automatic entitlement to committee representation in Chapter 11. But, if they can show that the debtor is not "hopelessly insolvent," the US Trustee may choose or the bankruptcy court may order the appointment of such a committee.

HOW WELL DO VALUATION MODELS WORK FOR DISTRESSED FIRMS?

Evidence on the Accuracy of Valuations

In a nondistressed M&A setting, Kaplan and Ruback (1995) find that the approaches described here yield relatively precise estimates of value for a sample of highly leveraged transactions. Valuation of distressed firms, however, may be more difficult for the reasons described above. Gilson, Hotchkiss and Ruback (2000) apply the methods described here to the cash flow forecasts from a sample of firms emerging from Chapter 11. The accuracy of these valuations is evaluated by comparing estimated values from comparable company and DCF models to the market value observed for the target company when it first trades following bankruptcy. They find that the estimated values are generally unbiased predictors of the realized market values, but they are not very precise. The sample ratio of estimated value to market value at emergence varies from less than 20% to greater than 250%. They argue that both the administrative process and the strategic use of values in the negotiation process explain the wide range of values. Demiroglu, Franks, and Lewis (2018) study 66 Chapter 11 bankruptcy reorganizations where the post-emergence stock is listed on an exchange; the average gap between the plan-based and postbankruptcy value of equity is 43.6%. Interestingly, the gap is significantly lower when bond prices are publicly observable during the bankruptcy case.

Ayotte and Morrison (2018) describe the methods used in 143 disputed Chapter 11 valuations, 1/3 of which arise in the context of plan confirmation. They point out the magnitude of these disputes, finding the percentage difference between high and low valuations has a mean of 47%.[20] Errors made by advisors are often self-serving, producing a valuation more favorable to their client. Diving into the methods and assumptions used, Ayotte and Morrison (2018) also demonstrate the sources of these disputes, such as disagreement over the choice of comparables (72% of cases), disagreement over the discount rate (46%), or disagreement over the projected cash flows (74%). They also criticize the significant usage of a "firm specific risk premium" which arbitrarily increases discount rates (decreasing estimated value) – as pointed out earlier in this chapter, there is no academic foundation for this type of ad hoc adjustment

Regardless of the method, estimated values are only as good as the cash flow forecasts and other assumptions on which they are based. Researchers have noticed over time that management projections are often biased upwards (Hotchkiss, 1995). Whether this reflects the debtor's preference to reorganize rather than liquidate, over-optimism regarding the firm's prospects, or other reasons is unclear.

Strategic Use of Valuation in Bankruptcy Negotiations

One explanation for the lack of precision in estimated values and the large magnitude of disagreements over value is the strategic use of these values as part of the plan negotiation process. When the incentives of the parties involved in negotiations conflict, stated estimates of value can reflect the biases of these parties. Case studies described by Gilson, Hotchkiss, and Ruback (2000) and the more recent examples presented in Figure 5.1 strongly suggest that stated positions on the value of the bankrupt firm can be self-serving.

Gilson, Hotchkiss and Ruback suggest several factors they expect to be related to these biases: the relative bargaining strength of competing (senior versus junior) stakeholders, management's equity ownership in the reorganized firm, the existence of outside bids to acquire or invest in the debtor, and senior management turnover. They develop empirical proxies for each of these factors, and show how they relate to whether the firm is over or undervalued relative to its market value at emergence from Chapter 11.

Provided that distributions under the plan approximately follow relative priority, basing the plan on a higher estimated value benefits junior classes by justifying a larger payout to their claims. Similarly, senior claimants benefit when the reorganization plan is premised on a low value. If firm value is low enough such that anyone below the senior claimholders is not entitled to any distribution, then typically the majority of the reorganized firm stock will be distributed to the senior claimants and those more junior will receive little or no distribution. If in fact the firm value after emerging is significantly higher than assumed in the plan, there would be a windfall to the senior claimants who received stock, who ex-post would have a recovery of greater than 100% (see the Cumulus example in Figure 5.8, under an assumed value of $2.2 billion). Any wealth gain that either group realizes ex post must come at the expense of the other group. Gilson, Hotchkiss, and Ruback find that values estimated from cash flow projections provided with the plan are higher (lower) when an investor holding junior (senior) claims has gained control of the reorganized firm.

Gilson, Hotchkiss and Ruback also find that the distribution of stock and/or options to managers of the reorganized firm is related to incentives to understate value. Options on the reorganized stock are typically granted "at-the-money", with an exercise price based on the presumed reorganization value. Presuming these managers are retained following emergence, they are incentivized to advance a low value for the reorganized firm because their post-emergence allocation of stock options immediately becomes in the money. Grants of equity also directly benefit from a windfall increase in value after emergence. In other words, if value is strategically understated at emergence, and the true intrinsic value is subsequently reflected in the market, there may be a substantial windfall to managers and others receiving the stock. Examples of windfalls in the value of management stock grants have notably been observed for the bankruptcy cases of National Gypsum (1990), and more recently for Visteon (2009); in the latter case, the stock received was worth twice its assumed value within six months of emergence, and management stock grants were worth $114 million.

More recently, Demiroglu, Franks, and Lewis (2018) find that enterprise values tend to be understated when the plan involves a rights offering of shares. They do find, however, that these effects are mitigated when there is publicly available information about prices of debt securities during the bankruptcy case.

Resolving debates over valuation remains a difficult issue. Interestingly, valuation disputes might also be resolved by issuing securities whose payoffs are explicitly tied to the firm's future market value (Bebchuk, 1988, 2002; Aghion, Hart, and Moore, 1992). Such securities provide a hedge against mistakes in valuation and are often used in corporate mergers. Although this is far from a new idea, relatively few Chapter 11 cases to date have used this type of mechanism to reduce the potential for large valuation errors. Resolving debates over valuation remains a difficult issue.

Is Estimated Value Consistent with Trading Prices of Claims?

Parties to a valuation dispute will sometimes refer to observed trading prices of claims to support their position on value. For example, it may be difficult to confirm a plan that provides little or no distribution to a class of claims, when those claims trade during the case at substantial positive values. Still, observed prices for the claims to a distressed or bankrupt firm may not be useful indicators of value for a number of reasons, including the following:

> *Transactions prices, even when observable, may be affected by other holdings of the buyer or seller.* An investor may have purchased distressed debt at a significant discount, or own debt from more than one class of claims which differ in priority; this can affect the investors incentives in subsequent trading decisions and as a participant in restructuring negotiations.

Holders of debt claims may have also purchased credit protection for their claims, the so-called "empty creditor" problem (Bolton and Oemke, 2011; Subramanyam, Tang, and Wang, 2014; Danis, 2016)

Prices paid for claims can depend on the investor's interest in gaining a controlling equity stake. An investor looking to purchase a "fulcrum" claim, meaning that its holders will become the ultimate equity owners of the restructured firm, may pay a premium for those claims (Feldhutter, Hotchkiss, Karakas, 2016) – yet determining which, if any, group of debt claims will be converted to equity itself depends on the value assigned to the restructured firm. Hotchkiss and Mooradian (1997) also show that investors may purchase claims, at increasing prices, up to the one third of the amount needed to potentially block any restructuring plan.

Prices paid for claims that are (or are near) out of the money reflect call option value rather than expected payoff values. For a company that is close to insolvency or is insolvent, the traded price of equity will reflect the deep out of the money call option value of the equity, rather than the expected value of the equity, which should be close to, or equal, to zero. Including the traded equity value in a sum-of-claims estimate of enterprise value will overstate that value. This criticism extends to any lower priority claim that is out of the money.

Prices paid may reflect the expected outcome of negotiations. To the extent one group has significant bargaining power, prices of claims may reflect expected settlement values differing from more objective valuations.

Market evidence is arguably more useful as one indicator of value in the context of a fraudulent conveyance dispute, where the objective is to determine solvency at a point in time prior to filing for bankruptcy.[21] It is possible to demonstrate that the trading prices of equity and debt claims are consistent with the belief that the firm was solvent at the time. Further, an insolvent firm would find it difficult to raise additional capital from unsecured creditors. Once in bankruptcy, however, several courts have rejected the idea that the trading price of debt or equity provides a useful estimate of value.[22]

A related issue involves whether offers to purchase the company should be used as an indicator of value. When a firm attempts to restructure, it may formally or informally solicit offers for some or all of the business. In addition, a group of claimants, for example the prior equity owners of the firm, may invest and gain (or retain) ownership of the firm.[23] A potential purchase price, or even the lack of an offer, may not however be relevant to the going concern value. As described earlier in this chapter, sales of distressed assets often occur at a substantial discount. Further, the lack of an offer for a company that may not even be up for sale is not clearly relevant for the reorganized firm. Lastly, while junior creditors or even equity holders may argue for a higher valuation, their inability to purchase the firm themselves at the value they put forth does not obviously negate a valuation based on the cash flow generating ability of the reorganized firm (cases involving such issues include *Chemtura* (Bankr. S.D.N.Y. 2009) and *Cumulus Media Inc.* (Bankr. S.D.N.Y. 2018)).

NOTES

1. Judge Michael D. Lynn, *In re Mirant Corp.*, 2005 WL 3471546 (Bankr. N.D. Tex., 12/9/05)
2. The market value of debt is calculated net of cash holdings, reflecting that cash and cashlike securities could be used to pay off debt.
3. Hotchkiss and Mooradian (1998) and Pulvino (1998, 1999).
4. For tests of their original models, see Fama and French (1993). For a recent discussion of multifactor models, and current debate as to best practice in applying these models out of sample, see Berk and DeMarzo, 4th ed., 2017, p. 474, and McLean and Pontiff (2016).
5. Delaware state courts have begun to recognize the Fama-French model as a substitute for the traditional CAPM. See Stan Bernstein, Susan H. Seabury, and Jack F. Williams, "Squaring Bankruptcy Valuation Practice with Daubert Demands," *ABI Law Review*, 2008 (see especially footnote 102).
6. A further example is implementation of the Black-Scholes model for pricing options, which is often applied to the valuation of warrants for restructuring companies.
7. See Stark, Siegel, and Weisfelner (2011) for a collection of readings regarding contested valuations in bankruptcy.
8. For certain industries such as oil or gas exploration and production (E&P) companies, a DCF model known as net asset value (NAV) typically incorporates the amount of reserves, which significantly affects value for companies which mine a finite resource.
9. Damodaran (1996), p. 304.
10. Figure 5.1 shows that the unsecured creditors committee supported a significantly higher value for Cumulus; based on their higher value, the debtor's plan could not be confirmed. This disagreement led to a valuation trial in April 2018.
11. Disclosure Statement for the First Amended Joint Plan Of Reorganization of Cumulus Media Inc. and Its Debtor Affiliates Pursuant to Chapter 11 of the Bankruptcy Code, filed February 12, 2018; Declaration of Evan A. Kubota in support of The First Amended Joint Plan of Reorganization of Cumulus Media Inc. and Its Debtor Affiliates Pursuant to Chapter 11 of the Bankruptcy Code, filed April 9, 2018. We use this example only to illustrate the methodology, and do not comment on the choice of comparables, the validity of cash flow forecasts, or other factors affecting conclusions about value.
12. As a further example, Chemtura Corporation's reorganization plan was opposed by prepetition equity holders. In the judge's decision regarding confirmation of the plan, he questioned whether a comparable transaction completed prior to September 2008 was relevant to Chemtura's valuation in 2010, given the deterioration in market conditions following Lehman Brother's bankruptcy filing.
13. Damodaran (1996).

14. Ruback (2002).
15. Morningstar provides such estimates, as does Damodaran Online (http://pages.stern .nyu.edu/~adamodar/).
16. The equity beta is also known as the "levered" beta. The relationship between the "levered" (β_L) and "unlevered" (β_U) beta is given as $\beta_L = \beta_U + (\beta_U - \beta_D)(1 - t)\frac{D}{E}$, where E equals the market value of the firm's equity and D equals net debt. If we assume the beta of debt (β_D) is equal to zero, this simplifies to the commonly used expression $\beta_L = \beta_U * \left[1 + (1 - t)\frac{D}{E} \right]$. Both of these specifications assume that the company will maintain a debt schedule such that it will not have a constant D/V ratio going forward, which is inconsistent with the assumed constant D/V ratio of the WACC. In calculating the WACC assuming a constant D/V ratio and nonzero debt beta, we use the relationship $\beta_L = \beta_U + (\beta_U - \beta_D)\frac{D}{E}$, which can also be written as $\beta_U = \left(\beta_L * \frac{E}{V} + \beta_D * \frac{D}{V} \right)$.
17. See Ruback (2002) for derivations. Groh and Gottschalg (2011) estimate betas for high-grade bonds of 0.296, and for low-grade bonds of 0.410.
18. Ruback (2002) shows that the capital cash flows approach is algebraically equivalent to discounting the firm's free cash flows by the WACC, assuming a constant projected debt-to-value ratio.
19. Also see Chapter 3 herein.
20. The authors define this difference as (high value − low value)/high value, excluding fraudulent transfer cases.
21. See: *In re Iridium Operating LLC* (Bankr. S.D.N.Y. 2007).
22. See: *In re MPM Silicones, LLC*, No. 14-22503-RDD (Bankr. S.D.N.Y. Mar. 9, 2018); *In re Cumulus Media Inc.* No. 17-13381 (Bankr. S.D.N.Y. May 4, 2018); and references therein.
23. Hotchkiss, Smith, and Stromberg (2016) show that private equity sponsors often inject capital and retain ownership of distressed portfolio companies.

CHAPTER 6

Corporate Governance in Distressed Firms

When a firm becomes financially distressed, almost every aspect of its governance is affected in some way. To start, managers and directors of a corporation owe fiduciary duties to the enterprise, encompassing a community of interests including creditors, shareholders, and other parties. For a distressed firm, the interests of creditors, shareholders, and other parties often conflict; the actions taken by managers in a restructuring can have far-reaching implications for who are the "winners" or "losers." The nature of compensation contracts needed to provide adequate incentives during a restructuring will change, as will labor relationships more broadly. Further, both management and board positions are likely to experience high turnover, particularly when the firm emerges from Chapter 11. Finally, many restructurings lead to large changes in ownership, with former creditors emerging as the new owners of the company. While such changes in control are common, the mechanisms through which they occur can be quite different than for nondistressed firms. This chapter discusses these aspects of governance and their impact on the incentives of managers and other participants in the restructuring process.

MANAGER, CREDITOR, AND SHAREHOLDER INTERESTS

Academics have long recognized that potential agency conflicts between debt and equity holders are exacerbated when leverage becomes high, as it does when firms near financial distress. Certain actions that may benefit shareholders can impose costs on other stakeholders. For example, managers have disincentives to raise additional equity capital if doing so dilutes shareholders' interests while it increases the value of existing debt claims by reducing default risk (the "debt overhang" problem; see Myers 1977). Firms may bypass positive net present

117

value investments if, again, the increase in firm value largely accrues to existing debtholders. At the same time, firms may prefer riskier (and potentially even negative net present value) investments, increasing the option value of equity but increasing default risk (the "risk shifting" problem; see Jensen and Meckling 1976). Economists have further pointed out the mechanisms by which creditors can protect themselves from opportunistic behavior by managers on behalf of shareholders, primarily through contractual provisions in debt contracts or through fiduciary duties owed by managers to creditors.[1]

As management evaluates the strategic options available to the distressed firm, some possible courses of action can favor the interests of one group of stakeholders over the recovery prospects of another. As explained in Chapters 3 and 4 herein, these options can include asset sales and/or an out-of-court restructuring. In some recent cases, managers have engaged in transactions where ownership of assets is moved within the corporate structure, arguably to the detriment of creditors, or modified or entered new financing contracts.[2] Management can attempt to delay bankruptcy filing, can negotiate a restructuring support agreement for a prenegotiated or prepackaged bankruptcy, or can file for bankruptcy with a stalking horse bidder already in place for a quick sale of the firm's assets.

Fiduciary Duties of Managers and Directors

The potential for a divergence of interests when a firm becomes distressed leads to a fundamental question – what determines who management should represent? And, who in practice does management appear to represent? Within bankruptcy, this is arguably more clear – the debtor has an obligation to maximize the value of the bankruptcy estate for the benefit of creditors.[3] What has been less clear over time is the fiduciary obligations of officers and directors as a firm nears or reaches insolvency.

Prior to bankruptcy, creditors may be protected in several ways. One is through enforcement of contractual provisions. Another is through the Uniform Fraudulent Transfer Act of 1984 (UFTA), which enables creditors to attempt to clawback funds transferred out of a failing businesses if those transfers were made with the "actual intention to hinder, delay or defraud" any creditor; the debtor did not receive "reasonable equivalent value" in return; or preference payments were made to an insider.[4] A third important mechanism is through common law regarding fiduciary duties. To the extent officers and directors have a fiduciary duty (to creditors, shareholders, or both), they face potential liability if they take actions that are not in the interest of whomever they owe these duties.

When a corporation is solvent, the managers and directors owe fiduciary duties to the corporation and its shareholders. Creditors are only entitled to protection as provided in the terms of their original contracts. A series of important case rulings consider how this relationship changes when the firm becomes insolvent. In the 1991 *Credit Lyonnaise v. Pathe Communications* bankruptcy case, the Delaware court held that the board of a firm "operating in the vicinity of insolvency" owed its fiduciary duty not just to shareholders but to the "corporate enterprise" as a whole. Directors could conceivably be held liable to creditors as well as shareholders. The trigger point for this expansion of duties was not a readily observable event, such as the filing of a Chapter 11 petition, but rather at the difficult to measure "zone of insolvency." Becker and Strömberg (2012) describe the market reactions to this case; consistent with the potential benefits of this ruling to creditor interests, they show empirically that firms closest to insolvency responded by increasing equity issues and investment and reducing investment risk, and also by increasing leverage and reducing reliance on covenants.

In two subsequent cases, the Delaware courts essentially reversed *Credit Lyonnais*, eliminating the idea that fiduciary duties shift in the zone of insolvency. In the 2007 *North American Catholic Educational Programming Foundation, Inc. v. Gheewalla (Gheewalla)* case,[5] the Delaware Supreme Court's decision limited the fiduciary duties that managers previously owed to creditors in times of financial distress. In 2014, the Delaware Chancery Court issued its opinion in *Quadrant Structured Products v. Vertin*, which further stripped creditors of the ability to claim a breach of fiduciary duty. The judge's opinion in this case reads as follows:

- There is no legally recognized "zone of insolvency" with implications for fiduciary duty claims. The only transition point that affects fiduciary duty analysis is insolvency itself.
- Regardless of whether a corporation is solvent or insolvent, creditors cannot bring direct claims for breach of fiduciary duty. After a corporation becomes insolvent, creditors gain standing to assert claims derivatively for breach of fiduciary duty.
- The directors of an insolvent firm do not owe any particular duties to creditors. They continue to owe fiduciary duties to the corporation for the benefit of all of its residual claimants, a category which now includes creditors. They do not have a duty to shut down the insolvent firm and marshal its assets for distribution to creditors, although they may make a business judgment that this is indeed the best route to maximize the firm's value.
- Directors can, as a matter of business judgment, favor certain noninsider creditors over others of similar priority without breaching their fiduciary duties.
- Delaware does not recognize the theory of "deepening insolvency." Directors cannot be held liable for continuing to operate an insolvent entity in the good faith belief that they may achieve profitability, even if their decisions ultimately lead to greater losses for creditors.

■ When directors of an insolvent corporation make decisions that increase or decrease the value of the firm as a whole and affect providers of capital differently only due to their relative priority in the capital stack, directors do not face a conflict of interest simply because they own common stock or owe duties to large common stockholders. Just as in a solvent corporation, common stock ownership standing alone does not give rise to a conflict of interest. The business judgment rule protects decisions that affect participants in the capital structure in accordance with the priority of their claims.

The last of these points speaks directly to the type of actions described at the beginning of this chapter, and represent a weakening of creditor protection via the channel of fiduciary law. The point of debate, therefore, is whether this specific channel is needed to protect creditors. These cases, and related discussion among legal scholars, argue that "contractual agreements, fraud and fraudulent transfer law, implied covenants of good faith and fair dealing [and] bankruptcy law ..." sufficiently protect creditors' interests (*Gheewalla*), particularly in the presence of large, sophisticated creditors (Baird and Rasmussen, 2006; Hu and Westbrook, 2007). In contrast, Ellias and Stark (2018) argue that the ex-ante design of debt contracts and bankruptcy law are insufficient to protect creditors during periods of distress. They provide case studies for several egregious examples of "corporate opportunism" at the expense of creditors, in the post-*Gheewalla* era of governance.

Debtor versus Creditor Biases in Bankruptcy

If a firm enters Chapter 11 bankruptcy, except in rare cases when creditors or other parties successfully petition the court to appoint a trustee, the debtor's management continues to operate the business. As described in Chapter 3 herein, management has an exclusive right to propose a reorganization plan during the first 120 days of bankruptcy, and during additional periods as approved by the court. These "prodebtor" features of the Bankruptcy Code yield considerable influence over the outcome of the restructuring to the incumbent management.

In the years following the 1978 bankruptcy reform, critics of the prodebtor orientation of Chapter 11 suggested that the process is too protective of incumbent management, and leads to excessive continuation of firms that are not viable. For example, Bradley & Rosensweig (1992) argued that, even when performance is so poor as to render a firm insolvent, incumbent managers go relatively unpunished because bankruptcy law allows them to retain control over corporate assets. The value of equity in an insolvent firm derives only from its option value – with time, equity claims may be back "in the money" should the firm sufficiently recover and return to solvency. Thus, continuation of the firm provides the only hope of restoring some value to these claims. A bias toward inefficient continuation can result when management is not adequately incentivized take actions leading to the highest valued use of the firm's assets (which may involve the sale or liquidation of assets) or when a priority of the debtor is to preserve employment.

Early research showed that numerous firms continue to perform poorly after emerging from reorganizations, often re-entering Chapter 11 (so called Chapter 22s). Critics faulted the prodebtor characteristics of the Bankruptcy Code for this behavior because it permitted debtors great leeway in satisfying the feasibility standard (see Chapter 3 herein). The high rate with which reorganized firms ultimately fail led to suggestions to replace Chapter 11 with a more market based auction process (Hart, 1999; Baird and Rasmussen, 2002). More recent examination of postbankruptcy performance suggests that concerns regarding the viability of post-emergence firms continues, with the caveat that much of this work is based on the performance of companies with publicly registered securities both before and after bankruptcy (see Chapter 7 herein).

The 2009 bankruptcy of General Growth Properties (GGP) illustrates the tensions between liquidation and reorganization. GGP, one of the largest mall operators and real estate companies in the United States, was solvent when it filed for bankruptcy, but faced large upcoming debt maturities through 2012. More than 160 of GGP's subsidiaries (special purpose entities which held the real estate assets of the company) were, to the surprise of many market participants, included in the Chapter 11 filings. Lenders to many of these subsidiaries moved to dismiss the SPEs' Chapter 11 cases, arguing that the filings were not in "good faith" because the SPEs were not insolvent and did not directly benefit from bankruptcy protection. The bankruptcy court denied these motions, and GGP went on to confirm a reorganization plan, emerging from bankruptcy with a significant recovery for shareholders. Thus, GGP successfully used Chapter 11 bankruptcy to avoid liquidation by creditors.

More recently, many observers have argued that the balance of negotiating power both before and during Chapter 11 has swung toward senior creditors, particularly with the rise of active distressed debt investors. When the upside to senior creditors is limited, they may prefer a quick resolution and return of their capital through a liquidation or sale of the business, even if the overall firm level recoveries would be greater in a reorganization. Or, senior creditors can participate in the upside potential of the reorganized firm if, based on a low presumed reorganization value, those creditor's claims are converted to stock in the emerging firm (see Chapter 5 herein). Ayotte and Morrison (2009) study a sample of large bankruptcy cases filed during 2001 and argue there is "pervasive" secured creditor control. Several important trends are arguably related to a shift in the balance of power in negotiations toward creditors of the firm. Specifically, deviations from absolute priority in favor of equity holders have become rare (consistent with a trend documented by Bharath, Panchapegesan, and Werner, 2010), and sales of the company are more likely when secured lenders are oversecured. Gilson, Hotchkiss, and Osborn (2016) also find that a higher proportion of secured debt at filing is associated with a greater incidence of asset sales in bankruptcy, though these sales generally preserve going concern value. To the extent assets are sold at "fire sale" discounts, there may be a redistribution of value from the former junior claimants of the company to the new owners of those assets.

In Chapter 11 bankruptcy, DIP financing has been criticized as the vehicle by which secured creditors can exert excessive control over the direction of the case. Earlier academic discussion speculated that the availability of additional funding to the debtor might exacerbate the overinvestment problem, allowing attempts at reorganization of unviable firms (Dahiya et al. 2003). As the debate shifted, however, the concern that the influence of secured creditors leads to excessive liquidation of viable companies has dominated the discussion. Although he is agnostic as to its benefits or harm, Skeel (2003a) presumes that specific provisions in DIP loan agreements provide secured creditors with the levers to control the debtor. For example, he notes that "lenders have increasingly used their postpetition financing agreements to shape the governance of the Chapter 11 case (p. 12)," and calls for "closer scrutiny of the provisions DIP lenders are currently trying to sneak into their lending agreements (p. 32)." Such provisions include "milestones" such as filing of a plan or the sale of certain assets, potentially giving the DIP lender undue control if these conditions are not met.[6]

CREDITOR CONTROL RIGHTS

Senior creditors have an important influence on the governance of firms well before defaults or bankruptcies. Academic research demonstrates two facts regarding the influence of senior creditors early in a firm's decline. First, when firms violate covenants in lending agreements, the control rights of senior lenders influence firm actions in ways that can increase firm value. Nini, Smith, and Sufi (2012) examine 3,500 incidences of financial covenant violations for the universe of US firms filing quarterly and annual financial reports with the SEC between 1997 and 2008. Their study shows that covenant violations are followed immediately by declines in acquisitions and capital expenditures, sharp reductions in leverage and shareholder payouts, and increases in CEO turnover. These changes coincide with amended credit agreements that contain stronger restrictions on firm decision making. Of critical importance, both operating performance and equity-market valuations improve following a covenant violation; in other words, strong creditor rights and the associated creditor intervention are associated with a turnaround in performance. This empirical evidence clearly shows that creditor influence on managerial decisions extends beyond states of default, and in particular that senior creditors begin to play an active role in corporate governance when firm performance first deteriorates.[7]

The second important observation is that contractual restrictions on the borrower increase when firm performance declines. Both Nini et al. (2012) and a survey by Ayotte, Hotchkiss, and Thorburn (2012) describe the terms of financing available to declining firms. They note that creditor control is exercised through secured lines of credit, which are extended to the firm both before and after it files a bankruptcy petition. Renegotiated credit agreements impose stronger contractual restrictions on the borrower, carry higher interest rate spreads, and are more likely to require collateral. Denis and Wang (2014) demonstrate the high incidence of loan renegotiation even absent a covenant default, which are likely workouts in advance of a potential default. These events indicate that even outside of observed defaults, creditors have strong influence over the borrower's operating and financial policies. Arguably, without these concessions, these lenders might not continue to extend credit to a distressed firm.

While senior creditors might be the first to intervene in a declining firm, control rights can also extend to more junior unsecured debt. Even when firms are not near distress, certain corporate actions specified in the bond indenture, such as changes to financings, asset sales, or acquisitions, can require the consent of a specified percentage of bondholders. If the firm violates these consent requirements, bondholders can accelerate payment of the debt. Hence, the threat of acceleration provides bondholders with a voice in what actions the firm can take, or, with the ability to negotiate a change in the terms of the debt that improves the value of the bond in exchange for consenting to these actions.[8]

Closer to or in default, control arguably matters the most. When a group of investors accumulates a significant stake in a class of debt, they can have a strong influence on the outcome of an out of court restructuring. For example, without their participation, a distressed exchange might fail or might not sufficiently reduce the firm's debt burden.[9] Empirically, Hotchkiss and Mooradian (1997) demonstrate that bond investors frequently purchase just over the "1/3 threshold" in a class of debt – a position sufficient to block any bankruptcy restructuring plan – often well in advance (or in the absence of) a subsequent bankruptcy. Bondholders, as well as other creditors, can form 'ad hoc' committees, which are "informal groups of sophisticated investors who pool resources to advance their common interests in out-of-court restructurings and bankruptcy cases" (Wilton and Wright 2011).

If a firm does file for Chapter 11, critical decisions – to which creditors can object – are made in the first days of the case. Such decisions involve financing, asset sales, rejection of contracts such as leases, formation of creditor committees, and so on (as described in Chapter 3 herein). Hence, debtholders can exercise substantial influence over both out of court and bankruptcy restructurings, and the corresponding decisions that are made in advance or to avoid such events.

CHANGES IN CONTROL AND IMPACT OF CLAIMS TRADING

When distressed firms restructure their debt, there are often significant changes in the ownership structure of the firms' residual claims. The original equity holders often receive little or no shares in the reorganized firm; stock is largely distributed to former creditors, who become the new owners of the company. Investors who specialize in investments in distressed firms frequently purchase claims from creditors looking to sell their stakes, which in some cases are converted into a sizeable or even a majority ownership position. There may also be an infusion of equity to the reorganized firm from an investor as part of the restructuring plan.

The earliest studies of ownership changes for distressed firms after the 1978 Bankruptcy Reform Act did not find a great deal of control activity for the firms studied. A possible explanation is that the prodebtor orientation early in the history of Chapter 11 discouraged acquisitions.[10] Still, equity distributions under reorganization plans led in many cases to a concentration of the firm's ownership in the hands of prior creditors. Gilson (1990) studies 61 Chapter 11 bankruptcies from the 1980s, finding that, on average, 80% of the common stock in the reorganized firm is distributed to creditors. The distribution of equity in the reorganized firm to prepetition creditors continues to be typical of Chapter 11 reorganizations; significant distributions to prepetition equity holders are the exception rather than the norm (see Chapter 7 herein for description of bankruptcy case outcomes).

While prepetition creditors often become shareholders of a successfully reorganized firm, two significant trends are important in understanding changes in control for firms in distress or bankruptcy. The first is the use of Section 363 of the Code to sell not just a portion, but "substantially all" assets of the firm. We discuss key aspects of these procedures in Chapters 3 and 7 herein. The second is the rise of the market for trading claims of distressed companies, which grew dramatically in the 1980s and early 1990s, accompanied by the rise of hedge fund and other investors in distressed debt (see Chapters 3 and 14 herein).

Generally, it is not possible to publicly observe the exact identity of owners of a firm's debt and other claims. An exception to this is a study by Ivashina, Iverson, and Smith (2016), who study 136 companies filing for Chapter 11 between 1998 and 2009. These researchers were able to construct a full snapshot of claims ownership at the firm level, both at the time of filing and at the time of voting on a reorganization plan, permitting them to document the types of owners and the consolidation of claims during the case, and to relate these changes to case outcomes.[11] Active investors, including hedge funds, are the largest net buyers of claims in bankruptcy. Perhaps the most important fact they document is that trading leads to more concentrated ownership, particularly of claims whose holders are eligible to vote on the bankruptcy plan and claims that represent the "fulcrum" security in the capital structure.

A common strategy for an activist investor is to acquire a large enough stake in the prereorganization claims to potentially block a plan of reorganization, giving the investor influence over the course of the restructuring. The investor can hold out or litigate for a greater distribution to their class of claims, or, depending on the final negotiated terms of the plan, potentially convert the claims into a majority equity stake in the reorganized firm. This activity provides an interesting contrast to control contests for nondistressed firms. When an investor acquires more than

5% of a firm's publicly traded stock, it must file a 13-D statement with the SEC disclosing its holdings and future intentions with respect to the company. For purchases of public debt, or other types of nonequity claims for that matter, there is no such requirement.

In an early study of the behavior of distressed debt investors, Hotchkiss and Mooradian (1997) find that "vulture" investors gain control of 16.3% of the sample of 288 firms defaulting on their debt between 1980 and 1993. The influence of these investors has become even more prevalent over time – Jiang, Li, and Wang (2012) show that close to 90% of their sample of 474 large Chapter 11 cases from 1996 to 2007 have publicly observable involvement by hedge funds. Li and Wang (2016) study the distinct effects when an active investor follows this strategy by becoming the DIP lender in bankruptcy; they find this occurs for 13% of their sample of 658 Chapter 11 cases of large public firms between 1996 and 2013 (up from 10% in the late 1990s to 20% a decade later). Whether the presence of these investors adds to or detracts from the efficiency of the restructuring process has been sharply debated, with anecdotal evidence supporting both sides.[12]

The negative view of distressed debt investors arises when these investors seek some payment by attempting to delay or "hold up" the process. For example, a hedge fund investing in junior claims that are not plausibly "in the money" might threaten litigation in order to negotiate a settlement payment in violation of absolute priority. Thus, the presence of such an investor makes the bargaining process in bankruptcy more difficult and costly. The negative view has also applied to hedge funds following a strategy of extending credit to a distressed firm before or during bankruptcy, with the potential to convert the loan into a controlling equity stake through a restructuring. This behavior has come to be known as a "loan-to-own" strategy. For example, Gilson (2014) writes about the case of School Specialty, in which a hedge fund provided the distressed company with a high-risk loan containing a massive "make whole" provision, which would be triggered by repayment of the loan before maturity. The hedge fund further agreed to provide DIP financing upon a bankruptcy filing, the terms of which required the firm to pursue a Section 363 sale of the firm with the hedge fund as the stalking horse bidder. Thus, the hedge fund's loans provided critical liquidity when banks were no longer willing to provide credit – but at the same time, the loans provided the hedge fund with substantial leverage over the debtor and the outcome of the bankruptcy case.

A more positive view is that an activist might push for a transaction that yields higher creditor recoveries. For example, litigation by hedge funds investing in the second lien debt of American Safety Razor led to the sale of the company to a strategic buyer for $301 million, dominating a planned credit bid of $244 by first-lien lenders (Ellias, 2016). Activist investors in a company's junior debt can also provide a check on incentives of senior lenders to undervalue the reorganized firm in an attempt to capture a greater share of the reorganized firm's value. Large sample academic studies point out the potential benefits of distressed debt investors. Hotchkiss and Mooradian (1997) find stronger postbankruptcy performance when hedge funds are active in the governance of reorganized firms; Jiang, Li, and Wang (2012) find that participation of hedge funds is associated with a higher probability of emergence and higher payoffs to junior claims; Lim (2015) finds that hedge fund involvement is associated with a higher probability of completing prepackaged restructurings, faster restructurings, and greater debt reduction. Ellias (2016) measures the intensity of the litigation campaigns embarked upon by investors in junior claims; he finds that "junior activism" is associated with higher values of reorganized firms, and that settlements junior activists receive are relatively small. The rise in defaulted debt of companies owned by private equity funds has further led to a tremendous increase in the activity of investors that also manage private equity funds (Hotchkiss, Strömberg, and Smith 2016). In some cases, though the PE fund's original equity stake is diluted or eliminated in the restructuring, the PE sponsor can maintain its control of the distressed firm via an equity infusion as part of the restructuring.

An important caveat to all of this research is that we cannot infer causality of the activists' behavior on case outcomes. Still, the positive role of these investors in corporate governance of distressed firms is consistent with broader research in corporate finance demonstrating that the market for corporate control provides an important discipline on management of underperforming firms (Jensen, 1986). This evidence is also consistent with literature that examines the impact of activist hedge fund investment in equity markets. This body of work generally demonstrates that hedge fund investments are associated with value gains for the target companies.[13]

CEOS, BOARDS, AND LABOR MARKETS

Management Turnover and Compensation

An important aspect of the governance changes that occur as part of a distressed restructuring, particularly when there is a change in control, are the changes in the firm's management and board of directors. Much of our understanding of management and board turnover dates back to Gilson (1989, 1990). From his sample of distressed firms restructuring in the 1980s, he finds that 52% of firms experience senior management turnover (individuals with the title of CEO, Chairman, or President) in a four-year period beginning two years before default. Turnover is significantly higher for firms that restructure in bankruptcy versus out of court. For 69 bankruptcy cases within his sample, 71% of these individuals are replaced as of two years after filing. Only 46% of incumbent directors remain following the restructuring. A number of recent studies continue to find extremely high turnover rates in bankruptcy. For 322 Chapter 11 filings of large public U.S. companies between 1996 and 2007, Eckbo, Thorburn, and Wang (2016) find that 86% of CEOs in place three years before filing leave by the end of the year following the bankruptcy restructuring. Li and Wang (2016) find that about two-thirds of firms that emerge from bankruptcy and continue to file with the SEC have experienced CEO turnover during the Chapter 11 restructuring, and 80% of board members at the time of filing retain their board seats after emerging. The loss of these positions is even higher when one also considers turnover before filing, and from firms that do not successfully emerge or which emerge as private companies.

As firms become financially distressed, a substantial commitment of time and attention is required of managers and directors to address the firms operating problems and develop a restructuring plan. In addition, increasing concerns about personal liability may limit the ability to attract high quality officers or directors (this has been suggested as one argument in favor of limiting director's fiduciary duties to creditors). Certain parties, such as those making investments in the distressed firm or groups concerned about protecting their interests in the restructuring, may desire to take board seats. Some, however, are unwilling to take board seats until the reorganization is complete if doing so would limit their ability to trade claims in the distressed firm. Finally, a potential conflict of interests exists when distressed debt investors (such as private equity funds) have simultaneous fiduciary responsibilities to shareholders of the reorganized firm and to investors in their own funds. In such cases, managers and directors involved with the company and its restructuring must take actions in the interest of the corporation "for the benefit of all of its residual claimants, a category which now includes creditors."[14]

The high turnover rate runs counter to the notion that managers are overly protected by the bankruptcy process. Another important component of governance changes are the compensation contract provided both to incumbent management as the firm becomes distressed, and to replacement managers. Gilson and Vetsuypens (1993) provide an early study of distressed firms' compensation policies, examining the managers in place as the firm enters financial distress as well as the managers who replace them. In addition to the high turnover previously documented, managers who remain generally take substantial cuts in their salary and bonus, and replacement CEOs earn on average 35% less than their predecessors. Outside replacements typically receive large grants of stock or options as part of their compensation, giving the CEO a high sensitivity of wealth to the firms' future stock price performance.

Compensation practices have significantly changes since the 1980s, however. Eckbo, Thorburn, and Wang (2016) compare compensation before and after bankruptcy, and follow the career paths of CEOs after departing the firm. Approximately two-thirds of departing CEOs leave the executive labor market, becoming nonexecutive directors, consultants, or retiring. For those that continue elsewhere as executives, they estimate the present value of the change in future compensation (relative to compensation at the year prior to departure) has a median of zero. Those that do not continue, however, experience substantial compensation losses, with a median present value of 4.6 times their predeparture compensation. These studies highlight the large potential personal costs to the CEO of a failing firm.[15] Unless managers are adequately incentivized to do otherwise, the risk of bearing these costs might incentive the CEO to take actions that hedge against bankruptcy at the expense of shareholder or firm value.

Setting the CEO's compensation contract, and the incentives it provides, can be problematic as firms move closer to default. An ideal contract balances the need to attract or retain qualified managers while not enabling entrenchment. For poorly performing firms, conditioning cash bonuses and the vesting of equity grants on achieving prespecified performance targets may be particularly useful. On the other hand, as the firm becomes financially distressed, performance metrics included in the contract may not be informative of the CEO's effort, and so would not adequately address agency issues. The use of incentive pay for firms facing financial distress may also impose too much risk on a risk-averse CEO, since missing performance targets in compensation contracts is tied to greater forced CEO turnover (Bennett, Bettis, Gopalan, and Milbourn 2017). Unlike criticism in the popular press that compensation practices for distressed firms are rigged in the CEO's favor, Carter, Hotchkiss, and Mohseni (2018) find that the use of performance based pay for distressed firms increases, and reliance on discretionary bonuses (those unrelated to performance) decreases. They also find that performance targets are set higher relative to past performance as the firm declines. Still, the press has brought attention to specific cases of large cash bonuses to executives in the days before a bankruptcy filing, even to executives who soon left the firm anyway. For example, Hostess Brands raised the salary of its CEO by approximately 300 percent (to over $2.5 million) and increased pay for other executives six months before filing; creditor pressure and negative press coverage led the company to cut back these raises after the firm filed for bankruptcy.[16] And, as Ellias (2018) points out, since bonuses are more heavily scrutinized should the firm file for bankruptcy, firms have incentives to make payments to executives before filing (subject to clawbacks – see Chapter 3 herein), or to wait until after their time in bankruptcy court.

Compensation in Bankruptcy: KERPs and KEIPs

Cash payouts may arguably be needed to retain management, and may be consistent with maximizing firm value by preserving firm-specific experience or relationships with customers. On the other hand, they may be seen as enriching the same managers responsible for the firm's decline.

In bankruptcy, payouts and contracts must be approved by the court, and can involve contentious debate as to whether compensation schemes provide incentives to ensure the success of the restructuring or are largely for retention.

Key Employee Retention Plans (KERPs) and Key Employee Incentive Plans (KEIPs) are compensation plans designed to award senior managers who are deemed critical to the restructuring and continuation of the business. KERPs award bonuses to key employees for staying with the debtor firm for a specific length of time or throughout the restructuring process. Firms advocate for such plans on the basis that management departures are highly disruptive to the successful and expeditious resolution of bankruptcy, and employee turnover results in a loss of continuity coupled with high search and training costs for replacement personnel. KEIPs tie bonuses to restructuring milestones and outcomes, including plan confirmation, speed of restructuring, sale of assets, debt recovery, improvement to EBITDA and enterprise value, and so on.

Following several high-profile companies paying large bonuses shortly after filing for bankruptcy, retention bonuses were essentially banned under an amendment of Section 503(c) of the bankruptcy code under the BAPCPA.[17] The bankruptcy court cannot authorize payments to induce an insider to stay with the debtor unless the manager can show proof of a competing job offer at the same or greater compensation, and it is deemed that the services provided by the manager are essential to successfully rehabilitate the firm. The new section also limits retention and severance payments to an amount not exceeding 10 times the mean amount given to nonmanagement employees in the same year and, if in the previous year no payments were paid to nonmanagement employees, cannot exceed 25% of the amount paid to the managers.

The new rule, however, does not impose payment restrictions on incentive bonuses. Figure 6.1, shows the annual frequency of use of KERPs and retention-only bonus plans by all Chapter 11 filings by U.S. public firms with book assets above $100 million (in constant 1980 dollars) from 1996 to 2013. It is clear that retention plans without incentive bonuses largely disappeared after the BAPCPA, while the overall adoption of KERPs and KEIPs remained around 35%.[18] Capkun and Ors (2016) suggest that firms conduct regulatory arbitrage

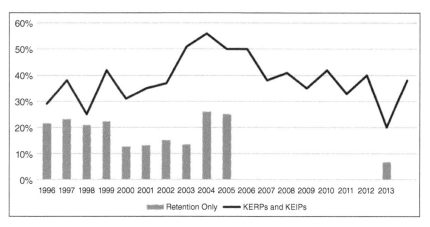

FIGURE 6.1 Adoption of KERPs and KEIPs
Source: Goyal and Wang (2017).

to avoid the BAPCPA's payment restrictions by disguising retention plans as incentive plans. Ellias (2018) further discusses the difficulties in distinguishing incentive from retention plans, and actions taken by firms that can serve to circumvent provisions of the Bankruptcy Code. For example, in recent cases debtors have paid out large retention payments to executives on the eve of bankruptcy, with a contractual obligation to give the money back if they quit before the retention term ends.[19] Adoption of compensation plans and their terms are frequently objected to and contested by the Creditors' Committee, the United States Trustee, unions, and other stakeholders.

Goyal and Wang (2017) take a comprehensive look at the features of KERPs and KEIPs and investigate the economic rationales for these plans. They find that approximately 40% of large Chapter 11 firms between 1996 and 2013 adopted either KERPs or KEIPs. In their sample, the plans typically include not only senior executives but also middle-level managers as key employees, and typically classify them into three to four tier groups with different pay ratios and bonus targets. Retention plans always tie bonus payments to the minimum length of stay and plan confirmation, while incentive plans most often tie bonuses to emergence, EBITDA targets, and asset sales. These plans cost the firm $8–9 million on average, representing about 0.4% of firms' book assets at filing. Their empirical results show that these plans are more likely to be adopted by firms with a complex operation and claim structure, strong creditor control, and that are located in stronger employment markets. Newly hired turnaround managers are more likely to be covered by these plans than incumbent CEOs.

COLLECTIVE BARGAIN AGREEMENTS (CBA) AND EMPLOYEE BENEFITS

For distressed firms, renegotiation of labor costs and employee benefits has often been an important part of the strategy to improve the chances of returning to profitability. Firms in industries with higher unionization rates than the broader economy, such as airlines and steel companies, have historically struggled to reduce labor costs and other employee benefit obligations such as health and pension plan contributions. In particular, as cash flows decline, many firms reduce contributions to retirement plans, resulting in underfunded pensions for firms with defined benefit plans. Although the share of defined benefit plans in large U.S. companies has significantly fallen over time, being replaced by defined contribution plans, a number of large companies face "legacy" costs from earlier plans.

Benmelech, Bergman, and Enriquez (2012) explore the role of financial distress in firms' ability to gain wage concessions. Using data from the airline industry, they find that firms that are distressed extract greater wage concessions from employees whose pensions are underfunded and not fully covered by the

Pension Benefit Guaranty Corporation (PBGC). Although most defined benefit pension plans are insured by the PBGC, coverage is limited, and higher paid employees stand to lose more. Thus, the risk of losing pension benefits should the pension be terminated in bankruptcy enables these firms to successfully use their financial condition to extract surplus from labor. Still, the ability to renegotiate labor agreements out of court is not sufficient for many of these firms to avoid a bankruptcy filing.[20]

In bankruptcy, the debtor can further renegotiate its prefiling labor contracts to enable it to emerge from bankruptcy with a more viable cost structure. As discussed in Chapter 3 herein, among the myriad of business decisions the debtor must make in Chapter 11 are decisions to assume or reject outstanding contracts (Section 365); damages to the counterparty of the contract become an unsecured claim. However, since the 1980s, collective bargaining agreements are governed by a different set of rules (Section 1113), which provides that rejection can happen only after the debtor has tried to negotiate with the union. The debtor must follow a set of steps showing the judge that bargaining has failed, and the debtor had no choice but to reject the contracts.[21] Rejecting labor contracts has yielded annual savings in billions of dollars in some major airline bankruptcies (e.g., UAL Corp. in 2002).

Union representatives can be appointed to the unsecured creditors committee when labor claims are large. Greater distributions negotiated on behalf of labor can therefore come at the expense of other unsecured and more junior claims. Consistent with the importance of unions to the outcome of restructurings, Campello, Gao, Qiu, and Zhang (2018) show that recovery rates to unsecured bondholders are lower when unions have more seats on the unsecured creditor's committee. In addition, firms with labor unions incur significantly higher expenses and fees paid in bankruptcy court, and have a higher likelihood of refiling after emergence from bankruptcy.

Bankruptcy can impose large losses to participants in both defined benefit (DB) and defined contribution pension plans. For DB plans, debtors in bankruptcy can petition the PBGC to terminate an underfunded pension if the company cannot pay its debt under a plan of reorganization and continue outside bankruptcy, or if the cost of providing pension benefits has become unreasonably burdensome as a result of declining covered employment. Once a pension plan is terminated, the PBGC assumes control of the plan's assets and pays benefits (at capped dollar limits) to participants out of these assets and its own funds (the PBGC has a large deficit itself). The PBGC's claim for termination liability is treated as a general unsecured claim. Not all firms in Chapter 11 attempt to, or successfully terminate DB plans: Duan, Hotchkiss, and Jiao (2015) find that among bankrupt firms with DB plans, only 20% of those cases in fact terminate the plan and transfer the assets to the PBGC. Defined contribution plans can potentially imposes large costs on employees of distressed firms when the plan has a high percentage of assets invested in the sponsor company's stock. The highly publicized lawsuits against

Enron by its 401(k) participants were driven by the plan's loss from Enron's stock, which accounted for more than 60% of plan assets in 2000 (*Wall Street Journal*, November 23, 2001). Similarly, employees of Lehman Brothers who owned stock through the company's Employee Stock Ownership Plan (ESOP) suffered millions of dollars in losses as the firm descended into bankruptcy.

ERISA provides only limited protection for other post-employment benefits (OPEBs), such as health coverage and life insurance. Section 1114 of the Code was added in 1988 to provide the court with guidance and protection of OPEBs against the threat of termination, in response to public outrage over LTV Corp's abrupt termination in bankruptcy of the benefits of 78,000 retirees without advance notice. Section 1114 provides a procedure, similar to Section 1113, for modifying debtor liabilities for OPEBs, and was further amended by the BAPCPA regarding benefit modifications prior to filing.

NOTES

1. Tirole (2001) describes the costs and benefits of various methods of protecting non-controlling stakeholders including covenants, exit options, and enlarged fiduciary duty. Gilje (2016) documents that firms reduce investment risk when they approach financial distress. His results are consistent with the argument that debt composition and financial covenants serve as important mechanisms to mitigate debt-equity conflicts that are not explicitly contracted on. See also Berk and DeMarzo (2017) (p. 565) for discussion of shareholder/debtholder conflicts.
2. For one such example, see *Forbes,* August 28, 2014: "Apollo and TPG pulled off a $30 billion leveraged buyout of the casino company in 2008 that has performed badly. In May, Apollo stripped protections on debt payments on some Caesars bonds and earlier Caesars transferred some casino assets to affiliates." See also Bloomberg, June 4, 2018: Both J. Crew Group Inc. and Claire's Stores Inc. created subsidiaries to hold assets including intellectual property, insulating them from creditors while freeing them up for use as collateral to back new debts. With retailers dominating lists of troubled issuers, lenders and analysts have been speculating about who'll be next to "pull a J. Crew."
3. See Toibb v. Radloff, 501 U.S. 157, 163 (1991): a fundamental Bankruptcy Code objective is "maximizing the value of the bankruptcy estate."
4. In bankruptcy, creditors are similarly protected under the Bankruptcy Code, Sections 544, 547 and 548. Ersahin, Irani, and Waldock (2017) summarize the history of fraudulent transfer law in the U.S.
5. See *North American Catholic Educational Programming Foundation, Inc. v. Gheewalla*, 930 A.2d 92, 99 (Del. 2007).
6. A review by the law firm Wilmer Hale, prepared for the Loan Syndications and Trading Association, documents the inclusion of milestones for a large sample of bankruptcies. 89/106 include a milestone related to the plan approval or effective date, which in effect limit the life of the loan. 90/107 include a milestone related to the closing of a sale or specifically to the bidding process for a Section 363 sale. Their descriptive analysis

points out that in many cases of liquidation or asset sales, reorganization was never anticipated at the onset of the case and a sale had been planned prior to filing.

7. A growing literature suggests that creditors directly influence corporate decisions when a firm violates debt covenants: Chava and Roberts 2008, Roberts and Sufi, 2009b, Nini, Smith, and Sufi, 2009, 2012 (investment and other decisions); Ferreira, Ferreira, and Mariano, 2017 (board composition); Balsam, Gu, and Mao (2018) and Akins, Bitting, De Angelis, and Gaulin, 2018 (CEO compensation contracts) and Zhang, 2018 (trade credits).

8. Kahan and Rock (2009) document cases in which bondholders accumulate a large enough position to engage in negotiations with management to improve the value of their claims. Feldhütter, Hotchkiss, and Karakaş (2016) show that control rights are priced in corporate bonds as firms approach default.

9. Gertner and Scharfstein (1991) suggest that large debtholders can negotiate directly with the firm to ensure the success of the offer.

10. While acquisitions in bankruptcy are a more recent phenomenon, equity infusions as part of a Chapter 11 restructuring are not a new. Gilson, Hotchkiss, and Ruback (2000) find that 12 out of 63 cases (17.5%) they study have such an investment, for which the median percentage of equity acquired is 54.2%. These changes in control were largely friendly, in the sense that incumbent management was retained.

11. Ownership structure is constructed using data from Chapter 11 claims agents, under a research project funded by the American Bankruptcy Institute (ABI) Endowment Fund.

12. The ABI Bankruptcy Reform Commission (see Chapter 3 herein) expressed concern that the growth of trading in distressed debt, and in particular the participation of hedge funds, has reduced the effectiveness of the Chapter 11 process.

13. A useful survey of this literature is provided by Brav et al. (2010), who summarize it by explaining that "the evidence generally supports the view that hedge fund activism creates value for shareholders by effectively influencing the governance, capital structure decisions, and operating performance of target firms."

14. See the Quadrant case discussed earlier in this chapter.

15. The costs of financial distress are also borne by nonmanagement employees. Graham et al. (2016) show that annual employee earnings deteriorate by 10% when a firm files for bankruptcy and remain below prebankruptcy earnings for at least six years. Brown and Matsa (2016) show that financial distress results in fewer and lower quality job applicants. Using detailed employer-employee matched data from Sweden, Baghai (2017) show that firms lose their most skilled workers as they become financially distressed.

16. Fortune, July 26, 2012.

17. Examples include: FAO Schwarz, 2003 (executive retention plan paying $1.1 million in bonuses 10 days after filing (NYT 2/15/2003); WorldCom, 2002 (NYT 12/17/2002).

18. Only one case adopted a retention-only plan through Section 503(c): GMX Resources, Inc. filed for Chapter 11 in the Western District of Oklahoma on April 1, 2013, and had its KERP approved on June 11, 2013.

19. For example, Exco Resources filed for Chapter 11 on January 15, 2018, shortly after granting "prepaid cash retention awards" to the five seniormost executives of the company described as "critical to the restructuring efforts." After filing, the firm sought court approval for KEIPs with terms satisfying the requirements of the bankruptcy code.

20. When American Airlines filed for Chapter 11 in 2011, its competitors had already reduced their most expensive union contracts and restructured their debt in bankruptcy. American tried to gain concessions from its unions out of court, but was left with higher unit labor costs than its competitors. The CEO explained the reason for filing: "It became increasingly clear that the cost gap between us and our biggest competitors was untenable." See "American Lands in Bankruptcy: Parent of No. 3 Airline Seeks Court Protection Amid High Fuel and Labor Costs," *Wall Street Journal*, November 30, 2011).

21. The court may approve an application to reject the CBA only if it finds that: (1) the debtor has, prior to the hearing, made a proposal that fulfills the requirements; (2) the authorized representative of the employees has refused to accept the proposal without good cause; and (3) the balance of the equities clearly favors rejecting such agreement.

CHAPTER **7**

Bankruptcy Outcomes

A key goal of Chapter 11 is to provide economically viable firms an opportunity to reorganize, while liquidating those that are not viable. There has been considerable debate as to whether the current bankruptcy code strikes the right balance between reorganization and liquidation, or whether it is biased toward allowing inefficient firms to reorganize. The number of failures following Chapter 11, as seen in the significant number of Chapter 22 filings, might be taken as evidence of a problem with the structure of Chapter 11. At the same time, we have seen some spectacular success stories upon emergence from bankruptcy, at least from the perspective of equity holders in the reorganized company. For example, Kmart's common stock traded at under $14 per share when the firm emerged from Chapter 11 in May 2003, with 53% of its shares owned by ESL Investments. The stock rose more than seven times to over $100 per share by November 2004, when its merger with Sears was announced (the longer term fate of the merged company was, of course, less stellar, leading to its bankruptcy filing in 2018). Six Flags emerged from Chapter 11 in 2010, with its stock trading at less than $10 per share; as the firm underwent a successful operational restructuring, the stock price approximately doubled in the first postbankruptcy year and continued to increase thereafter.

There are several ways in which one might evaluate the success of Chapter 11. The simplest measure of a "successful" restructuring might be whether the firm in fact emerges from the process as a going concern. For firms that do emerge, researchers have examined several measures of postbankruptcy success. This chapter describes recent evidence on bankruptcy outcomes and postbankruptcy performance, and their relevance to the debate over the efficiency of Chapter 11.

OUTCOMES OF CHAPTER 11 CASES

When a firm enters Chapter 11, the expected outcome in most cases is to confirm a plan under which the firm is reorganized (or sometimes sold) as a going concern. However, only a fraction of firms that enter Chapter 11 emerge as independent companies. The Executive Office for United States Trustees provides comprehensive national analysis of the confirmation rates of Chapter 11 cases, and periodically publishes analyses of case outcomes. The most striking facts emerging from their analysis are the following:

- A large proportion of cases are closed without confirmation, or closed as a no asset Chapter 7 case.
- Estimated confirmation rates for cases do not exceed 45% (and are likely substantially less) in any year since 1990.
- Many confirmed plans are liquidating plans.
- There is a strong correlation between the amount of assets listed by the debtor and the confirmation rate.

Confirmation rates for all national Chapter 11 cases since 2008 are provided in Figure 7.1.

These figures include all national filings, not only those of publicly registered companies. A caveat in interpreting these figures, however, is that large companies entering Chapter 11 often file a number of bankruptcy petitions for the various entities within their firm, and each individual case is treated as a separate

Fiscal Year Filed	Total Cases Files	Total Confirmed	Percent Confirmed[1]
2008	8,869	2,934	33.1%
2009	14,816	5,892	39.8%
2010	14,296	5,110	35.7%
2011	12,115	4,186	34.6%
2012	10,678	3,902	36.5%
2013	9,679	3,573	36.9%
2014	7,691	2,434	31.6%
2015	7,110	1,918	27.0%
2016	7,494	1,515	20.2%
2017[2]	7,100	572	8.1%

[1]Confirmation rates may be slightly understated because cases that were dismissed or converted after confirmation but before closing are not counted as confirmations.
[2]As of September 30, 2017.

FIGURE 7.1   National Chapter 11 Filing and Confirmation Figures by Year since 2008
Source: Federal Judicial Center, based on reports from the Administrative Office of the U.S. Courts (https://www.fjc.gov/research/idb).

observation in this analysis. For example, in early 2004, Footstar Inc. filed in the Southern District of New York. This case included approximately 2,510 separate filings—about 20 percent of all of the Chapter 11 cases filed nationwide that year. While the number of related cases is generally not nearly as extreme as for Footstar, this problem does lead to an overstatement of confirmation rates, since bigger cases with a number of related filings are more likely than average to reach confirmation. Still, it is clear that well over 60% of cases are closed without confirmation. A likely explanation is that the direct and indirect costs of distress and bankruptcy make reorganization in bankruptcy infeasible for many smaller firms (Bris, Welsh, and Zhu 2004).

Confirmation rates are particularly useful in understanding the outcomes of cases for smaller firms. For larger firms entering Chapter 11, further classifications of case outcomes describe of how assets are redeployed. Broadly, there are four economic outcomes to a Chapter 11 bankruptcy case:

- *Emerging from Chapter* 11. A plan of reorganization is confirmed, and the firm exits bankruptcy as an independently operating private or publicly registered company. A subset of the publicly registered firms successfully list their stock on an exchange sometime following emergence.
- *Acquired in Chapter* 11. The firm is merged with another operating company as part of a Chapter 11 plan,[1] or is sold as a going concern to another operating company or to a financial buyer, generally under Section 363 of the Bankruptcy Code.
- *Liquidated*. This includes firms whose case is converted to a Chapter 7 liquidation, or which are liquidated piecemeal (not as a going concern) under a liquidating Chapter 11 plan.
- *Dismissed*. The bankruptcy case is dismissed shortly after filing for a small number of firms. More recently, a number of cases have utilized a "structured dismissal," particularly when the debtor has sold most assets under Section 363 sale but cannot confirm a plan because it is unable to pay administrative claims in full.[2] A "structured dismissal" provides for the exit from Chapter 11 but can require distributions to certain creditors, grant third-party releases, or preserve transactions undertaken during the case.

Much of what we know about Chapter 11 case outcomes is, by necessity, based on studies of public companies that have entered Chapter 11. An exception is Waldock (2017), who uses information from PACER to examine 2,621 Chapter 11 cases between January 2004 and July 2014 with assets of at least $10 million (based on the bankruptcy petition). Of the firms in her study, 33.3% of firms are reorganized, 15.8% acquired, 21.5% liquidated, and 27.1% dismissed. The relatively high rates of liquidation and dismissals are likely due to the inclusion of smaller firms than previous studies based on firms reporting financial statements (from SEC filings, as reported to Compustat) prior to filing.

In a study of over 3,000 public companies entering Chapter 11 between 1981 and 2013, Altman (2014) reports that 65% of firms either emerge as a continuing entity or are acquired. Similarly, Iverson, Madsen, Wang, and Xu (2018) find that 57% of firms filing from 1980 to 2012, with assets of at least $50 million, emerge from Chapter 11. Earlier studies report similar statistics, showing that larger public companies have a significantly greater likelihood of emerging or being sold as a going concern. Firms that are liquidated after failed efforts to reorganize are typically smaller firms, but do include some well-known large failures such as Circuit City. These firms spend several months or longer in Chapter 11 before they move to Chapter 7 or a liquidating Chapter 11 plan.

A relatively small number of firms merge with another operating company under a plan of reorganization. Hotchkiss and Mooradian (1998) find that mergers are an effective mechanism for redeploying the assets of Chapter 11 firms, in the sense that the combined cash flows of the merged company after Chapter 11 increase by more than is observed for similar nonbankrupt transactions. It is significantly more likely that acquisitions occur using Section 363 of the bankruptcy code (see Chapter 3 here in and discussion below).

Several researchers have tried to identify factors that are related to the probability a firm successfully emerges from Chapter 11. One of the first such attempts was Hotchkiss (1993). The overwhelmingly most important firm characteristic related to whether firms successfully reorganized rather than liquidated was firm size, measured by the prepetition assets of the company. Hotchkiss shows that many of the emerging firms have considerably downsized while in Chapter 11, so that the ability to divest assets and use the proceeds to fund remaining operations is likely to be important in understanding why these firms are more likely to survive Chapter 11. Dahiya, John, Puri, and Ramirez (2003) argue that the availability of debtor-in-possession (DIP) financing to large companies is an important determinant of the reorganization versus liquidation outcome. Consistent with these studies, Denis and Rodgers (2007) focus on events during the Chapter 11 process and show that firms that both reduce their assets and liabilities while in Chapter 11 are more likely to emerge as an independent firm.

POSTBANKRUPTCY PERFORMANCE

If Chapter 11 does in fact suffer from economically important biases toward continuation of unprofitable firms, poor investment decisions will be reflected in the postbankruptcy performance of firms emerging from the process. Therefore, researchers have found it useful to examine several dimensions of post-emergence performance:

- Accounting measures of profitability
- Ability to meet cash flow projections on which the reorganization plan is based

- Incidence of subsequent distressed restructurings (including Chapter 22 filings)
- Post-emergence stock performance

These studies are, by necessity, based on firms that survive Chapter 11 as publicly registered companies. Existing studies of performance of firms emerged from Chapter 11 are summarized in Figure 7.2, and described in the remainder of this section.

1. *Accounting measures of profitability*

The first comprehensive analysis of postbankruptcy operating performance was Hotchkiss (1995), who examined firms that emerged as public companies from Chapter 11 by 1989. These firms had an average book value of assets prior to filing of $285 million, were generally insolvent at the time of filing, and spent on average 1.7 years in bankruptcy. The financial performance of each firm was traced for up to five years following the time of emergence from bankruptcy.

This early analysis produced some striking results. Over 40% of the firms emerging from bankruptcy continued to experience operating losses in the three years following bankruptcy. Based on accounting ratios such as return on assets and profit margins, performance was substantially lower than for matched groups of firms in similar industries. For example, in the year following emergence from Chapter 11, almost three quarters of the sample firms have a ratio of operating income to sales that is lower than observed for nonbankrupt firms in the same industry. The firms showed some positive growth in revenues, assets, and number of employees in the postbankruptcy period, but showed little improvement in profitability, especially in comparison to industry groups. Performance varied little over the five-year postbankruptcy period, which suggests the firms did not simply need more time to recover.

This analysis has been confirmed for a subsequent time period by Hotchkiss and Mooradian (2004). The most significant difference from the earlier study is that larger firms, which become more prominent in later sample years, have somewhat better performance based on these accounting measures. Still, for even the larger firms in the updated sample, more than two-thirds of the firms underperform industry peers for up to five years following bankruptcy, and over 18% of the sample firms have negative operating income in the year following emergence. Denis and Rodgers (2007) also study postbankruptcy operating performance, and show that operational restructuring during bankruptcy is associated not only with a greater likelihood of emergence, but a higher industry adjusted operating performance following emergence.

An important concern in interpreting any analysis of postbankruptcy performance based only on firms that survive Chapter 11 is the fact that firms' asset composition changes significantly before and during bankruptcy. Maksimovic

	Operating Performance	Ability to Meet Cash Flow Projections	Stock Performance	Sample
Hotchkiss (1995)	√	√	√	197 firms emerging by 1989
Hotchkiss and Mooradian (2004)	√		√	620 firms emerging by 2004
Maksimovic and Phillips (1998)	√			Plant level data for 302 manufacturing firms in Chapter 11 1978–1989
Alderson and Betker (1999)	√			89 firms emerging from Chapter 11 1983–1993 (includes 62 emerging 1990–1993)
Hotchkiss and Mooradian (1997)	√			288 firms defaulting on public debt 1980–1993 (166 are reorganized in Chapter 11)
Michel, Shaked, and McHugh (1998)		√		35 firms filing for Chapter 11 from 1989 to 1991
Betker, Ferris, and Lawless (1999)		√		69 firms emerging from Chapter 11 1984–1994
Eberhart, Altman, and Aggarwal (1999)			√	131 firms emerging from Chapter 11 as a public firm from 1980 to 1993
Goyal, Kahl, and Torous (2003)			√	Firms distressed between 1980 and 1983; 35 firms in first year after resolution of distress to 25 firms five years after
Jiang and Wang (2019)			√	266 Chapter 11 cases filed by U.S. firms with at least $50m assets at filing and that emerged as a public firm from 1982 to 2013

FIGURE 7.2   Academic Studies of Postbankruptcy Performance

and Phillips (1998) examine this issue by studying plant-level operating data for manufacturing firms in Chapter 11 between 1978 and 1989. They examine measures of productivity of capital, as well as operating cash flow, for 1,195 plants of 302 bankrupt firms, as well as plants of nonbankrupt counterparts. Since they are able to track the productivity of individual plants, regardless of whether these plants are redeployed to new owners or are closed, Maksimovic and Phillips are able to examine asset performance even for firms that are liquidated or emerge private from Chapter 11, thus avoiding survivorship bias.

For the manufacturing firms they study, changes in bankrupt firms' performance can be explained for the most part by asset sales and closures, not by changes in the efficiency of retained assets. Bankrupt firms in high growth industries are more likely to sell assets than bankrupt firms in declining industries. The plants that are not sold by these firms have lower productivity compared to those that are sold off. In contrast, for nonbankrupt firms in the same industry, plants retained have significantly higher productivity than those sold. This result provides an alternative explanation as to why operating performance of emerging firms does not improve from prebankruptcy levels, namely that some firms have retained their least profitable assets. Further, in high growth industries, the productivity of the assets sold increases under new ownership. This evidence is consistent with the efficient redeployment of assets to more productive uses.

A key insight of the Maksimovic and Phillips study is that industry conditions are an important determinant not just of the frequency of bankruptcy, but of economic decisions such as asset redeployment in bankruptcy. In contrast to higher growth industries, in declining industries, the productivity of plants in Chapter 11 and subsequent to emerging does not significantly differ from their industry counterparts. This finding remains even when controlling for the changing asset composition of bankrupt firms as they make decisions to retain, sell, or close plants.

A final issue in evaluating whether firms return to profitability after Chapter 11 concerns the ownership and governance of the postbankruptcy firm. Hotchkiss and Mooradian (1997) find that the involvement of distressed debt investors, which rose dramatically starting in the early 1990s (see Chapters 3 and 14 herein), is strongly related to postbankruptcy success. Their study is based on a sample of 288 firms that defaulted on public debt between 1980 and 1993. The percentage of firms experiencing negative operating income in the year following bankruptcy is 31.9% for firms with no evidence of vulture involvement, versus 11.7% when a vulture has been involved in the restructuring. Strikingly, when the investor remains active in the governance of the firm post–Chapter 11, the percentage of firms experiencing operating problems drops to 8.1%. Improvements in performance relative to predefault levels are greater when the investor joins the board, becomes the CEO or Chairman, or gains control of the firm. When there is evidence of vulture involvement but the vulture is not subsequently active in the restructured company, performance appears no better than for those firms with no evidence of vulture involvement. Thus, the presence of these investors in the governance of the restructured firm is strongly related to postbankruptcy success for the sample studied. Distressed debt investors, largely hedge funds, has become the norm in most large Chapter 11 cases, and a number of firms emerging from Chapter 11 become portfolio companies of some of these investors. The post-emergence performance and changes in value of these portfolio companies, which are privately held following the bankruptcy, remains a question for further study.

2. *Ability to meet cash flow projections provided at the time of reorganization*

In order for a plan of reorganization to be confirmed by the court, the debtor must show that the plan is feasible. To meet this requirement, many firms provide cash flow forecasts, generally prepared by management or their financial advisors, when the plan is submitted to creditors and the court. The ability to meet these projections provides another measure of postbankruptcy success.

Cash flow projections are typically provided in the firm's disclosure statement, as part of the effort to gain creditor approval of the reorganization plan. Although the court has reviewed and approved these statements, it can still be difficult for outsiders to assess the validity of projections. First, the quantity and quality of financial information produced for firms in Chapter 11 may be reduced relative to prebankruptcy levels. For example, security analysts have often reduced coverage of these firms, and some firms cease to report audited financial statements. Second, the various constituencies involved in the case including management and various creditor groups can have divergent interests; the cash flow forecasts and the values they imply can arise either from negotiations among these parties, or from the group that largely controls the process.

Ex-post comparisons of projected versus realized cash flows were examined in earlier studies, each of which found that firms, on average, failed to achieve their projections. In her study of postbankruptcy performance, Hotchkiss (1995) shows that projections were on average overly optimistic. For example, operating income was lower than projected for 75% of the 72 sample firms for which cash flow projections are available. However, shortfalls between projected and actual performance were significantly greater when prebankruptcy managers were still in office at the time the plan was submitted. If management is concerned with the firm's survival, they may need to convince creditors and the court that the firm value is high enough to justify reorganization rather than liquidation. A shareholder-oriented management might also overstate forecasts in order to justify giving a greater share of the reorganized stock to prepetition equity holders.

Similar evidence demonstrating that firms on average are unable to meet cash flow projections is provided by Michel, McHugh, and Shaked (1998) and Betker, Ferris, and Lawless (1999). These researchers conclude that these forecasts have a systematic, optimistic and inaccurate bias in favor of reorganization.

Developments in the activity of distressed debt investors since the time of these studies have changed the dynamics of negotiations between stakeholders, and there have been notable instances of firms being undervalued at emergence based on projected cash flows. Gilson, Hotchkiss, and Ruback (2000) discuss incentives for over- (under-) valuation, based on the successful influence of junior (senior) stakeholders. Thus, one might expect a greater number of cases of understated cash flow projections in some recent cases.

3. *Incidence of subsequent distressed restructurings*

The high rate of Chapter 22 filings is not a new phenomenon, and has persisted since the early years of Chapter 11. Many repeat filers first entered bankruptcy in the wave of the early 1990s, and reentered in the bankruptcy waves of the early and late 2000s. Several early studies of postbankruptcy performance demonstrated the striking incidence of distressed restructuring taking place after firms have left bankruptcy. For example, Hotchkiss (1995) finds 32% of her sample firms restructure again either through a private workout, a second bankruptcy, or an out-of-court liquidation. Similar findings have been provided by LoPucki and Whitford (1993), Gilson (1997), and Halford, Lemmon, Ma, and Tashjian (2017). As shown in Chapter 1 herein, through 2017, an estimated 20% of all firms emerging from Chapter 11 as a going concern have subsequently refiled for bankruptcy.

The high rate of subsequent failures occurs despite requirements of the Bankruptcy Code that, in order for a reorganization plan to be confirmed, the company must demonstrate that further reorganization is not likely to be needed. Section 1129(a)(11) of the Code specifies: *Confirmation of the plan is not likely to be followed by the liquidation, or the need for further financial reorganization of the debtor or any successor to the debtor under the plan, unless such liquidation or reorganization is proposed in the plan.* High recidivism arguably runs counter to this requirement. Altman (2014) suggests that default prediction models, such as those explained in Chapter 11 herein, can be helpful in identifying firms with high risk of a subsequent failure.

There are several potential explanations for this high rate of subsequent failures. Many uncontrollable events, such as a major economic downturn, can lead to recidivism. Among the Chapter 22 or 33 filers are a number of airline, steel and textile companies, clearly reflecting difficult industry conditions. Still, subsequent failures have frequently occurred within two to three years after exiting the first Chapter 11, and additional factors likely contribute. For example, former Fortune 500 company Pillowtex (manufacturer of Fieldcrest sheets and towels) reorganized and emerged from its first bankruptcy in May 2002, then refiled for Chapter 11 in July 2003, and immediately announced it would liquidate. The firm immediately failed to meet the operating projections that the first bankruptcy's reorganization plan was based on, and it became clear that reorganized firm was not viable.

One factor explaining recidivism is that many firms have not sufficiently reduced their debt under their first restructuring plan. Supporting this idea, Gilson

(1997) finds that firms remain highly levered after emerging from Chapter 11, though not as high as for those firms which complete an out of court restructuring: the median ratio of long term debt to the sum of long term debt and common shareholders' equity is 0.47. Another potential explanation is that the debtor's management is overly optimistic about the prospects for the reorganized firm, as reflected in overly optimistic cash flow forecasts (Hotchkiss, 1995).

4. *Postbankruptcy stock performance*

While only a portion of firms entering Chapter 11 emerge as an independent, publicly registered company, an even smaller fraction of these firms successfully relist their stock following emergence. For example, of the 197 emerging firms studied by Hotchkiss (1995), only 60% had their stock relisted on NYSE/Amex or Nasdaq following emergence. She finds positive unadjusted but negative market adjusted stock returns for the year following emergence from bankruptcy. A large proportion of emerging firm stocks trade only on the OTC Bulletin Board or Pink Sheets, and so are not reflected in these studies. If the ability to relist the company's stock is a reflection of its postbankruptcy success, this might bias studies of postbankruptcy stock performance toward better performing firms.

Still, studying the performance of this group of stocks is interesting for several reasons. Most generally, it allows us to test the efficiency of the market for stocks of emerging firms, and is of particular interest to potential investors. More specific to our evaluation of the Chapter 11 process, it allows us to assess the accuracy of valuations of firms as projected in reorganization plans, by comparing the plan based valuation to the actual traded market value of the emerging firm. Further, creditors who receive stock as part of a reorganization plan but do not plan to hold the stock long term have a need to understand this market.

The first comprehensive academic study to date of the equity performance of firms emerging from bankruptcy is Eberhart, Altman, and Aggarwal (1999). Examining a sample of 131 firms emerging from Chapter 11 between 1980 and 1993, their key result is that there are large positive excess returns in the 200 days following emergence. A key issue in estimating these returns is the benchmark comparison or "expected return" from which to calculate abnormal performance. However, their results are robust with respect to different methods of estimating expected returns. Under their most conservative estimates, the average cumulative abnormal return over the first 200 days following emergence is 24.6% (median is 6.3%). Overall, they conclude that while returns in the first two days following emergence are not clearly significant, there are large positive and significant abnormal returns in the year following emergence.

More recent evidence on the stock performance post–Chapter 11 is provided by Jiang and Wang (2019), who report returns for 266 stocks (22% of their initial sample of Chapter 11 filings) of firms whose stock is traded on an exchange or over-the-counter after emergence. Median abnormal returns, relative to various benchmarks, reach 8% to 9% for horizons of one to three years after emergence, though are smaller and only significant for two- and three-year horizons once stocks trading OTC are excluded. Thus, their results are largely consistent with the earlier findings of Eberhart, Altman, and Aggarwal (1999).

While the results of these studies are clearly of interest to potential investors, the implications for evaluating Chapter 11 are less clear. Direct comparison with studies of operating performance are difficult because of sample selection issues, with only some firms emerging with publicly traded stock. Regardless, in contrast to studies showing poor operating performance, the stock performance indicates that firms do better than the market had expected at the time of emergence. These results may also be related to management incentives to issue relatively low firm valuations as part of the plan confirmation process.

IMPLICATIONS FOR EFFICIENCY OF CHAPTER 11

Early critics of the Chapter 11 process have primarily argued that the current U.S. bankruptcy system is biased toward allowing inefficient firms to reorganize. These critics have focused either on specific provisions of the Bankruptcy Code, which are characterized as "prodebtor" (such as management's ability to remain in control and initially propose a plan of reorganization), or on the behavior of particular courts which have been characterized as too prodebtor. More recently, the balance of power is argued to have swung in the direction of senior creditors (see Chapter 6 herein).

The high incidence of failures subsequent to leaving Chapter 11, which has been described in detail in this chapter, is certainly consistent with the view of excessive continuation. However, it is also clear that the governance structure of the emerging firm is importantly related to postbankruptcy performance; as more firms exit Chapter 11 under the control of senior creditors, the problem of recidivism may decline. The fact that Chapter 22s have continued even recently still suggests, however, that a number of firms emerging from bankruptcy continue to be overlevered or have poor operating prospects.

NOTES

1. An example is the 2013 merger of American Airlines (filing for bankruptcy in 2011 in SDNY) and US Airways.
2. In the 2017 case of Czyzewski v. Jevic Holding Corp., the Supreme Court held that structured dismissals providing for distributions that do not follow the normal priority rules and do not have the consent of affected creditors are impermissible.

CHAPTER 8

International Evidence

Insolvency laws vary considerably across countries, as do protections for creditors and enforceability of contracts outside of insolvency procedures.[1] The country-specific legal system drives the treatment of claimants to the firm's assets, the use of court supervised restructuring versus restructuring out of court, and ultimately the prospects for reorganization versus liquidation of distressed firms in those countries.

While the specific provisions of insolvency laws differ by country, several key characteristics can be used to broadly characterize the system as debtor or creditor friendly. Legal regimes oriented toward protecting the debtor, such as the U.S. system, provide more opportunity for rehabilitation. Creditor-based regimes, such as those modeled off UK common law, more often lead to or even require sale of the business assets or liquidation. Answering a set of eight key questions at the country level is helpful in characterizing insolvency procedures, as well as the overall usage or avoidance of in court restructurings in a given country:[2]

1. Do classes of creditors have unilateral rights to seek court protection or appoint responsible parties to handle affairs of a business in default?
2. Are there any restrictions, such as creditor consent or an insolvency test, for a debtor to seek court protection?
3. Are officers and directors of the company subject to civil or criminal proceedings for operating the business while insolvent?
4. Does the debtor continue to manage the firm or are they replaced by a court-appointed administrator (or liquidator)?
5. Are creditors prohibited from seizing collateral, through an automatic stay?
6. Do secured creditors rank first in priority for distributions, for example above claims of the government or workers?
7. Can the debtor obtain postfiling super-priority credit?
8. Is there cooperation with foreign jurisdictions in the administration of cross border restructurings?

On these dimensions, the United States appears relatively debtor friendly. As described in Chapter 3 herein, the debtor in possession retains control of ongoing business activities, though transactions considered as beyond ordinary activities, such as sales of significant assets, are conducted under court supervision and approval. In stark contrast, upon filing for bankruptcy in Scandinavian countries including Sweden, an administrator is immediately appointed to oversee a process that frequently leads to a sale of the firm as a going concern. In Germany, firms are required to file for court proceedings upon a determination of insolvency, and directors can be subject to criminal action for failing to do so and continuing to operate an insolvent firm.

Questions 2, 5, and 6 from the list above are used by La Porta, Lopez-de-Silanes, Shleifer, and Vishny (LLSV,1998) and Djankov, McLiesh, and Shleifer (2007) to construct a country level index of creditor rights that has been used extensively in subsequent academic research.[3] La Porta, Lopez-de-Silanes, Shleifer, and Vishny combine this index with a classification of the origins of company law or Commercial Code of each country (English, French, German, Scandinavian, and Socialist) and analysis of the quality of enforcement of laws to examine the relationship between investors' legal protection and firms' observed ownership concentration in those countries. Djankov, McLiesh, and Shleifer further show the importance of this measure of creditor power in explaining cross-country differences in the development of private credit markets.

Related research suggests that creditor rights and the efficiency of enforcement of debt contracts in bankruptcy have other ex-ante effects within a country. For example, empirical research demonstrates that ex-ante effects of stronger creditor power include lower corporate leverage (Acharya, Sundaram, and John 2011; Acharya, Amihud, and Litov 2011), reduced corporate risk taking (Acharya, Amihud, and Litov 2011), and less corporate innovation and patenting (Acharya and Subramanian 2009). In addition to enforcement of contracts, the predictability of legal regimes is also important in enabling investors, lenders, and others to assign a risk of repayment or repatriation to a regime.

The costs imposed by an insolvency regime determine claimholders' incentives to voluntarily restructure claims outside of formal in-court procedures. Claessens and Klapper (2005) find that filing rates are generally higher in countries with an efficient judicial system. Moreover, controlling for judicial efficiency, out of court restructurings are less frequent in countries where creditors have weaker rights under insolvency procedures. Thus, when comparing outcomes under different systems, one should keep in mind the caveat that the characteristics of firms under court supervision and their financial condition may differ when comparing across countries. Further, to the extent the process imposes greater costs on managers, for example, requiring replacement by a court-appointed administrator, managers can have greater incentives to delay filing as the firm value declines.

Understanding the differences in legal procedures across countries provides insights into the relative efficiency of different mechanisms for resolving distress. Nevertheless, evidence on restructuring outcomes outside the United States is sparse. In the next section of this chapter, we provide brief descriptions of the key features of insolvency codes for a set of countries explored in prior academic research – France, Germany, Japan, Sweden, and the UK – to demonstrate the varying degrees of creditor orientation. As available from this research, we describe outcomes (reorganization versus liquidation), costs and time in restructuring, and recovery rates.[4] We also summarize the current framework in two economies with significant growth in the restructuring industry, China and India. As bankruptcy laws in these countries are constantly evolving, our discussion reflects these systems as of 2017. Where an insolvency code provides multiple procedures, we focus on classifications based on the reorganization procedure. Figure 8.1 summarizes characteristics for these countries.

Country	China	France	Germany	India	Japan	Sweden	United Kingdom	United States
Legal origins	German	French	German	English	German	Nordic	English	English
1. Creditors have unilateral rights to seek court protection or appoint parties to manage the business in default.	1	1	1	1	0	0	1	0
2. There are restrictions (creditor consent or insolvency test) to enter court proceedings.	1	0	0	1	0	1	1	0
3. Officers and directors face liability for operating the business while insolvent.	1	1	1	0	1	1	1	0

FIGURE 8.1 Characteristics of Insolvency Codes in Eight Representative Countries

Country	China	France	Germany	India	Japan	Sweden	United Kingdom	United States
4. The debtor continues to manage the firm pending the resolution of the case.	0	0	1	0	0	1	1	0
5. Secured creditors can seize their collateral, i.e. there is no "automatic stay."	0	0	1	0	0	0	1	0
6. Secured creditors rank first in priority (above other creditors such as government or workers).	1	0	1	1	1	0	1	1
7. The debtor can obtain post-filing super-priority credit.	0	1	0	0	0	1	0	1
8. There are provisions for cooperation between domestic and foreign courts for cross border insolvency cases.	0	EU only	EU only	0	1	0	1	1

FIGURE 8.1   (*Continued*)

CHINA

The 2007 Enterprise Bankruptcy Law represented the first comprehensive bankruptcy code that covers liquidation, compromise, and corporate reorganization, the latter of which was previously only available to state-owned enterprises. There have been no major amendments made to the 2007 Law as of 2018, but the Supreme People's Court (SPC) from time to time has issued judicial interpretations (2011, 2013, 2015) with respect to bankruptcy petitions and acceptance, and to better facilitate filing and identifying the debtor's property.[5]

Both liquidation and reorganization procedures are available under the 2007 Law, with applications filed in the People's court where the corporation is domiciled (e.g. place of head office or place of main business). The bankruptcy cases are often heard by the civil or economic division of the court.[6] Either a debtor or creditor can initiate a filing; the court has discretion over whether to convert the case to reorganization if originally filed as a liquidation. Often it is the local government or shareholders that initiate the filing. After filing, the court has 15 days to consider whether to "accept" the application, and uses its discretion in applying standards for whether or not to accept the case. An automatic stay becomes effective only after the court's acceptance of a case. The debtor can apply to continue to manage the business, but under the supervision of an appointed administrator. In fact, the old management has limited impact on the reorganization or liquidation process. The administrator proposes a reorganization plan, which often requires the support of the local government. Voting requirements and conditions for a cramdown closely resemble those of the United States. Litigation for fraudulent transfers and preferences are rarely filed, even though asset transfers before bankruptcy are common.

Jiang (2014) notes that ironically, following enactment of the 2007 Law, the number of accepted bankruptcy applications in China fell. Low utilization of reorganization procedures may be due to the broad power of the court and the debtor's risk of losing control of the business, as well as management exposure to civil liabilities. Thus, while many of the provisions of China's law resemble those of the U.S. Chapter 11, distressed firms in China tend to avoid filing for reorganization. Moreover, the low number of filings can be also attributed to a lack of understanding of the 2007 Law by courts, which often refuse to accept bankruptcy petitions. Most creditors choose to sue and take immediate enforcement actions to attempt to recover their claims rather than filing for bankruptcy petitions.

Historically, most bankruptcy cases filed after the enactment of the 2007 Law were by small and medium-sized private enterprises, and they were often liquidated. Recent years have seen a greater number of cases, including many large state-owned enterprises (SOEs), filing for reorganization or liquidation. The number of filings for both reorganization and liquidation has risen quickly, with 2017 seeing a record number of bankruptcy cases (about 9,000) since the introduction of the "old" Bankruptcy Law (effective 1988).

As of yet, there are no published economic studies of bankruptcy outcomes in China. There are, however, predictive models for financial distress among listed Chinese firms. For example, Bhattacharjee and Han (2014) estimate hazard models for the incidence of financial distress in the period from 1995 to 2006. They note that firms are shielded to varying degrees by active state protection, such that the incidence of firms ceasing to operate is very low. Their models demonstrate the importance of firm specific factors such as firm age, size, and cash flow, but also the institutional effects of state protection (e.g., large former SOEs are observed to have lower hazard rates of financial distress).

With the ongoing structural economic reform, the number of large SOEs filing for bankruptcy, and the establishment of an information platform for bankruptcy cases and an online asset auction platform, we expect to see a growing research examining the efficiency of the bankruptcy process in China and its implications for Chinese financial markets.

FRANCE

French law, as reformed in 2015, provides that a firm which is not yet insolvent according to an insolvency test can ask the court to appoint an insolvency practitioner to assist in negotiating a voluntary restructuring with some or all creditors. The firm can also apply to the court for the opening of solvent reorganization proceedings, called *Safeguard proceedings*. The objectives of the procedure are, in order of priority, to continue the firm's operations, to maintain employment, and to pay back creditors. The administrator can raise new super-priority financing[7] (with the exception of claims of employees and insolvency expenses). During the first six (and up to 18 months), there is a stay on payments as the company negotiates a restructuring plan. The court appoints a judicial administrator to supervise or assist management in developing the plan, and a creditors' committee may present an alternative.

An insolvent debtor must apply for the opening of insolvency proceedings (liquidation) within 45 days of the occurrence of insolvency, unless it has already requested an insolvency practitioner to assist in a voluntary restructuring or has requested liquidation proceedings. A court-appointed administrator investigates the debtor and makes proposals for liquidation or for a sale of the going concern. Priority is given to saving the business and jobs, and employee and tax creditors rank ahead of other creditors.

The French code, with its explicit objective to maintain operations and preserve jobs, has a predisposition to allow continuation of inefficient firms. At the same time, strong employee rights and entitlements upon separation can increase the costs to shed a money losing business (for example to close an unprofitable plant without lengthy bargaining) and potentially hinder attempts to successfully restructure. Prior empirical evidence indicates that relatively few firms survive in-court proceedings in France. Kaiser (1996) reports that only 15% of filing firms continue to operate as a going concern after bankruptcy. In a broader sample comprised of both in-court and private restructurings, Davydenko and Franks (2006) find that 62% of French firms are liquidated piecemeal, a higher fraction than they find for the UK.

The low survival rates in France translate into relatively low creditor recovery rates. Davydenko and Franks (2006) document an average bank recovery rate in

French proceedings of 47% (median 39%), which is much lower than the recovery rates they report for UK banks. The median reorganization takes three years.

Overall, firm survival rates and creditor recoveries in France compare poorly with evidence from the UK and the United States. It is possible that the costs associated with the creditor-friendly French insolvency procedures ultimately are borne by the claimants of distressed firms. A more positive view is that the French legal code provides firms with incentive to restructure their debt prior to default in order to avoid being subject to insolvency laws, leaving only the lemons to the bankruptcy procedure.

GERMANY

The German Insolvency Act specifies that directors must file with the court within three weeks of becoming insolvent, based either on illiquidity or over-indebtedness. In addition, the debtor, but not creditors, can initiate a reorganization when illiquidity threatens. The debtor can continue to operate the firm but under the supervision of a court-appointed administrator. There is a three-month observation period, during which claims are stayed, and following which insolvency proceedings are formally opened and a permanent administrator with no prior connection to the debtor or creditors is appointed. The debtor (or administrator) submits a plan for either the liquidation or restructuring of the firm, including a potential sale of the firm as a going concern. A reorganization plan must receive creditor approval before it can be implemented.

Davydenko and Franks (2008) document that 57% of distressed German firms that defaulted on their debt between 1984 and 2003 are liquidated piecemeal, which is higher than liquidation rates reported for Sweden and the UK and lower than liquidation rates in France. The median duration of the reorganization procedure in Germany is 3.8 years, and banks recover on average 59% (median 61%) of their claims.

In Germany, the debt is typically concentrated with a house bank that often also has an equity interest. As a result, one should expect coordination failures that would prevent an out of court restructuring to be relatively rare. According to Kaiser (1996), most German firms with a chance of survival are reorganized in an out-of-court workout. Davydenko and Franks (2008), however, find that 78% of the distressed firms in their sample enter formal bankruptcy, with the remaining 22% of sample firms working things out with creditors informally.

While the German procedure has some resemblance to the U.S. Chapter 11, it imposes a strict three-month limit on the reorganization. This period risks being too short to allow a firm with complex operations and capital structure to carefully develop a reorganization plan. The empirical evidence, however, predates implementation of the current code.

INDIA

Historically, companies in India relied on different laws for bankruptcy reorganization and liquidation. These included the Sick Industrial Company Act (SICA), the Recovery of Debtors due to Banks and Financial Institutions Act (RDDBFI), the Securitization and Reconstruction of Financial Assets and Enforcement of Security Interests Act (SARFAESI), and the Companies Act. The Indian parliament enacted the Insolvency and Bankruptcy Code (IBC) in December 2016, substantially reforming processes for handling corporate insolvency. The IBC was a welcome overhaul of all prior framework dealing with insolvencies of corporations, partnerships, and individuals.[8]

Under the revised system, a plea for insolvency is submitted to the National Company Law Tribunal (NCLT), which accepts or rejects the plea within 14 days. If accepted, the NCLT appoints an insolvency resolution professional as an interim resolution professional to take over the management and control of the insolvent company. The interim resolution professional invites claims against the insolvent company and constitutes a committee of creditors comprising of all financial creditors. The committee of creditors appoints the interim resolution professional or some other insolvency professional as the resolution professional to manage the insolvency resolution process. Operational creditors holding large amounts of claims have the right to attend committee of creditors meetings but have no voting rights.

A resolution plan must be drafted within 180 days extendable by 90 days to total 270 days, failing which the debtor will undergo liquidation. The plan must be placed before the committee of creditors for approval, which requires a vote of not less than 75% of the voting share of the financial creditors. The resolution plan is then placed before the NCLT for final approval. Upon approval, the plan becomes binding on all key stakeholders including creditors, guarantors, employees, and others. The board of directors is suspended, although the appointed professional can rely on management to continue operations. The IBC also provides for voluntary liquidation and an unambiguous framework for a liquidation waterfall.

The new system replaces one in which a quasi-judicial body known as the Board for Industrial and Financial Reconstruction (BIFR) oversaw potential reorganizations. Under the prior system, a determination of negative net worth triggered a mandatory referral of industrial firms to the BIFR, which assessed the financial condition of the firm and its viability. The company was required to develop a restructuring plan in consultation with lenders. During this time period, management continued to control the company's assets, and an automatic stay protected the firm from enforcement of creditors' claims. Thus, the prior code was substantially more debtor friendly than the current system, which is substantially more oriented toward creditors' control.

Vig (2013) was among the first researchers to argue that changes in India's insolvency procedures strengthened the rights of secured creditors, had an important influence on capital structures, and decreased the use of secured debt financing. More recently, Gopalan, Martin, and Srinivasan (2017) use data for all firms filing with the BIFR from 1990 to 2013; they show that prior to bankruptcy, firms use accounting accruals to manage earnings and depress their net worth. Given the debtor-friendly provisions of the code during this period, an unintended consequence of using accounting book values to determine eligibility for bankruptcy protection was that firms could time their filing to opportunistically take advantage of a safe haven from creditors. Gormley, Gupta, and Jha (2018) also provide evidence consistent with the view that the earlier system offered debtors protection from creditors. Their study shows that increased banking competition induced by foreign bank entry was associated with more firms seeking bankruptcy protection.

JAPAN

Japan's insolvency code historically provided creditor-oriented procedures, often dominated by large keiretsu banks. Under these procedures, managers would typically lose their jobs, and creditors would control the outcome of the proceedings. However, banks and companies were known to collaborate to delay an ultimate day of reckoning, in line with a long standing cultural bias against taking away someone's honor (to "lose face"); this resulted in relatively less use of formal in-court proceedings.

Over the past 15 years, however, Japan has undertaken a series of revisions of its insolvency procedures aimed at strengthening the provisions for restructuring distressed firms as ongoing concerns. Current law provides for a *bankruptcy* procedure under which a trustee is appointed to oversee a liquidation. It also provides two procedures for reorganization. Under the *civil rehabilitation* procedure, further payments to debtholders are stayed, but secured claims can be enforced outside of the proceedings. Although the debtor continues to run the business, the court also appoints a supervisor which must approve actions such as transfers of assets and further borrowing. The second *corporate reorganization* procedure enables stock companies to develop a reorganization plan with approval by a majority of creditors. Both secured and unsecured claims are stayed. These proceedings are generally used for reorganization of larger debtors or complex cases. In 2009 a "quasi-DIP" practice was introduced under which the debtor's director or counsel is appointed as trustee. A court appointed examiner supervises the debtor's business management and disposition of assets.

Research on restructurings in Japan predates these reforms and largely examines the role of industrial groups known as *keiretsus*, which were dominant in the Japanese economy for the second half of the twentieth century. At the core of a keiretsu are banks, which finance much of the industrial operations, both as creditors and equity holders of the firms affiliated with the group. Hoshi, Kashyap, and

Scharfstein (1990) examine the role of a keiretsu affiliation for a sample of 125 publicly traded firms that become financially distressed. They find that distressed firms associated with a keiretsu invest more and sell more than nonkeiretsu firms in the years following the onset of financial distress. This suggests that keiretsu banks helped to relax financial constraints (consistent with the cultural bias noted above), possibly mitigating the costs of financial distress.[9] Helwege and Packer (2003) study the role of keiretsu banks for the outcome of procedures for 172 troubled Japanese firms. They report that the probability of liquidation is higher for firms affiliated with keiretsu banks than for nonkeiretsu firms, controlling for firm size. However, since there is no discernible difference in the profitability of the liquidated firms, they conclude that there is no evidence that keiretsu banks forced excessive liquidations.

SWEDEN

In Sweden, a debtor is considered insolvent when it is unable to pay its debts as they are due, and this inability is not temporary. Besides a voluntary petition of a debtor, a creditor may file a petition but must prove its claim regarding insolvency. Upon a filing, the court appoints an administrator with fiduciary responsibility to all creditors, who organizes the sale of the firm in an auction. The winning bidder determines whether the firm is liquidated piecemeal or continues to operate as an ongoing concern. Payment must be in cash, and creditors are paid strictly according to the absolute priority of their claims. Debt payments are stayed, and collateral cannot be repossessed. Moreover, trade credits and other debt raised while in bankruptcy get super-priority. These provisions help protect the operations until the firm is auctioned off.

Swedish law also provides a procedure for reorganization, where the debtor continues to operate the firm but must consult with the administrator. Secured debt and priority claims (taxes and wages) must be offered full repayment, and junior creditors at least 25% of their claim. These high thresholds make composition unfeasible for the vast majority of distressed firms. A new reorganization law was enacted in 1996, but Buttwill and Wihlborg (2004) argue that the new law shares many of the weaknesses of the old composition procedures and is rarely used. Thus, in Sweden, court-supervised renegotiation of the firm's debt contracts is effectively not an alternative to auction bankruptcy.

Thorburn (2000) examines a sample of 263 small, private Swedish firms filing between 1991 and 1998. Her evidence counters widespread fears that auctions tend to excessively force liquidation. She demonstrates that three-quarters of firms continue as a going concern under the buyer's reign, with the remaining one-quarter of firms being liquidated piecemeal.[10] To gauge the quality of the continuation decision, Eckbo and Thorburn (2003) examine the operating profitability of the emerging firms. In contrast to U.S. evidence for firms emerging from Chapter 11, they show that auctioned firms perform at par with industry competitors.

Thorburn (2000) also estimates the costs of Swedish insolvency proceedings. She reports direct costs of on average 6% of prefiling book value of assets, with an average of 4% for the one-third largest firms in her sample. When measured as a fraction of the market value of assets in bankruptcy, costs average 19%, with a median of 13%. The direct costs decrease with firm size and increase with measures of industry distress, suggesting that trustees may increase their sales effort in periods when auction demand is relatively low. Importantly, the auction is speedy, with an average time from filing to sale of the assets of only two months, implying relatively low indirect costs.

The value of the assets remaining at the end of the process reflects all the different costs imposed on the financially distressed firm. This value is split between the firm's creditors. The higher the total costs of bankruptcy, the lower are creditor recovery rates. Under the Swedish system, creditors' claims are paid with the cash generated in the auction. Thorburn (2000) reports average recovery rates of 35% (median 34%). Recovery rates are higher in going-concern sales (mean 39%) than in piecemeal liquidations (mean 27%). Secured creditors receive on average 69% (median 83%).

Overall, the Swedish evidence suggests that mandatory auctions provide a relatively efficient mechanism for restructuring financially distressed firms. Because of protections for the going concern, the auction process leads to the survival of many firms, not unlike the outcomes of a reorganization. Survival rates, direct costs, and recovery rates compare well with extant evidence from the United States and the UK.

THE UNITED KINGDOM (ENGLAND AND WALES)

UK insolvency law is based on the Insolvency Act of 1986 and Insolvency Rules of 1986, as replaced by the Insolvency Rules of 2016. Current law provides for three court-administered reorganization procedures. A *Company Voluntary Arrangement* (CVA) permits directors to reach a restructuring agreement with creditors. The procedure is supervised by an insolvency practitioner, and secured creditors' claims cannot be reduced without their consent. A moratorium on the collection of secured debts is only available for small private companies. Therefore, it is more likely that firms undergo *Administration*, under which a secured creditor appoints (or directs the board to appoint) an administrative receiver to represent their interests. If the claim is secured by floating charge collateral, this administrator has full control over the firm and has wide discretion regarding the decision whether to reorganize the firm or sell assets.[11] With court permission, creditors secured with fixed liens on particular assets have the right to repossess their collateral, even if the assets are vital for the firm's operations. Any excess balance is distributed to remaining claimholders according to the absolute priority of their claims. Since the Enterprise Act of 2002, firms are often sold soon after entering administration in a "pre-packaged administration," often to the company's directors. Finally, a

Scheme of Arrangement can be used to effect a balance sheet restructuring through a compromise with certain classes of creditors, sometimes in combination with a CVA or administration.

UK law further provides for *Liquidation*, which can be *Voluntary* or forced (*Compulsory*) by creditors. An administrator also has the power to move a company into liquidation. Overall, UK insolvency procedures are considered creditor-oriented. The liquidation decision is typically left to representatives of secured creditors, who may lack incentives to generate proceeds above the value of their claim.

Empirical evidence predates more recent changes to these procedures. Since secured creditors fare relatively well in formal court proceedings, one would predict voluntary workouts to be relatively rare in the UK. Davydenko and Franks (2008) find that 75% of small firms that default on their debt restructure in court, with the remaining 25% of firms reorganizing out of court.

Despite the strong protection of secured creditors, empirical evidence indicates that a large fraction of distressed UK firms survive as ongoing concerns. Franks and Sussman (2005) examine 542 small to medium-sized financially distressed UK firms, and find that 60% of firms filing for the UK receivership code continue to operate as going concerns after emergence. In a sample of UK firms filing for administrative receivership, Kaiser (1996) finds that almost half are sold as going concerns. Similarly, Davydenko and Franks (2008) show that 43% of small UK firms that default on their debt are liquidated piecemeal.

To the extent the UK procedures are speedy, direct costs are expected to be low. Still, Franks and Sussman (2005) report direct costs averaging 33% of asset values. Davydenko and Franks (2008) report a median time in reorganization of 1.4 years.

Secured creditors fare relatively well in the UK procedures, as expected. Franks and Sussman (2005) document average bank recovery rates of 75%, with a median of 100%. Interestingly, banks tend to liquidate collateral at prices close to the face value of the secured claim, possibly because secured creditors have few incentives to generate additional proceeds for junior claimants. Similarly, Davydenko and Franks (2008) report an average bank recovery rate of 69% (median 82%).

Overall, the weak protection of the firm's operations and the strong rights allocated to secured creditors in the UK may raise concerns of excessive liquidations. Nevertheless, firm survival and recovery rates in the UK compare well to the U.S. Chapter 11 despite the strong creditor orientation.

CROSS-BORDER INSOLVENCIES

Even with recent reforms in a number of countries, significant differences persist in the ability to enforce creditor rights in bankruptcy and in provisions that increase the likelihood of a reorganization. This presents significant difficulties for firms operating in multiple international jurisdictions. At the broadest level, judges in common law countries generally have wider discretionary authority than those in civil law countries. Disparities or conflicts between specific insolvency codes present further challenges for cross-border cases. An illustrative case is the 2009 bankruptcy of Nortel Networks, Ltd., involving operations in 140 jurisdictions. The parent company filed in Canada under the Companies' Creditors Arrangement Act, the U.S. subsidiaries filed under Chapter 11 in the United States, and certain European entities filed in the UK under the Insolvency Act of 1986. One dispute arose as to whether the UK pension regulator could liquidate the amount of the group's pension liabilities in that country, in violation of automatic stays in the United States and Canada. Another difficulty involved claims to the proceeds from asset sales, since auctions for Nortel's assets were conducted based on business lines across multiple jurisdictions rather than for geographically separate entities.

Given that few multilateral treaties have been successfully negotiated between countries, some international initiatives have arisen to aid in the harmonization of proceedings for cross-border cases. These include the Cross-Border Insolvency Concordat, which has been utilized in a number of cases involving the United States and Canada, and most importantly the United Nations Commission on International Trade Law (UNCITRAL) Model Law on Cross-Border Insolvency.[12] Among other benefits of the Model Law, it is expected that the greater certainty in matters of creditor and debtor rights will assist international trade, commerce and the availability of capital in less-developed countries.

The Model Law provides a legal framework for cross-border cooperation and communication between courts in different countries. It is important to note that it does not specify how that cooperation should be achieved. The objective of the Model Law is to establish a set of uniform principles to deal with the requirements a corporate entity needs to meet in order to have access to the courts of other countries involved in the case. This framework provides the basis for the execution of a "cross-border insolvency agreement," sometimes referred to as a "protocol," that can guide the management of the case. These agreements are designed to assist in the management of procedural rather than substantive issues between the jurisdictions involved. Still, they can be a necessary component in resolving issues in certain complex cases. As an extreme example, "The disputes involving the affiliates of Lehman Brothers, which involved 75 distinct bankruptcy proceedings relating to its more than 7,000 subsidiary entities in over 40 countries, were even more protracted. It took the insolvency administrators of the 18 major foreign subsidiaries of Lehman Brothers seven months to work out a protocol that contained general principals of coordination and cooperation, and in which the administrators agreed to cooperate in attempting to calculate the inter-company claims among the group" (Gropper, 2012).

Foreign insolvency proceedings under the Model Law are divided into two categories – "main" and "nonmain" proceedings. The main proceeding (or "COMI," the center of main interest) takes place in the country where a debtor has its primary operations; normal requirements of that country's bankruptcy law then proceed. The main country court can appoint someone to communicate between the relevant courts and coordinate the administration of the debtor's assets and affairs in the main and nonmain jurisdictions.

Over 70 countries and international organizations participated in the development of the Model Law. As of 2017, 45 nations or territories have passed laws based on this framework, 20 of them in 2015 and most recently Singapore in 2017. The U.S. Congress adopted the Model Law as part of the Bankruptcy Act of 2005 (BAPCPA), which became Chapter 15 of the new Bankruptcy Act. Adoption of the Model Law in European countries has been limited, but the EU Insolvency Regulation establishes a cross-border insolvency regime within the EU. Cross-border insolvency regimes remain largely absent, however, in major emerging economies.

NOTES

1. In many countries, "bankruptcy" refers to the legal provisions for liquidation. More broadly, the legal framework outside the United States is referred to as "insolvency" laws.
2. "Restructuring & Insolvency 2018" published by Getting the Deal Through provides a set of answers to 50 standardized questions that characterize the insolvency codes of 44 countries and the EU.
3. See LLSV Table 4, p. 1136; see also Djankov, Hart, McLiesh, and Shleifer (2008), who characterize country level debt enforcement procedures based survey responses regarding the handling of a hypothetical default case from local insolvency professionals in 88 countries.
4. See also Hotchkiss, John, Mooradian, and Thorburn (2008) for a review of earlier evidence regarding restructuring outcomes in these countries. Other studies of the impact of country-specific bankruptcy reform include Rodano et al. (2016), who study the 2005 and 2006 changes to bankruptcy law in Italy, and Schoenherr (2017), who studies the 2006 reform in Korea.
5. A special report by INSOL International explains the Chinese Enterprise Bankruptcy Law; see "PRC Enterprise Bankruptcy Law and Practice in China: 2007 to a record-breaking 2017" by Wang, Xu, and Tang, January 2018.
6. The SPC promoted the setting up of specialist bankruptcy courts and the training of professional bankruptcy judges. On June 21, 2016, the SPC formally issued "Work Plan of the Supreme People's Court on Establishing Liquidation and Bankruptcy Tribunals in Intermediate People's Courts." As of the end of 2017, the number of national tribunals specializing in bankruptcy proceedings increased to 97 from 5 at the beginning of 2015.
7. See discussion of super-priority debt financing in Chapter 3 herein.
8. See the special report by INSOL International: "The New Insolvency & Bankruptcy Code in India: Impact on the Distressed Debt Market," by V. Bajaj, April 2018.
9. Claessens, Djankov, and Klapper (2003) examine the use of bankruptcy in the restructuring of 644 financial distressed firms across East Asia. Bankruptcy filing is less likely for firms owned by banks or affiliated with a business group, and in countries with weak creditor rights and low judicial efficiency.

10. Prior to 1993, Finnish bankruptcy also mandated a sale of the firm. In a sample of 72 small firms filing under the old Finnish code, Ravid and Sundgren (1998) find that only 29% of the firms are sold as a going concern. They report average direct costs of 8% of pre-filing book value of asset, and average recovery rates of 34% in going-concern sales and 36% in piecemeal liquidations.

11. The collateral of a floating charge claim includes inventory, accounts receivables, working capital, and intangible assets.

12. The Model Law on Cross-Border insolvencies was promulgated at the thirtieth session of UNCITRAL on May 12–30, 1997. The official text of the Model Law is available at www.uncitral.org. Also available is the UNCITRAL Practice Guide on Cross-Border Insolvency Cooperation, which illustrates how cross border insolvency agreements can be used to facilitate resolution of conflicts between jurisdictions.

High-Yield Debt, Prediction of Corporate Distress, and Distress Investing

The High-Yield Bond Market

Risks and Returns for Investors and Analysts

"**H**igh-yield junk bonds, they are finished." This was not an uncommon refrain heard from various pundits on Wall Street and in Washington in the wake of the corporate default surge in 1990 and 1991 and after the bankruptcy of the market's leading underwriter of these non-investment-grade bonds (Drexel Burnham Lambert) and the criminal indictment of the market's leading architect, Michael Milken. We argued then (Altman 1992, 1993), and in every other subsequent instance of a major domestic or international credit crisis, that high-yield bonds are a legitimate and effective way for firms that have an uncertain credit future to raise money. One should expect periodic times of relatively high defaults commensurate with the risk premiums that issues need to offer investors to lend money. These investors, primarily from institutions like mutual funds and pension funds, seek higher fixed income returns than are available from safer, corporate investment-grade and government bonds.

Figure 9.1 displays the rating hierarchy of credit and default risk from the leading bond and bank loan rating agencies – Fitch Ratings, Moody's Investors Service, and Standard & Poor's. Note the now familiar distinction between the relatively safe investment grade bonds (AAA to BBB–, or Aaa to Baa3) and the more speculative, non-investment-grade, or high-yield, securities (below BBB– or Baa3).

In 2018, almost one-quarter of the total North American dollar-denominated corporate bond market was comprised of lower-graded bonds (Figure 9.2). And the market for high-yield bonds had grown dramatically to a total amount outstanding at 2017 (Q2) of over $1.6 trillion! (See Figure 9.3.) Note the consistent growth of the size of the market from 1996, when it totaled under $300 billion, to the impressive total in 2017.

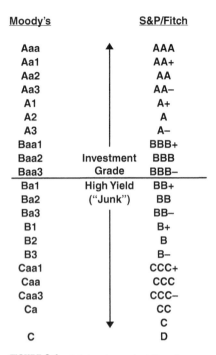

Moody's		S&P/Fitch
Aaa		AAA
Aa1		AA+
Aa2		AA
Aa3		AA–
A1		A+
A2		A
A3		A–
Baa1		BBB+
Baa2	Investment	BBB
Baa3	Grade	BBB–
Ba1	High Yield	BB+
Ba2	("Junk")	BB
Ba3		BB–
B1		B+
B2		B
B3		B–
Caa1		CCC+
Caa		CCC
Caa3		CCC–
Ca		CC
		C
C		D

FIGURE 9.1 Major Agencies' Bond Rating Categories

Size of Market	
U.S. Corporate Investment-Grade (IG)	$5,899B
U.S. High-Yield (HY)	$1,656B
Total	$7,555B

Distribution of Market (in $ billions)		
	IG	HY
Financials	$1,727	$152
Industrials	$3,590	$1,407
Utilities	$582	$97
Totals	$5,899	$1,656

FIGURE 9.2 Size of the North American Corporate Bond Market, June 30, 2017
Sources: S&P Global Ratings and NYU Salomon Center.

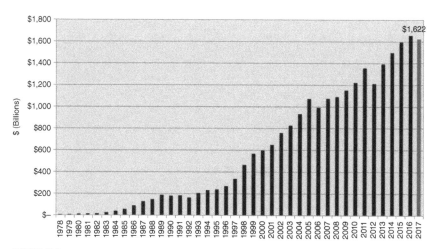

FIGURE 9.3   The Size of the North American High-Yield Bond Market, 1978–2017 (Mid-year, $ Billions)
Source: NYU Salomon Center estimates using Credit Suisse, S&P, and Citi data.

The growth and sustainability of the U.S. high-yield bond market is reflected in the record amount of new issuance since the great financial crisis of 2008–2009, as shown in Figure 9.4, and the also impressive growth in the European market, as shown in Figure 9.5. Indeed, according to Bank of America, Merrill, statistics, the market has averaged significant annual new issuance exceeding, on average, $230 billion per year from 2010 to 2017. European high-yield bonds also began its own impressive new issuance run in 2010 and its total amount outstanding, according to Credit Suisse, in 2017 (Q1), totaled €477 billion ($563 billion). Europe's growth reflects the substitution of capital market bonds for the more traditional commercial bank financing, as banks were reluctant to lend to subinvestment grade companies due to higher capital requirements under Basel II, and the quest for yield by investors in Europe, and other parts of the world.

High-yield bond estimates from Asia and South America (not shown in our figures) are in the range of $150 billion in each region as of 2017. So, all together, the global high-yield bond market was about $2.5 trillion in 2017, indicating it as a legitimate diversified asset class.

High-Yield Bond Market Developments in Russia and Mini-Bonds in Italy   We are now aware of some recent efforts to promote high-yield bond issuance and trading in Russia and expand the role of a new local rating agency, the Analytical Credit Rating Agency (ACRA). It has been estimated that in recent years there have been about 70 original-issue HY bonds issued in Russia, aggregating about $5 billion. The size distribution in rubles of these bonds are all the way from less

Annual	Total	Ratings				
		BB	B	CCC	(CCC % H.Y.)	NR
2005	81,541.8	18,615.0	45,941.2	15,750.9	(19.3%)	1,234.7
2006	131,915.9	37,761.2	67,377.3	25,319.2	(19.2%)	1,458.2
2007	132,689.1	23,713.2	55,830.8	49,627.6	(37.4%)	3,517.5
2008	50,747.2	12,165.0	25,093.1	11,034.4	(21.7%)	2,454.6
2009	127,419.3	54,273.5	62,277.4	10,248.4	(8.0%)	620.0
2010	229,307.4	74,189.9	116,854.7	35,046.8	(15.3%)	3,216.1
2011	184,571.0	54,533.8	105,640.4	21,375.0	(11.6%)	3,021.8
2012	280,450.3	71,852.1	153,611.1	48,690.2	(17.4%)	6,297.0
2013 (1Q)	73,492.3	31,953.1	29,534.2	11,480.0	(15.6%)	525.0
(2Q)	62,135.0	24,380.0	23,665.0	13,790.0	(22.2%)	300.0
(3Q)	73,770.8	22,964.2	32,610.0	18,196.6	(24.7%)	0.0
(4Q)	60,936.8	24,050.0	22,686.8	14,175.0	(23.3%)	25.0
2013 Totals	270,334.8	103,347.3	108,495.9	57,641.6	(21.3%)	850.0
2014 (1Q)	51,634.7	17,585.0	25,792.2	7,842.5	(15.2%)	415.0
(2Q)	74,629.6	23,893.7	30,852.3	19,363.6	(25.9%)	520.0
(3Q)	59,777.3	25,537.3	22,550.0	10,875.0	(18.2%)	815.0
(4Q)	52,721.1	21,975.0	28,906.1	1,840.0	(3.5%)	0.0
2014 Totals	238,762.7	88,991.0	108,100.6	39,921.1	(16.7%)	1,750.0
2015 (1Q)	76,059.5	23,184.2	44,785.3	8,090.0	(10.6%)	0.0
(2Q)	73,428.4	21,219.0	40,037.3	12,052.1	(16.4%)	120.0
(3Q)	31,740.0	14,770.0	12,675.0	4,295.0	(13.5%)	0.0
(4Q)	34,584.1	20,500.9	13,100.0	660.0	(1.9%)	323.2
2015 Totals	215,812.0	79,674.2	110,597.5	25,097.1	(11.6%)	443.2
2016 (1Q)	34,665.0	19,325.0	11,070.0	4,270.0	(12.3%)	0.0
(2Q)	63,490.4	31,427.4	26,043.0	6,020.0	(9.5%)	0.0
(3Q)	56,403.6	21,614.6	28,788.9	5,650.0	(10.0%)	0.0
(4Q)	40,509.1	13,610.7	20,763.4	4,960.0	(12.2%)	1,175.0
2016 Totals	194,718.0	85,977.7	86,665.3	20,900.0	(10.7%)	1,175.0
2017 (1Q)	72,673.9	33,327.5	29,802.4	9,544.0	(13.1%)	0.0
(2Q)	50,342.4	18,531.3	23,366.0	7,895.1	(15.7%)	550.0
(3Q)	57,157.3	19,639.7	27,699.3	9,818.3	(17.2%)	0.0
(4Q)	61,415.0	23,877.5	27,050.0	9,587.5	(15.6%)	0.0
2017 Totals	241,588.7	95,376.0	107,917.8	36,844.9	(15.3%)	550.0

FIGURE 9.4 New Issuance: U.S. High-Yield Bond Market ($ millions), 2005–2017
Source: Based on data from Bank of America, ML.

than $20 million to about $200 million. Most of these bonds have been issued to private individuals seeking higher yields, with only a small amount of issuance to financial institutions. In October 2018, two conferences were held for the first time on Modeling Bond Ratings and one on the high-yield bond market, sponsored by ACRA and at least one investment bank, SOLID, respectively.

This parallels efforts in Italy to issue and promote SME bond issuance, known as *mini-bonds*. Since 2013, more than €10 billion of these bonds, mostly less that €25 million in size have been issued. These bonds, yielding from 5 to 7%, have mostly been bought by institutions, such as foundations, and individuals, and some structured into CBOs. One of our authors, Altman, has developed

Annual	Total	Ratings					Currency		
		BB	B	CCC	(CCC % HY)	NR	USD	EUR	GBP
2005	**19,935.6**	1,563.3	11,901.0	5,936.6	(29.8%)	534.8	2,861.0	15,080.3	1,668.3
2006	**27,714.6**	5,696.2	16,292.1	5,020.5	(18.1%)	705.9	7,657.8	19,935.7	121.1
2007	**18,796.7**	5,935.3	11,378.5	562.0	(3.0%)	920.9	4,785.5	12,120.9	1,890.3
2008	**1,250.0**	1,250.0	25,093.1				1,250.0		
2009	**41,510.3**	18,489.4	16,697.4	4,771.3	(11.5%)	1,552.2	12,315.0	28,696.9	498.3
2010	**57,636.5**	22,751.3	29,050.5	2,170.7	(3.8%)	3,663.9	12,775.0	43,147.7	1,403.3
2011	**60,435.8**	24,728.9	29,919.7	4,108.7	(6.8%)	1,678.6	16,720.0	33,758.0	8,842.4
2012	**65,516.1**	27,001.7	29,013.0	7,186.7	(11.0%)	2,314.6	28,198.0	32,270.4	2,929.3
2013 (1Q)	27,954.5	6,783.8	15,008.4	5,160.6		1,001.7	10,050.0	12,380.7	4,837.4
(2Q)	30,335.3	6,860.2	19,295.1	3,724.1		455.9	9,913.0	14,149.9	6,074.0
(3Q)	16,558.4	3,375.3	9,609.6	2,721.8		851.7	5,310.0	8,644.0	2,604.4
(4Q)	16,655.9	2,588.0	10,657.6	2,366.4		1,043.9	5,210.0	9,086.5	2,359.4
2013 Totals	**91,504.1**	19,607.3	54,435.2	13,972.9	(15.3%)	3,353.2	30,483.0	44,125.6	15,875.3
2014 (1Q)	27,169.2	12,565.7	11,685.2	1,230.0	(4.5%)	1,688.3	7,315.0	16,352.8	3,501.4
(2Q)	65,671.4	13,730.1	45,808.3	4,111.1	(6.2%)	2,021.9	23,150.0	36,009.0	6,096.7
(3Q)	15,980.5	3,586.3	10,593.2	1,241.3	(7.8%)	559.7	2,750.0	8,216.2	4,744.6
(4Q)	10,646.9	3,893.7	4,288.8	654.5	(6.1%)	1,810.0	6,305.0	4,341.9	
2014 Totals	**119,468.0**	**33,775.8**	**72,375.4**	**7,236.9**	**(5.1%)**	**6,080.0**	**39,520.0**	**64,919.9**	**14,342.7**
2015 (1Q)	30,535.5	15,387.8	10,054.6	938.7	(3.1%)	4,154.3	10,225.0	17,149.0	2,622.0
(2Q)	26,458.3	11,282.6	12,253.2	2,334.6	(8.8%)	587.8	12,465.0	11,744.4	1,782.2
(3Q)	12,605.5	2,068.1	10,125.9	411.5	(3.2%)		5,850.0	5,170.1	1,585.4
(4Q)	5,645.5	3,032.2	2,350.9	262.4	(4.6%)		2,050.0	3,169.1	426.4
2015 Totals	**75,244.9**	**31,770.8**	**34,784.6**	**3,947.3**	**(5.2%)**	**4,742.1**	**30,590.0**	**37,232.6**	**6,416.0**
2016 (1Q)	2,771.6	334.7	2,280.6	156.2	(5.6%)		1,675.0	1,096.6	
(2Q)	22,337.5	6,553.8	13,629.8	2,153.8	(9.6%)		14,590.0	7,115.4	569.8
(3Q)	23,206.6	12,240.5	8,745.2	2,220.8	(9.6%)		10,565.0	10,590.8	2,050.8
(4Q)	9,131.6	600.0	7,316.5	1,215.1	(13.3%)		2,595.0	6,268.6	268.0
2016 Totals	**57,447.2**	**19,729.1**	**31,972.2**	**5,745.9**	**(10.0%)**		**29,425.0**	**25,071.3**	**2,888.6**
2017 (1Q)	**18,283.5**	6,094.4	7,805.9	4,133.3	(22.6%)	250.0	7,665.0	7,074.9	3,543.7
(2Q)	17,632.6	9,117.5	6,702.3	1,812.7	(10.3%)		4,920.0	8,513.8	3,854.3
(3Q)	17,903.1	5,868.7	9,002.3	2,278.4	(12.7%)	753.7	4,200.0	11,879.6	1,823.5
(4Q)	31,845.9	9,888.6	19,448.9	2,508.4	(7.9%)		6,355.0	23,913.6	1,381.9
2017 Totals	**85,665.1**	**30,969.2**	**42,959.5**	**10,372.8**	**(12.5%)**	**1,003.7**	**23,140.0**	**51,381.9**	**10,603.3**

FIGURE 9.5   New Issuance: European High-Yield Bond Market (Face Values in US $ millions), 2005–2017
Source: Based on data from Bank of America, ML.

a credit risk model, called the Italian SME Z-Score model (Altman, Sabato, and Esentato 2016), to assist investors and issuers in both original issuance and secondary market trading on Borsa Italiana.

Size and Growth of High-Yield Bond Market   The yields on various debt securities are determined by the market's assessment of three major risks in purchasing and holding a given issue: (1) its sensitivity to changes in interest rates, (2) its liquidity or lack thereof, and (3) its probability of default. Such yields are set by the market to provide investors with promised yields that increase with the level of these three risks. It is the third category of risk – the probability of default – that defines the high-yield bond market and periodically has provided substantial amounts of "raw-material" securities that make up the distressed and defaulted debt market – a focus of this book.

High-yield bonds are comprised of, basically, two types of issuing companies. About 20–25% of the market in recent years is made up of so-called fallen angels – securities that at one time (usually at issuance) were investment grade, but, like most of us, get uglier as they aged, and migrated down to non-investment-grade or so-called junk level status. When the modern-age high-yield market started in the late 1970s, just about 100% of the very small market was comprised of these fallen angels. As an example, in the early-to-mid 2000s, one of the icons of American industry, General Motors (GM), was being scrutinized by at least one of the major rating agencies as a possible fallen angel candidate. And, in May 2005, GM was downgraded to noninvestment grade by S&P, soon to be followed by Fitch and then by Moody's. Ford Motor Company was also downgraded to high-yield status by S&P at the same time.

The other source of high-yield bonds is original-issue securities that receive a non-investment-grade rating at birth. Today, all major investment banks have teams of bankers, analysts, and sales/trading personnel dedicated to this dynamic and growing speculative grade market. Since many investment-banking divisions are part of larger commercial bank organizations, there is usually a close relationship between the low-grade bonds of an issuer and its corporate loan analogue, the so-called leveraged loan market (see Chapter 3 herein). The latter are loans either issued by non-investment-grade companies or that require a risk premium, or yield spread, over the London Interbank Offered Rate (LIBOR) of at least 150 basis points.

The continuing growth of both the U.S. corporate high-yield bond and the leveraged loan markets was punctuated by the record amount of new issuance in both markets of late. As we will show clearly, the most important risk area, that of default rates, has seen a relatively large number of years when the annual rate of default has approached and even exceeded 10%. Indeed, since 1978, there has been five years with a default rate of about 10%, or more (Figure 9.6). Yet, the most recent high-default period of 10.7% in 2009 has been shrugged off by many in the market as ancient history with an impressive rebound. More and more analysts and regulators, including one of the authors, have sounded serious alarms of late, like he did about one decade earlier; see Altman (2007).

What Is a Default?

We, and the credit rating agencies, define "default" in the bond or loan markets as an issue or issuer experiencing one of the following three events:

1. A bankruptcy petition is filed and accepted, either a Chapter 7 (liquidation) or Chapter 11 (reorganization). In this case, all liabilities of the bankrupt entity are in default.
2. A missed interest payment or principal repayment, whereby the payment-default is not "cured" within the grace period (usually 30 days), or not within a "forbearance" period agreed upon by the debtor and specific creditors whose debt is affected.

Year	Par Value Outstanding[a] ($)	Par Value Defaults ($)	Default Rates (%)
2017	1,622,365	29,301	1.806
2016	1,656,176	68,066	4.110
2015	1,595,839	45,122	2.827
2014	1,496,814	31,589	2.110
2013	1,392,212	14,539	1.044
2012	1,212,362	19,647	1.621
2011	1,354,649	17,963	1.326
2010	1,221,569	13,809	1.130
2009	1,152,952	123,878	10.744
2008	1,091,000	50,763	4.653
2007	1,075,400	5,473	0.509
2006	993,600	7,559	0.761
2005	1,073,000	36,209	3.375
2004	933,100	11,657	1.249
2003	825,000	38,451	4.661
2002	757,000	96,855	12.795
2001	649,000	63,609	9.801
2000	597,200	30,295	5.073
1999	567,400	23,532	4.147
1998	465,500	7,464	1.603
1997	335,400	4,200	1.252
1996	271,000	3,336	1.231
1995	240,000	4,551	1.896
1994	235,000	3,418	1.454
1993	206,907	2,287	1.105
1992	163,000	5,545	3.402
1991	183,600	18,862	10.273
1990	181,000	18,354	10.140

Year	Par Value Outstanding[a] ($)	Par Value Defaults ($)	Default Rates (%)
1989	189,258	8,110	4.285
1988	148,187	3,944	2.662
1987	129,557	7,486	5.778
1986	90,243	3,156	3.497
1985	58,088	992	1.708
1984	40,939	344	0.840
1983	27,492	301	1.095
1982	18,109	577	3.186
1981	17,115	27	0.158
1980	14,935	224	1.500
1979	10,356	20	0.193
1978	8,946	119	1.330
1977	8,157	381	4.671
1976	7,735	30	0.388
1975	7,471	204	2.731
1974	10,894	123	1.129
1973	7,824	49	0.626
1972	6,928	193	2.786
1971	6,602	82	1.242

		Standard Deviation (%)
Arithmetic Average Default Rate (%)		
1971 to 2017	3.104	3.006
1978 to 2017	3.308	3.160
1985 to 2017	3.759	3.280
Weighted Average Default Rate (%)[a]		
1971 to 2017	3.378	
1978 to 2017	3.381	
1985 to 2017	3.394	
Median Annual Default Rate (%)		
1971 to 2017	1.806	

[a]Excluding Defaulted Issues from Par Value Outstanding (US $ millions).

FIGURE 9.6 Historical High-Yield Bond Default Rates, Straight Bonds Only, 1971–2017

Source: E.I. Altman and B. Huehne (2018a), Solomon Center.

3. A distressed exchange (DE), through which the debtor presents a tender-offer to creditor group or groups, whereby some swap is accepted. The swap, or exchange, usually involves one or more of the following: (a) cash for debt, (b) new debt for old debt, or (c) equity for debt. Note that at least one of the rating agencies, S&P, records the tender-offer as a default, regardless if it is accepted by the creditors. See Altman and Karlin (2009) for a discussion and analysis of DEs.

Defaults and Returns (1971–2017)

As noted, the key risk variable for investors in high-yield, and high-risk, corporate bonds is default and default loss expectations and results. This paramount issue has always been a major determinant of the success, or not, for investors, issuers, traders, and commentators about the return/risk tradeoff and popularity of "junk" bonds. This was especially important in the embryonic stage of the modern high-yield bond market in the early 1980s, when the market was being introduced to newly issued, non-investment-grade debt securities. Altman and Nammacher (1985, 1987) and Altman (1987) were the first academics to explore the default rate calculation in a rigorous fashion, although *Drexel* and other investment banks had been commenting and reporting their own default results, primarily only on original issue junk, for several years. We felt then, as we do now, that the appropriate market size of high-yield bonds to measure default rates was the entire population of non-investment-grade bonds, both fallen angels and original-issue speculative grade. Currently, everyone has adopted our approach.

Back in the mid-1980s, believe it or not, the average annual high-yield bond default rate was a miniscule 1.4% per year (see Altman and Nammacher 1985), which market "players" actually argued was too high! Over the years, default rates, measured in terms of dollar amounts, fluctuated according to the business cycle, industry trends, and corporate leveraged strategies, such as leveraged buyouts. Indeed, as shown in Figure 9.6, the average annual default rate for the period 1971–2017 was about 3.4% per year, weighted each year by the amount outstanding. The standard error was 3.01%, so a two-standard deviation year implies a rate of about 10% on the high side and near zero on the low side (of course, default rates cannot go below zero). Default rate measures using the number of issuers in the market as the base, as reported by S&P and Moody's, has averaged a bit higher, at close to 4.0% per year (e.g., S&P Global Ratings, 2017). While we understand the argument for using number of issuers as the base in default rate calculations, we remain convinced that most market participants identify more with rates based on dollar amounts, which weight larger defaults more heavily than smaller ones. Indeed, most of the rating agencies now include dollar-denominated rates, as well as issuer-denominated default rates.

For the relationship between default rates and economic recessions, we can observe in Figure 9.7 that corporate default rates tend to peak at the end of a

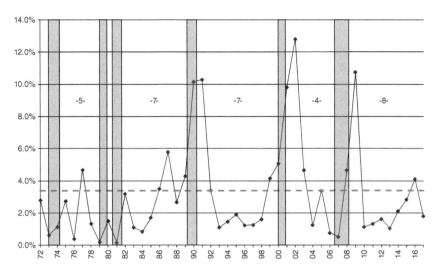

FIGURE 9.7 Historical Default Rates and Recession Periods in the United States,*
High-Yield Bond Market, 1972–2017
*All rates annual.
Periods of Recession: 11/73–3/75, 1/80–7/80, 7/81–11/82, 7/90–3/91, 4/01–12/01,
12/07–6/09
Sources: E. Altman (NYU Salomon Center) and National Bureau of Economic Research.

recession, or, in most cases, soon after the recession ends (shaded areas represent recession periods). What is also clear, however, is that in many recession-stressed periods, default rates actually begin to rise two to three years before the recession starts! This was especially the case in two of the three most recent recessions; the exception being the most recent recession period in 2008–2009 when increasing defaults in the mortgage-backed debt market preceded the recession, while corporate defaults were essentially coincident with economic declines. Note in Figure 9.7, that the most recent benign credit cycle, where default rates are below the historic average, was into its eighth year, a record number of years for this low-stressed benign period. We will comment on the definition and measurement issues of benign and stressed credit cycles shortly.

Another particularly important variable to consider when analyzing corporate defaults is the relationship between nonfinancial corporate debts as a percentage of GDP versus default rates. Figure 9.8 shows that in the past three recessionary-stressed cycles, when default rates spiked to double-digits, U.S. nonfinancial corporate debt to GDP peaked within one year or less, of the peak in high-yield bond default rates. An ominous signal existed in 2017 when it appeared that the Debt/GDP ratio was peaking. But, the market for risky debt was in no mood to be concerned and the low-interest rate and low-spread environment continued through most of the summer of 2017. As global political tensions

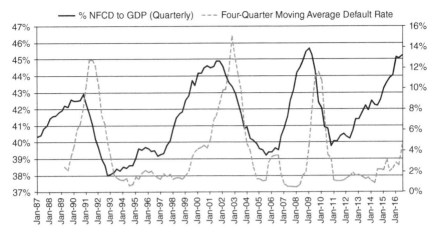

FIGURE 9.8   U.S. Nonfinancial Corporate Debt to GDP: Comparison to Four-Quarter Moving Average Default Rate
Sources: FRED, Federal Reserve Bank of St. Louis, and Altman/Kuehne High-Yield Default Rate data.

escalated and certain industries (e.g. retail) showed distinct signs of stress, spreads and volatility did increase toward summer's end, but still did not signal a major change in investor sentiment.

Default rates on leveraged loans also remained in a benign state into 2017 (and in 2018), mirroring the trends we observed for high-yield bonds. Similar to bonds, defaults escalated temporarily in 2015–2016 due to energy defaults and the distressed ratio had a momentary spike upward in late 2015 as this sector "produced" many potential defaults. As data from S&P Global's LCD group showed, as oil prices turned upward in mid-2016 and into 2017, pressures in the energy industry eased and Recovery Rates (see below) on new and existing defaulted issues increased significantly to over 60% in 2017, from as low as 30% one year earlier (see Figure 9.9).

	Default Rate	Overall Recovery Rate	Energy/Mining Recovery Rate	All Other Recovery Rate
2014	2.11%	63.19	n/a	63.19
2015	2.83%	33.91	25.64	46.78
2016	4.11%	29.99	27.33	40.03
2017	1.81%	56.68	57.97	57.47
Weighted Average Default Rate (1971–2017)		3.38%		
Arithmetic Average Recovery Rate (1978–2017)		45.88		

FIGURE 9.9   Default and Recovery Rates for High-Yield Bond Defaults, 2014–2017

Recovery Rates

The subject of recovery rates on high-yield bond and leveraged loan defaults will be addressed in Chapter 16 of this volume in more detail, but the topic deserves some specific mention at this point, as we move from defaults to loss given defaults (LGD). We define the recovery rates on defaulted securities as the price, in the market, just after default.

The simple relationship between estimating PDs and LGDs is through the recovery rate measurement and its correlation, or not, with PD. Altman, Brady, Resti, and Sironi (2005) presented a straightforward, but powerful, study that showed that over time, the coincident relationship between default rates and recovery rates (price at default) was negative and highly significant (see Figure 9.10). Indeed, when default rates in the high-yield bond market were high in such years as 1990/1991, 2001/2002, and 2009, recovery rates were quite low, between 25% and 35%, compared to the historical average of about 45%. Likewise, when default rates were low (see top-left quadrant of Figure 9.10), recovery rates were far above average. So, with just one explanatory variable (default rate), we can explain as much as between 58% and 62% (depending upon the regression model) of the level and variance of our key dependent variable: recovery rates (see Altman, Brady, Resti, and Sironi, 2005, for explanation and various regressions).

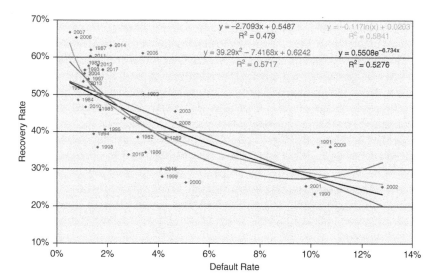

FIGURE 9.10 Recovery Rate/Default Rate Association: Dollar-Weighted Average Recovery Rates to Dollar Weighted Average Default Rates, 1982–2017
Source: Altman, Brady, Resti, and Sironi (2005).

Over time, high-yield bond investors have lost about 2.4% per year from default losses, considering default rates, recoveries, and lost-coupon receipts (see Altman and Kuehne 2017). Compared to average promised yield spreads of 5.21% from 1978 to 2016 (Figure 9.11), the *expected* return spread was about 2.8% (5.2% – 2.4%). This simple, but important expected relationship was remarkably close to *actual* return spread results of an average 2.4–2.8% (lower left-hand part of Figure 9.11) per year! See Altman and Bencivenga, (1995) for an in-depth discussion of the powerful association between promised yield spreads, expected losses from defaults, lost coupon payments from defaults, and expected return spreads.

What Is a Benign Credit Cycle?

We are often asked what differentiates a benign credit cycle from one that is stressed. There really is no formal definition, but we have found it helpful to focus on four variables that help to describe a credit cycle, namely:

1. Default rates on risky debt.
2. Recovery rates on defaulted debt.
3. Yields and spreads on risk debt compared to "riskless" benchmark rates.
4. Liquidity of the market.

These four determinants, and related factors, are shown in Figure 9.12.

1. It is instructive to view again Figures 9.6 and 9.7 and focus on the historical default rate and the duration of lower-than-average default rates; one clear definition of a benign credit cycle. We can observe that the duration of the periods of lower-than-average default rates between stressed periods (double-digit rates), had lasted between 4 and 7 years, with an average duration of 5.5 years, during the history of the modern high-yield bond market. Indeed, in the most recent lower-than-average default cycle, we are already (at the end of 2018) in the ninth year of below-average high-yield bond default rates since the "lofty" rate of 10.7% in 2009. The duration of default rates that are below the historical average annual rates of 3.5% is one of the most important barometers of a benign cycle. Investors are extremely sensitive to "current" defaults when analyzing required yields and spreads for investing in risky debt and the duration of below, or above, average rates, is viewed carefully by market participants.
2. Related to default rates is the other "half" of the default loss calculation, recovery rates. As we saw earlier, during periods of extremely high or low default rates, we have observed the opposite for recoveries. So, a second determinant of a benign cycle is certainly recoveries and default losses. The latter is much above average in stressed periods.

Year	Return (%)			Promised Yield (%)		
	HY	Treas	Spread	HY	Treas	Spread
2017	7.05	2.13	4.92	6.35	2.41	3.95
2016	17.83	(0.14)	17.96	6.55	2.43	4.12
2015	(5.56)	0.90	(6.46)	9.27	2.27	7.00
2014	1.83	10.72	(8.89)	7.17	2.17	5.00
2013	7.22	(7.85)	15.06	6.45[b]	3.01	3.45
2012	15.17	4.23	10.95	6.80	1.74[b]	5.06
2011	5.52	16.99	(11.47)	8.41	1.88	6.54
2010	14.32	8.10	6.22	7.87	3.29	4.58
2009	55.19	(9.92)	65.11	8.97	3.84	5.14
2008	(25.91)	20.30	(46.21)	19.53	2.22	17.31
2007	1.83	9.77	(7.95)	9.69	4.03	5.66
2006	11.85	1.37	10.47	7.82	4.70	3.11
2005	2.08	2.04	0.04	8.44	4.39	4.05
2004	10.79	4.87	5.92	7.35	4.21	3.14
2003	30.62	1.25	29.37	8.00	4.26	3.74
2002	(1.53)	14.66	(16.19)	12.38	3.82	8.56
2001	5.44	4.01	1.43	12.31	5.04	7.27
2000	(5.68)	14.45	(20.13)	14.56	5.12	9.44
1999	1.73	(8.41)	10.14	11.41	6.44	4.97
1998	4.04	12.77	(8.73)	10.04	4.65	5.39
1997	14.27	11.16	3.11	9.20	5.75	3.45
1996	11.24	0.04	11.20	9.58	6.42	3.16
1995	22.40	23.58	(1.18)	9.76	5.58	4.18
1994	(2.55)	(8.29)	5.74	11.50	7.83	3.67
1993	18.33	12.08	6.25	9.08	5.80	3.28
1992	18.29	6.50	11.79	10.44	6.69	3.75
1991	43.23	17.18	26.05	12.56	6.70	5.86
1990	(8.46)	6.88	(15.34)	18.57	8.07	10.50
1989	1.98	16.72	(14.74)	15.17	7.93	7.24
1988	15.25	6.34	8.91	13.70	9.15	4.55
1987	4.57	(2.67)	7.24	13.89	8.83	5.06
1986	16.50	24.08	(7.58)	12.67	7.21	5.46
1985	26.08	31.54	(5.46)	13.50	8.99	4.51
1984	8.50	14.82	(6.32)	14.97	11.87	3.10
1983	21.80	2.23	19.57	15.74	10.70	5.04
1982	32.45	42.08	(9.63)	17.84	13.86	3.98
1981	7.56	0.48	7.08	15.97	12.08	3.89
1980	(1.00)	(2.96)	1.96	13.46	10.23	3.23
1979	3.69	(0.86)	4.55	12.07	9.13	2.94
1978	7.57	(1.11)	8.68	10.92	8.11	2.81
Arithmetic Annual Average						
1978–2017	10.39	7.55	2.84	11.25	6.07	5.18
Compound Annual Average						
1978–2017	9.51	7.02	2.49			

[a]End-of-year yields.
[b]Lowest yield in time series.

FIGURE 9.11 Annual Returns, Yield, and Spreads on 10-Year Treasury (Treas) and High-Yield (HY) Bonds,[a] 1978–2017
Source: Citi's High-Yield Composite Index.

1. Length of Benign Credit Cycles
2. Key Variables: Default Rates and (Forecast), Recovery Rates, Yields, and Liquidity
3. Coincidence with Recessions
4. Level of Nonfinancial Debt as a Percentage of GDP
5. Comparative Health of High-Yield Firms (2007 versus 2016)
6. High-Yield and CCC New Issuance
7. Impact of Maturity Profile?
8. LBO Statistics and Trends
9. Liquidity Concerns (Market and Market-Makers)
10. Possible Timing of the Bubble Burst (Short-Term versus Longer-Term

FIGURE 9.12   What Is a Benign Credit Cycle?

3. One of the most closely watched statistics in the high-yield bond market is the yield and spreads over Treasuries currently being required, that is., accepted by the market. The more benign are market conditions, the lower the required yield and spreads, either the yield-to-maturity or option-adjusted spread, that investors are comfortable with. From Figure 9.11, we see that the historical average yield-to-maturity spread was 5.18% as of December 31, 2017. Spreads at year-end have been as low as 2.81% in 1978, and dropped to 2.6% in June 2007 (Figure 9.13). On the upper-side, aside from the unprecedented level as of year-end 2008 of 17.3% (spreads actually exceeded 20% in mid-December 2008), when the nominal yield spiked to well over 20%, the highest yield spread was 10.5% in 1990, when default rates also exceeded 10% for the first of two years in succession (Figure 9.6). Of course, Treasury yields impact the level of required yields and spreads. The spread as of year-end 2017 of just 3.9% partially reflects the low 10-year T-Bond rate of 2.4%, but also reflects the benign status of the HY bond market at that time. So, a 3.9% spread as of December 31, 2017 is not as low as a similar spread of 4.0% in 2005, when 10-year Treasuries were higher than 2.3%, at 4.4%.

4. The liquidity of a market, be it fixed-income or equity, is perhaps the most challenging variable to define precisely. At the same time, the perceived liquidity is always a major factor in a credit cycle and sometimes, for instance during highly stressed periods, is the dominant variable in defining a credit cycle's phase. While some traditional measures as bid-ask spreads, size of issues, and market volume are related to liquidity, we tend to look more closely at the volume of newly issued debt, and the proportion of this debt issued at the lowest credit rating category (i.e., CCC/Caa). If very risky firms with this lowest category of credit worthiness are able to easily access debt markets for refinancing or growth, this is definitely a sign that markets are not concerned much about credit risk and liquidity is high. Figure 9.14 shows the percentage of all high-yield bond new issuance that was CCC in both

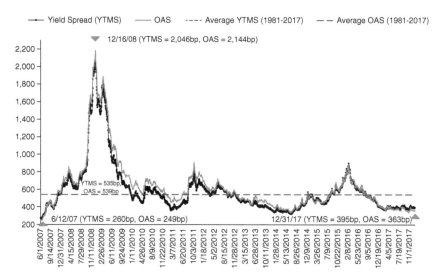

FIGURE 9.13 YTM and Option-Adjusted Spreads between High-Yield Markets and U.S. Treasury Notes, June 1, 2007–December 31, 2017
Sources: Citigroup Yieldbook Index Data and Bank of America Merrill Lynch.

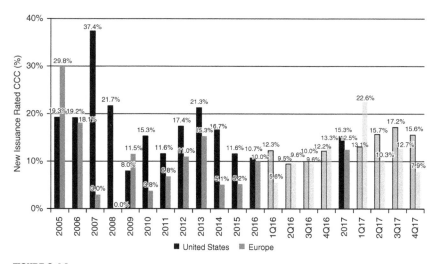

FIGURE 9.14 U.S. and European High-Yield Bond Market: CCC-Rated New Issuance (%), 2005–2017
Source: Based on data from S&P Global.

the United States and European markets. Note that the percentage reached an incredibly high 37.4% in 2007, just before the credit crisis hit and at the end of that cycle's benign state. Since 2009, CCC percentages ranged in the United States from as low as 8.0% in 2009 to over 21% in 2013, with more recent periods of 10.0–16%. It does appear that liquidity remains high with new HY issuance in 2017 of about $242 billion. So, CCC issuers seemingly had relatively easy access to the market at attractive rates and the overall high-yield market's ability to attract investors and new issuers was continuing its easy-money environment.

The last bullet of Figure 9.12 notes that when a market has enjoyed a benign credit cycle for many years and new issuance, combined with investor enthusiasm, seems perhaps excessive, a careful analysis must ask the question of a possible bubble growing and what might precipitate its bursting. Our conclusion in 2017 is that the high-yield bond market had several of the bubble characteristics of the frothy and excessive market in 2007, but also a few important mitigating factors which helped to provide confidence to investors that the benign cycle would continue, at least for the short-term future. There was no question that the four determinants, described above, were all pointing to a continuance of the benign cycle.

Where we were concerned was in several risk factors that could change the market's sentiment fairly quickly and precipitate a more risky environment. These included:

- The unprecedented duration already of the benign cycle (Figure 9.7)
- Level of nonfinancial corporate debt/GDP (Figure 9.8)
- Comparative health of high-yield issuing firms in 2016 compared to 2007 (see Chapter 10 herein on Z-Scores for these periods)
- High-yield and CCC new issuance (Figures 9.4, 9.5, 9.14)
- LBO statistics and trends (discussed further on)

LBO Risk Factors

Throughout the history of the high-yield bond market, one pervasive supply factor has been the use of these leveraged-financed instruments, including leveraged-loans, to finance takeovers of companies by private-equity (PE) firms, either in a friendly or hostile transaction. When the demand for company buyouts is particularly strong, either for reasons of low valuation of potential buyout candidates, low interest rates, and easy access to credit, or enormous stockpiles of cash acquired by PE funds, or several or all of the above, the competition will drive up the purchase cost and result in larger transactions at higher levels of debt than when demand is not as excessive. Figure 9.15 shows the Purchase Price Multiple paid by private-equity firms in their LBO transactions from

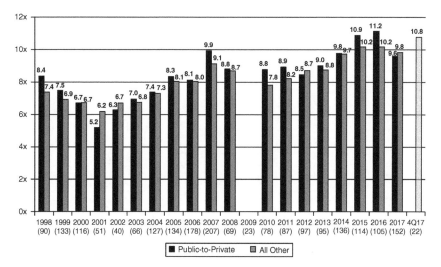

FIGURE 9.15 Purchase Price Multiples, Excluding Fees for LBO Transactions, 1998–2017
Source: Based on data from S&P Global.

1998 to 2017, for public-to-private or "other" (secondary buyouts) transactions. The public-to-private multiples have ranged from as low as 5.2 times in 2001 to 11.2 times recently in 2016. This latter average multiple was the highest recording that we have ever seen. This level was even higher than the frantic and catastrophic buyouts binge in the late 1980s that eventually led to a huge increase in defaults in 1990–1992, which was, in our opinion, one of the main catalysts of the economic recession in 1991 and record level of high-yield bond and leveraged loan default rates.

Skeptics of this concern about very recent record levels of purchase price multiples will cite the low current interest rate environment and the greater proportion of equity invested in current deals by PE firms compared to the average 80–90% debt to total capital transactions in 1987–1989. And, these skeptic factors are correct. LBO deals are more conservatively financed in recent years than back in the late 1980s and debt levels have averaged only 60–70% of late. Still, when debt is measured relative to EBITDA (Figure 9.16), we do see that the median levels of late are just below the dangerous level of six times. And, a median level of, say, 5.8 times indicates that half of the LBO transactions had debt/EBITDA ratios above 5.8 times – an ominous signal, even at low interest rates. One final point involves the significant amount of floating rate-leveraged loans in these transactions. So, if interest rates rise considerably in the future, LBO firms will face much higher hurdles to survive the high level of debt.

It should be noted that our concern with LBO purchase price multiples and Debt levels is mitigated by the PE firm industry's impressive track record with

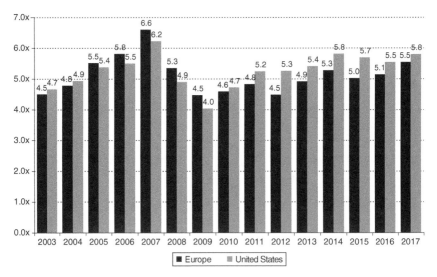

FIGURE 9.16 Average Total Debt Leverage Ratio for LBOs: Europe and United States with EBITDA of €/$50M or More, 2003–2017
Source: Based on data from S&P Global.

respect to avoiding defaults, either through superior portfolio management of their stockpile of firms and/or their network of alternative financing sources to provide liquidity should a crisis manifest. Indeed, our research has shown (not published) that, controlling for credit risk levels, PE-financed LBOs default significantly less frequently than do comparable risk firms, which are not part of a PE firm portfolio. Also, an element of late is the increased availability of financing from "shadow-banking" funds, which were not evident nearly as much in prior LBO cycles. This access-to-capital is more prominent for PE-owned firms than stand-alone entities.

Correlations between HY Bonds and Stock Returns

One final ingredient in our discussion of risk and returns in the HY bond market involves its critical relationship to another risky security market – equity prices. Classical financial-economic theory of the 1960s discussed the "pecking-order" theory of various types of capital financing for companies, firms' asymmetric information with respect to choices of capital, and their ability to maintain a target capital structure, debt/equity ratio, but also to take advantage of periodic opportunities (tradeoff theory) either to increase debt levels or to cut down on leverage when levels become excessive. This alternative financing theory works very well when firms have the ability to relatively seamlessly substitute equity for debt when it desires to, at costs which are not prohibitive. And, such was the case when it was easy to switch and the cost of equity was not highly correlated with the cost of debt. Figure 9.17 shows that the correlation between high-yield bonds and the S&P 500 Stock Index was reasonably low at 59% for the entire sample period 1987–2017 and even lower during the stressed cycles of 1990–1991 and 2001–2002 (48% and 54%, respectively). So, when the risky debt market "tanked" in those two cycles, equities fell also, but not nearly as much as what occurred in the 2008–2009 financial crisis (73% correlation) and, indeed, for the subsequent most recent period 2010–2017, when the correlation was 72%. Hence, it appeared that high-yield issuing or fallen-angel companies did have the opportunity to switch to equity in benign credit cycle markets, as it was in mid-2017, if they were concerned with the debt buildup since 2010. But, if they wait until the bond market shows unmistakable signs of stress to switch to equity, it may be too late to do so at reasonable

		Citi HY Index	S&P 500 Stock Index
Stressed Cycle I[a] 01/1990–12/1991 (24 obs.)	Defaulted Bond Index	68%	12%
	S&P 500 Stock Index	48%	
Stressed Cycle II[b] 01/2001–12/2002 (24 obs.)	Defaulted Bond and Bank Loan Index	76%	23%
	S&P 500 Stock Index	54%	
Stressed Cycle III 01/2008–03/2009 (15 obs.)	Defaulted Bond and Bank Loan Index	80%	73%
	S&P 500 Stock Index	73%	
Recovery Cycle 04/2009–04/2011 (25 obs.)	Defaulted Bond and Bank Loan Index	71%	65%
	S&P 500 Stock Index	67%	
Full Sample Period 01/1987–012/2017 (372 obs.)	Defaulted Bond and Bank Loan Index[c]	62%	38%
	S&P 500 Stock Index	59%	
Most Recent Period 01/2010–12/2017 (96 obs.)	Defaulted Bond and Bank Loan Index	52%	35%
	S&P 500 Stock Index	72%	

[a]Correlation between Defaulted Bond Index and S&P 500 was –16% during recovery period.
[b]Correlation between Defaulted Bond and Bank Loan Index and S&P 500 was 43% during recovery period.
[c]Based on only the Defaulted Bond Index from 01/1987 to 12/1995.

FIGURE 9.17 Total Monthly Return Correlations on Various Asset Class Indexes during Stressed and Recovery Credit Cycles
Source: E. Altman and B. Kuehne (2017), NYU Salomon Center.

costs, or at all! The temptation to continue to take advantage of low interest rates during a benign cycle may dominate the more conservative strategy to increase equity when the stock market is also at historic high record levels.

Forecasting Default Rates and Recoveries

Forecasting market default and recovery rates is a tricky exercise at best, that can be based on a "bottom-up" approach on individual issues and issuers or a macro, "top-down" approach – or both. For practical and track-record reasons, we have chosen the top-down approach using several techniques (models) that include (1) aggregate amounts of new issuance over the past decade stratified by the major ratings categories (mortality statistics). We also analyze the information content of market-based measures, based on (2) yield spreads and (3) distress ratios, to forecast the near-term default incidence of the market. Distress ratios are based on the percentage of high-yield bonds selling at least at 10% above the benchmark government bond rate (see Chapters 14 and 15 on distressed debt). In the past, these three techniques have been averaged to arrive at our single default rate estimate, although the range of possible outcomes can be observed as well. We now believe that our two market-based measures overweight the final default rate forecast, so we also present our forecast based on equal weights between the mortality rate method and average of the two market-based estimates. Our default rate estimates are then used as inputs to form the basis for estimates of aggregate recovery rates on corporate high-yield bond defaults. For more details on our default rate forecasting methodologies and their estimates over time, see our annual High-Yield bond reports, for example, Altman and Kuehne (2017).

Comparing Cumulative Default Rates across Sources

One can observe in Figure 9.18 a marked difference in one-year default rates for many of the bond rating classes, if you compare our mortality results in Figures 9.19 and 9.20 (Altman 1989, Altman and Kuehne, 2017) with those cumulative default rates of the major rating agencies (Moody's and S&P). The different methodologies for computing marginal and cumulative rates can explain these seemingly confusing results. The differences can be explained by several factors:

- Dollar-weighted (Altman) versus issuer-weighted (rating agencies) data.
- Original issuance ratings (Altman) versus grouping by rating regardless of age (rating agencies).
- Mortality approach (Altman) versus default rates based on original cohort size (rating agencies).
- Sample periods.

	1	2	3	4	5	6	7	8	9	10
AAA/Aaa										
Altman	0.00%	0.00%	0.00%	0.00%	0.01%	0.03%	0.04%	0.04%	0.04%	0.04%
Moody's	0.00%	0.00%	0.00%	0.03%	0.13%	0.22%	0.31%	0.41%	0.51%	0.62%
S&P	0.00%	0.04%	0.17%	0.29%	0.42%	0.54%	0.59%	0.67%	0.76%	0.85%
AA/Aa										
Altman	0.00%	0.00%	0.20%	0.26%	0.28%	0.29%	0.30%	0.31%	0.33%	0.34%
Moody's	0.01%	0.02%	0.08%	0.21%	0.35%	0.47%	0.57%	0.66%	0.74%	0.82%
S&P	0.03%	0.08%	0.18%	0.31%	0.45%	0.60%	0.74%	0.86%	0.96%	1.07%
A/A										
Altman	0.01%	0.04%	0.15%	0.27%	0.36%	0.41%	0.43%	0.67%	0.74%	0.78%
Moody's	0.03%	0.13%	0.30%	0.46%	0.64%	0.86%	1.09%	1.34%	1.60%	1.85%
S&P	0.07%	0.20%	0.36%	0.54%	0.73%	0.95%	1.19%	1.41%	1.65%	1.89%
BBB/Baa										
Altman	0.32%	2.65%	3.86%	4.80%	5.27%	5.48%	5.71%	5.86%	6.02%	6.33%
Moody's	0.15%	0.43%	0.78%	1.23%	1.68%	2.14%	2.59%	3.06%	3.59%	4.18%
S&P	0.22%	0.58%	0.99%	1.50%	2.05%	2.60%	3.09%	3.58%	4.07%	4.55%
BB/Ba										
Altman	0.92%	2.94%	6.68%	8.50%	10.71%	12.11%	13.37%	14.32%	15.53%	18.16%
Moody's	1.07%	2.93%	5.08%	7.27%	9.25%	11.13%	12.80%	14.46%	16.16%	17.94%
S&P	0.80%	2.52%	4.57%	6.57%	8.38%	10.14%	11.62%	12.98%	14.17%	15.25%
B/B										
Altman	2.86%	10.31%	17.29%	23.70%	28.08%	31.29%	33.76%	35.12%	36.24%	36.72%
Moody's	3.90%	9.07%	14.31%	18.97%	23.25%	27.16%	30.70%	33.81%	36.64%	39.10%
S&P	3.92%	9.00%	13.43%	16.88%	19.57%	21.76%	23.56%	24.98%	26.24%	27.42%
CCC/Caa										
Altman	8.11%	19.50%	33.79%	44.55%	47.27%	53.40%	55.91%	58.01%	58.28%	60.05%
Moody's	10.37%	18.29%	24.97%	30.52%	35.25%	38.77%	41.96%	45.31%	48.89%	51.59%
S&P	28.85%	39.23%	44.94%	48.55%	51.31%	52.53%	53.95%	55.00%	55.96%	56.66%

Sources:

- Altman, Market value weights, by number of years from original Standard & Poor's issuance ratings, 1971–2016; see Altman (1989) and Altman and Kuehne (2017)
- Moody's Average Cumulative Default Rates, all issues (1970–2016), Moody's Investors Services, 2017
- S&P 2016 Annual Global Corporate Default Study and Rating Transitions: Average Cumulative Default Rates for Corporates (1981–2016)

FIGURE 9.18 Cumulative Default Rate (Mortality Rate) Comparison (in Percent for Up to 10 Years)

$$\text{MMR}_{(r,t)} = \frac{\textit{total value of defaulting debt from rating } (r) \textit{ in year } (t)}{\textit{total value of the population at the start of the year } (t)}$$

MMR $=$ marginal mortality rate

One can measure the cumulative mortality rate (CMR) over a specific time period $(1, 2, \ldots, t$ years) by subtracting the product of the surviving populations of each of the previous years from one (1.0), that is,

$$CMR_{(r,t)} = 1 - \Pi SR_{(r,t)}$$
$$t = 1 \rightarrow N$$
$$r = \text{AAA} \rightarrow \text{CCC}$$

here

CMR $_{(r,t)}$ = cumulative mortality rate of (r) in $_{(t)}$,
SR $_{(r,t)}$ = survival rate in $_{(r,t)}$, $1 - \text{MMR}_{(r,t)}$

FIGURE 9.19 Marginal and Cumulative Mortality Rate Actuarial Approach

		\multicolumn{10}{c}{Years After Issuance}									
		1	2	3	4	5	6	7	8	9	10
AAA	Marginal	0.00%	0.00%	0.00%	0.00%	0.01%	0.02%	0.01%	0.00%	0.00%	0.00%
	Cumulative	0.00%	0.00%	0.00%	0.00%	0.01%	0.03%	0.04%	0.04%	0.04%	0.04%
AA	Marginal	0.00%	0.00%	0.19%	0.05%	0.02%	0.01%	0.01%	0.01%	0.01%	0.01%
	Cumulative	0.00%	0.00%	0.19%	0.24%	0.26%	0.27%	0.28%	0.29%	0.30%	0.31%
A	Marginal	0.01%	0.03%	0.10%	0.11%	0.08%	0.04%	0.02%	0.23%	0.06%	0.03%
	Cumulative	0.01%	0.04%	0.14%	0.25%	0.33%	0.37%	0.39%	0.62%	0.68%	0.71%
BBB	Marginal	0.31%	2.34%	1.23%	0.97%	0.48%	0.21%	0.24%	0.15%	0.16%	0.32%
	Cumulative	0.31%	2.64%	3.84%	4.77%	5.23%	5.43%	5.66%	5.80%	5.95%	6.25%
BB	Marginal	0.91%	2.03%	3.83%	1.96%	2.40%	1.54%	1.43%	1.08%	1.40%	3.09%
	Cumulative	0.91%	2.92%	6.64%	8.47%	10.67%	12.04%	13.30%	14.24%	15.44%	18.05%
B	Marginal	2.85%	7.65%	7.72%	7.74%	5.72%	4.45%	3.60%	2.04%	1.71%	0.73%
	Cumulative	2.85%	10.28%	17.21%	23.62%	27.99%	31.19%	33.67%	35.02%	36.13%	36.60%
CCC	Marginal	8.09%	12.40%	17.71%	16.22%	4.88%	11.60%	5.39%	4.73%	0.62%	4.23%
	Cumulative	8.09%	19.49%	33.75%	44.49%	47.20%	53.33%	55.84%	57.93%	58.19%	59.96%

*Rated by S&P at issuance.
Based on 3,359 issues.

FIGURE 9.20 Mortality Rates by Original Rating, All Rated Corporate Bonds,*
1971–2017
Source: Standard & Poor's (New York) and author's compilation.

In our opinion, by far the most important reason is the mortality rate vs. cumulative default rates, whereby we use the rating of an issue, and its size, when the bond was first issued (second and third bullets above). The rating agencies all use a basket of bonds with the same rating at a point in time, regardless of how long they have been outstanding and what the original rating was. So we observe substantially lower default rates in the first few years in the mortality rate results than we do in the rating agency data. For example, the mortality rate for single-B rated issues in the first year is 2.86%, whereas the one-year default rate for Moody's (5.0%) is substantially higher. Note that the differences between mortality and cumulative rates persist until the fourth or fifth year, when the aging effect is diminished and all methods give fairly similar results (Figure 9.18).

These differences in results are immensely important for the bank or investor, who needs to estimate the one-year expected default rate for Basel II purposes (banks) or for expected defaults and loss reserves (all users). We believe that the age distribution of the portfolio under analysis should dictate the method used and the data referenced. Certainly, when making a new loan or investing in a newly issued bond, the mortality rate approach would be logical to use for estimating cash flows, net of defaults, and other purposes. For more mature portfolios, there is perhaps more logic in the rating agencies' cumulative-default approach.

Conclusion

In conclusion, the high-yield bond market, both in the United States and in several other geographical areas of the world, is now a mature, universally accepted means of financing lower-rated companies for a great variety of purposes, and its securities are a legitimate asset class for investors. We will in Chapter 14 and Chapter 15 of this volume explore a "derivative" asset class to high-yield bonds and leveraged loans, specifically distressed and defaulted securities of companies that may, or actually do, default but whose securities continue to trade as the firm attempts to restructure, either out-of-court or while in a Chapter 11 reorganization. Keep in mind that while high-yield bonds are a legitimate risk-return asset class, especially for institutional investors, they are also the potential inventory for distressed investors.

CHAPTER 10

A 50-Year Retrospective on Credit Risk Models, the Altman Z-Score Family of Models, and Their Applications to Financial Markets and Managerial Strategies

THE EVOLUTION OF CORPORATE CREDIT SCORING SYSTEMS

Credit scoring systems for identifying determinants of a firm's repayment likelihood probably go back to the days of the Crusades – when travelers needed "loans" to finance their travels – and certainly were used much later in the United States as companies and entrepreneurs helped to grow the economy, especially in its westward expansion. Primitive financial information was usually evaluated by lending institutions in the 1800s with the primary types of information required being subjective, or qualitative in nature, revolving around ownership and management variables and collateral (see Figure 10.1). It was not until the early 1900s that rating agencies and some more financially oriented corporate entities, for example, the DuPont System of corporate ROE growth, introduced univariate accounting measures and industry peer-group comparisons, with rating designations (Figure 10.2). The key aspect of these "revolutionary" techniques was the ability of the analyst to compare an individual corporate entity's financial performance metrics to a reference database of time-series (same entity) and cross-section (industry) data. Then, and as even more so today, data and databases were the key element for meaningful diagnostics. There is no doubt that in the credit scoring field, data is "king" and models to capture the probability of default ultimately succeed, or not, are based on its ability to be applied to databases of various size and relevance.

- Qualitative (Subjective) – 1800s
- Univariate (Accounting/Market Measures)
  - Rating Agency – for example, Moody's (1909), S&P (1916), and Corporate (e.g., DuPont) Systems (early 1900s)
- Multivariate (Accounting with Market Measures) – Late 1960s (Z-Score) → Present
  - Discriminant, Logit, Probit Models (Linear, Quadratic)
  - Nonlinear and "Black-Box" Models (e.g., Recursive Partitioning, Neural Networks, 1990s)
- Discriminant and Logit Models Used for
  - Consumer Models – *Fair Isaacs* (FICO Scores)
  - Manufacturing Public (U.S.) Firms (1968) – Z-Scores
  - Extensions and Innovations for Specific Industries and Countries (1970s → Present)
  - ZETA Score – Industrials (1977)
  - Private Firm Models (e.g., Z''-score (1983), Z'' Score(1995)
  - EM Score – Emerging Markets (1995)
  - Bank Specialized Systems (1990s) Basel 2 impetus
  - SMEs – for example, Edminster (1972); Altman and Sabato (2007); and Altman, Sabato, and Wilson (2017)
- Option/Contingent Claims Models (1970s → Present)
  - Risk of Ruin (Wilcox, 1971)
  - *KMVs* Credit Monitor Model (1993) – Extensions of Merton (1974) Structural Framework
- Artificial Intelligence Systems (1990s → Present)
  - Expert Systems
  - Neural Networks
  - Machine Learning
  - Recursive Partitioning (Frydman, Altman, and Kao, 1985)
- Blended Ratio/Market Value Models
  - Altman Z-Score (Fundamental Ratios and Market Values) – 1968
  - Bond Score (Credit Sights, 2000; RiskCalc Moody's, 2000)
  - Hazard (Shumway, 2001)
  - Kamakura's Reduced Form, Term Structure Model (2002)
  - Z-Metrics (Altman et al., RiskMetrics, 2010)
- Reintroduction of Qualitative Factors and Real-Time Data (FinTech)
  - Stand-alone Metrics, for example, Invoices, Payment History
  - Multiple Factors – Data Mining (Big Data Payments, Governance, Time spent on individual firm reports [e.g., *CreditRiskMonitor's* revised FRISK Scores, 2017], etc.)
  - Enhanced Blended Models (2000s)

FIGURE 10.1   Corporate Scoring Systems over Time

The original Altman Z-Score model (Altman, 1968) was based on a sample of 66 manufacturing companies in two groups, bankrupt and nonbankrupt firms, and a holdout sample of fewer (50) companies. In those "primitive" days, there were no electronic databases and the researcher/analyst had to construct his own database from primary (annual report) or secondary (Moody's and S&P Industrial manuals and reports) sources. To this day, instructors and researchers oftentimes ask me for my original 66-firm database, mainly for instructional or reference exercises. It is not unheard of today for researchers to have access to databases of

Moody's		S&P/Fitch
Aaa		AAA
Aa1		AA+
Aa2		AA
Aa3		AA–
A1		A+
A2		A
A3		A–
Baa1		BBB+
Baa2	Investment	BBB
Baa3	Grade	BBB–
Ba1	High Yield	BB+
Ba2	("Junk")	BB
Ba3		BB–
B1		B+
B2		B
B3		B–
Caa1		CCC+
Caa		CCC
Caa3		CCC–
Ca		CC
		C
C		D

FIGURE 10.2 Major Agencies' Bond Rating Categories

thousands, even millions of firms (especially in countries where all firms must file their financial statements in a public database, e.g., in the UK). To illustrate the importance of databases, Moody's, Inc. purchased in 2017 the extensive data on 200 million firms and customer access of Bureau van Dijk Electronic Publishing (EQT) for $3.3 billion and S&P-purchased SNL's financial institution's extensive database, management structure, and customer-book for $2.2 billion in 2015. As indicated in Figure 10.2, the three major rating agencies established a hierarchy of creditworthiness that was descriptive, but not quantified, in its depiction of the likelihood of default. The determination of these ratings was based on a combination of (1) financial statement ratio analytics, usually on a univariate, one ratio-at-a-time basis; (2) industry health discussion; and (3) qualitative factors evaluating the firm's management plans and capabilities, strategic directions and other, perhaps "inside-information," gleaned from its interviews with senior management and experience of the team that was assigned to the rating decision. To this day, the decision process of rating agencies remains essentially the same with the ultimate rating decision made based on the firm's likelihood of default, and in some cases the loss given default, based on expected recovery. These inputs were analyzed on a "through-the-business-cycle" basis, often based on a "stressed" historical analysis. While the stressed scenario basis for evaluating a firm's solvency is still an important input, rating agencies no longer embrace the business cycle as the key determinant as to whether to change a rating.

What can we say about this process and its evaluative results? Here are our opinions:

1. Since the process has been standardized and carried out fairly consistently over time, it can provide important reference points for the market and is well understood as an "international language of credit." This makes the database of assigned original ratings and rating changes an incredibly important source of data for both researchers and practitioners on an ongoing basis.

2. Original rating assignments are done carefully with adequate resources and a strong desire to assess the repayment potential of the firm on specific issues of bonds, loans, and commercial paper (the so-called "plain vanilla" issuances of firms) very accurately. The rating assignments do not provide specific quantitative estimates of the probability of default, but do provide important benchmarks for comparing the actual incidence of default on millions of bond issues for long periods of times and to assess the bond-rating-equivalent of nonrated firms and securities in order to eventually provide a PD (probability of default) of corporate debt issuances. Hence, we will show this capability in our own "mortality rate" determination based on our original work (Altman 1989) subsequently updated annually, as well as similar analytics introduced by Moody's and S&P in the early 1990s called cumulative default rates and rating transitions tables.[1]

3. We have found, however, that the track record of excellent original rating assignments by the rating agencies is not matched in their performance by the *timeliness of rating changes*, that is, transitions as the firm's financial and business performance evolves. Studies such as Altman and Rijken (2004) show clearly that agency ratings are generally slower to react to changes, primarily deteriorations in performance, than are established models based on "point-in-time" estimations, for example, Z-Score type models or KMV structural estimates. Indeed, rating agencies openly admit that stability of ratings is a very important attribute of their systems and volatile changes are to be avoided. So, it is no surprise that when rating changes do occur, they are slower than what an objective, unemotional model would produce, and these changes, principally downgrades, are typically smaller (i.e., less notches) than what a model would have produced. The latter implies that if another rating change would follow an initial downgrade, it is highly likely that the second change would be in the *same direction* as the first – that is, strong auto-correlation of rating downgrades (see Altman and Kao 1992). We have not encountered much, if any, denial from the rating agencies on this observation. After all, the Agencies' clients (firms issuing debt) are more comfortable with a system that provides more stable ratings than one which changes, especially

negatively, frequently. And, those using the service, like pension and mutual funds, also prefer stability to volatile ratings.

4. These observations illustrate the ongoing discussion and heated arguments as to the objectivity and potential bias of ratings based on the agencies' business model that the same entity that is being rated (firms) also pays for the rating. Critics of rating agencies point to this potential conflict of interest and call for other structures, such as the "investor-pay" model, or government agencies providing ratings. These have been floated but have not seemed to resonate well with the main protagonist in the rating industry, that is, the users of ratings, primarily investors. And, investors, in some cases, prefer stability of ratings over short-term volatility, especially if the changes involve a change from investment grade to noninvestment grade, or vice versa. Hence, despite efforts by regulators to encourage alternative systems for estimating PDs, like internally generated or vendor models, ratings from the major rating agencies continue to be an important source of third-party assessment for the market. We feel that models, like the Altman Z-Score family, can still play a very important role in the investment process despite the continued prominence of Agency ratings.

Multivariate Accounting/Market Measures

Continuing the evolutionary history of credit scoring beyond univariate systems, (such as those followed by rating agencies and prominent scholarly research studies by numerous academics, such as Beaver 1966) we now move to the first multivariate study to attack the bankruptcy prediction subjects; the initial Z-Score model. Utilizing one of the first Discriminant Analysis models applied to the economic-financial social sciences, Altman (1968) and later Deakin (1972), combined traditional financial statement variables with new and more powerful statistical techniques, and aided by early editions of main-frame computers, constructed the original 1968 Z-Score model. Consisting of five financial indicators, four of which required only one year of financial statements and one requiring equity market values, the original model (Figure 10.3) demonstrated outstanding original and holdout sample accuracies of Type I (predicting bankruptcy) and Type II (predicting nonbankruptcy) based on a derived cutoff-score approach (discussed later) and from financial data from one annual statement prior to bankruptcy. The original sample of firms utilized *only manufacturing companies*, that filed for bankruptcy-reorganization under the "old" system called Chapter X or XI (now combined under Chapter 11). All firms were publicly held and, given the economic environment in the United States prior to 1966, all had assets under $25 million. The sample sizes were small, only 33 in each grouping, which is remarkable in that the model is still being used extensibly 50 years after its introduction on firms of all sizes, including those with billions of dollars of assets.

Variable	Definition	Weighting Factor
X_1 - - - -	$\dfrac{\text{Working Capital}}{\text{Total Assets}}$	1.2
X_2 - - - -	$\dfrac{\text{Retained Earnings}}{\text{Total Assets}}$	1.4
X_3 - - - -	$\dfrac{\text{EBIT}}{\text{Total Assets}}$	3.3
X_4 - - - -	$\dfrac{\text{Market Value of Equity}}{\text{Book Value of Total Liabilities}}$	0.6
X_5 - - - -	$\dfrac{\text{Sales}}{\text{Total Assets}}$	1.0

FIGURE 10.3 Original Z-Score Component Definitions and Weightings

The original Z-Score model was linear and did not require more than one set of financial statements. Subsequent to its introduction, similar models utilizing linear and nonlinear variable structures and different classification techniques, such as quadratic, logit, probit, and hazard model structures were introduced to attempt not only to classify a firm as bankrupt or not, but also to express the outcome in terms of the probability of default based on the characteristics of the sample of firms used in the model's development. An alternative approach for developing PDs, based on the Altman Bond-Rating Equivalent (BRE) method, combined with empirically derived estimates of default incidence for long horizons (e.g., 1–10 years), will be discussed shortly.

These Discriminant, or Logit, models were applied to consumer credit applications (e.g., Fair Isaac's FICO scores), nonmanufacturers (e.g., ZETA scores; see Altman, Haldeman, and Narayanan (1977), to private firms as well as publicly owned ones, in many other countries (built over several decades and continuing to be derived even in current years), emerging markets (e.g., Altman, Hartzell, and Peck 1995), for internal rating systems (IRBs) of banks (starting in the mid-1990s and especially since Basel II was first introduced for discussion in 1999) and for various industries and sizes of firms, including models specifically derived for small and medium sized firms (SMEs), for example, Edmister (1972), Altman and Sabato (2007), Altman, Sabato, & Wilson (2010), and, most recently for mini-bond issuers in Italy, by Altman, Esentato, and Sabato (2016).

Many other exotic statistical and mathematical techniques have been applied to the bankruptcy/default prediction field including expert systems, neural-networks, genetic algorithms, recursive partitioning, and so on, and the latest attempts using sophisticated machine-learning methods, motivated by the existence of massive databases and the introduction of nonfinancial data. While these techniques usually surpass the now "primitive" discriminant financial statement based models in terms of prediction-accuracy tests on original and sometimes holdout samples, the more complex the algorithm and specialized data sources, the less likely that the model will be understood and be able to be replicated by other researchers and by practitioners in its real-world applications.

One class of model that attained both scholarly and practitioner acceptance and usage is the so-called structural models built after the introduction of Merton's (1974) contingent-claims approach for valuing risky-debt and later the commercialization of the Merton model by KMV's Credit Monitor System. The latter was and still is (marketed by Moody's since 2001) based on a very large sample of defaults derived from companies on a global basis. The result is a PD estimate derived from a distance-to-default calculation, relying primarily on firm market values, historical market volatility measures and levels of debt. Academic researchers and several consultants have replicated the Merton-structural approach and have oftentimes compared it to their own models, as well as more traditional models, like Z-Scores and Kamakura's reduced-form approach, see Chava and Jarrow (2004), with results that are not always clear as to which approach was superior; see, for example, Das, Hanouna, and Sarin (2009), Bharath and Shumway (2008), or Campbell, Hilcher, and Szilagi (2008).

The most recent attempts at building both accurate and practically acceptable models have utilized what we call a "blended" Ratio/Market Value/Macro Variable approach, with some attempts to also include nonfinancial variables, where data exists. These blended models, for example, Z-Metrics (Altman, Rijken, Watt, Balan, Forero, and Mina 2010), introduced by RiskMetrics, are probably the ones that consultants and many financial lenders are either considering or utilizing today, at least in comparison to more traditionally derived models, sometimes with judgmental adjustments by lending officers. And, finally, the FinTech innovations of late are exploring the use of "Big-Data" and nontraditional metrics, like invoice-analysis, payable-history, and governance attributes, "clicks" on negative information events and data,[2] and social-media inputs, in order to capture, on a real-time basis, changes in credit quality of firms and individuals.

Machine-Learning Methods

As for machine-learning and "big-data" techniques, we remain somewhat skeptical whether many practitioners will accept "black-box" methods for assessing credit risk of counterparties. Yes, it is undeniable that the current surge in the application of such techniques has captured the interest of many academics and several startups in the FinTech space. Indeed, I collaborated with some colleagues (Barboza, Kumar, and Altman 2017) using several machine-learning models, for example, support vector machines (SVM), boosting, random forest, and so on, to predict bankruptcy from one year prior to the event and compared the results to discriminant analysis, logistical regression, and neural network methods. Using data from 1985–2013, we found substantial improvement in prediction accuracy (of about 10%) using machine learning techniques, especially when, in addition to the five Z-Score variables, six additional indicators were included. These results add one more study to the growing debate in the past few years (2014–2017) about the superiority of SVM versus other machine-learning methods. Almost all of the machine-learning credit models have been published in expert systems and computational journals, most prominent found in *Expert Systems with Applications* (see the reference list in Barboza et al., 2017).

FROM A SCORING MODEL TO DEFAULT PREDICTION

The construction of a credit-scoring model is relatively straightforward with an adequate and appropriate database of default and nondefault securities, or firms, and accurate predictive variables. In the case of our first model, the Z-Score method (named in association with statistical Z-measures and also chosen because it is the last letter in the English alphabet), the classification as to whether a corporate entity was "likely" to go bankrupt, or not, was done based on cutoff scores between "Safe" versus "Distress" zones, with an intermediate "Gray" zone (Figure 10.4).

These zones were selected based solely on the results of the original, admittedly smallish samples of 33 firms in each of the two groupings (bankrupt and nonbankrupt) from manufacturing firms and their financial statement and equity markets values from the 1960s. Any firm whose Z-Score was below 1.8 (Distressed Zone) was classified as "bankrupt" and did, in fact, go bankrupt within one year (100% accuracy) and firms whose score was greater than 2.99 did not go bankrupt (also 100% accuracy), at least until the end of the study period in 1966. There were a few errors in classification for firms with scores between 1.81 and 2.99 (Gray Zone – 3 errors out of 66; see Figure 10.5).

Keep in mind that these cutoff-scores were based solely on the original sample of firms. But, because the zones were clear, unambiguous and consistently accurate in their subsequent predictions of greater than 85%, based on data from one year prior to bankruptcy (Type I) (Figure 10.5), these designations remain to this day as accepted and useful to market practitioners. While flattering to this writer, this is unfortunate as it is obvious that the dynamics and trends in credit-worthiness have changed significantly over the past 50 years. And, the classification as to "Bankrupt" or "Nonbankrupt" is no longer sufficient for many applications of the Z-Score model. Indeed, there is very little difference between a firm whose score is 1.81 versus one whose score is 1.79, yet, the zones are different. In addition,

Z > 2.99 – "Safe" Zone

1.81 > Z < 2.99 – "Grey" Zone

Z < 1.81 – "Distress" Zone

FIGURE 10.4   Zones of Discrimination: Original Z-Score Model (1968)

Year Prior to Failure	Original Sample (33)	Holdout Sample (25)	1969–1975 Predictive Sample (86)	1976–1995 Predictive Sample (110)	1997–1999 Predictive Sample (120)
1	94% (88%)	96% (72%)	86% (75%)	85% (78%)	94% (84%)
2	72%	80%	68%	75%	74%
3	48%	–	–	–	–
4	29%	–	–	–	–
5	36%	–	–	–	–

FIGURE 10.5 Classification and Prediction Accuracy Z-Score (1968) Failure Model*
*Using 2.67 as cutoff (1.81 cutoff accuracy in parenthesis).

certainly the "holy grail" in credit assessment, namely the probability of default (PD) and when the default associated with the probability is to take place, are not specified clearly by a certain credit score. We now examine how credit dynamics have changed over our relevant time periods and how have we moved on to precise PD and timing of default estimates.

Time-Series Impact on Corporate Z-Scores

When we built the original Z-Score model in the mid-1960s, financial credit markets were much simpler, some might say primitive, compared to today's highly complex, multistructured environment. Such innovations as high-yield bonds, leverage-loans, structured financial products, credit derivatives, such as CDS (credit default swaps), and "shadow-banking" loans were nonexistent then, and riskier companies had little financing alternatives outside of traditional bank loans and trade-debt. For example, we see from Figure 10.6 that the North American high-yield bond market had not been "discovered" until the late 1970s when the only participants were the so-called "fallen-angel" companies that raised debt originally when they were investment grade (IG). In 2017, the size of the high-yield "junk-bond" market has grown from about $10 billion of fallen angels in 1978 to about $1.7 trillion, with over 85% of this market consisting of "original-issue" high-yield issues. In addition, the pace of newly issued high-yield bonds and leverage-loans, has accelerated since the great financial crisis (GFC) of 2008/2009, with more than $200 billion of new bond issues each year since 2010, fueled by a benign credit-cycle, which has continued now into its ninth year in 2018. Not to be outdone, loans to noninvestment grade companies have again become numerous and available at attractive rates as interest rates have fallen, in general, and banks, despite regulatory oversight guidelines (e.g., Debt/EBITDA ratios not to exceed a certain level, e.g., 6), have competed with the public markets in the United States and Europe. These leveraged loans' new issues of several hundred billions per year ($651 billion in 2017 in the United States alone) have swelled corporate debt ratios to an unprecedented level as firms have exploited the easy-money, low-interest-rate environment and lenders seek yield, even for

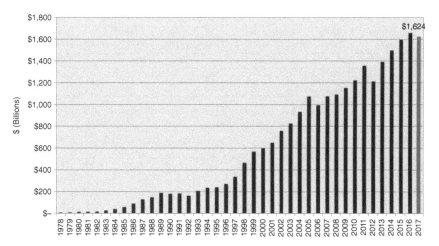

FIGURE 10.6   Size of the U.S. High-Yield Bond Market 1978–2017 (Mid-year US$ Billions)
Source: NYU Salomon Center estimates using Credit Suisse, S&P, and Chili data.

the most risky corporate entities. For example, CCC new issues have averaged 15.0% of all high-yield bond new issues over the period 2010–2017 (2Q).

Other factors that have reduced the average credit-worthiness of companies as the Z-Score model has matured in the 50-year period since its inception are global competitive factors, the enormous power of market dominating firms in certain industries, such as Walmart and Amazon in the retail space, and the amazing susceptibility of larger companies to financial distress and bankruptcy. Indeed, when we built the Z-Score model in the 1960s, the largest bankrupt firm in our sample had total assets of less than $25 million (about $125 million, inflation adjusted) compared to an environment with a median annual number of 14 firms each year since 1990 with liabilities (and assets) of more than $1 billion.

To demonstrate the implied deterioration in corporate creditworthiness over the past 50 years, one can observe our median Z-Score statistics by S&P credit rating for various sample years shown in Figure 10.7. First, the number of AAA ratings has dwindled to just two in 2017 (Microsoft and Johnson & Johnson) from 15–20 years ago and 98 in 1992! Hence, we now combine AAA- and AA-rated companies to analyze average Z-Scores and that median has decreased from a high of 5.20 in the 1996–2001 period to 4.30 in 2017. Of even more importance, is the steady deterioration of median Z-Scores for single-B companies from 1.87 in 1992–1995 to 1.65 in 2017. Recall that a score of below 1.8 in 1966 was classified as a firm in the Distress Zone and a strong bankruptcy prediction. However, in the past 15 years, or so, the dominant and largest percentage of issuance in the high-yield market was single-B, and surely all single-Bs do not default! Yes, a median single-B has a distribution with 50% of the issues higher than 1.65,

Rating	2017 (No.)	2013 (No.)	2004–2010	1996–2001	1992–1995
AAA/AA	4.30 (14)	4.13 (15)	4.18	5.20	5.80*
A	4.01 (47)	4.00 (64)	3.71	4.22	3.87
BBB	3.17 (120)	3.01 (131)	3.26	3.74	2.75
BB	2.48 (136)	2.69 (119)	2.48	2.81	2.25
B	1.65 (79)	1.66 (80)	1.74	1.80	1.87
CCC/CC	0.90 (6)	0.33 (3)	0.46	0.33	0.40
D	−0.10 (9)[1]	0.01 (33)[2]	−0.04	−0.20	0.05

*AAA Only; No. = Number of firms in the sample.
[1]From 1/2014 to 11/2017;
[2]From 1/2011 to 12/2013.

FIGURE 10.7 Median Z-Score by S&P Bond Rating for U.S. Manufacturing Firms, 1992–2017
Source: Compustat database, mainly S&P 500 firms, compilation by E. Altman, NYU Salomon Center, Stern School of Business.

but the probability of all Bs that default within five years of issuance is approximately "only" 28% (see our mortality-rate discussion shortly). Finally, the median "D" (Default-rated company) had a Z-Score of –0.10 in 2017, while the median Z-Score in 1966 for bankruptcy entities was +0.65 (see Altman 1968 or Altman and Hotchkiss 2006). In all time periods of late, the median defaulted firm's Z-Score was zero or below (Figure 10.7). Hence, we suggest that a score below zero is consistent with a "Defaulted" company. The cutoff of 1.8, based on our original sample, will place an increasing number, perhaps as much as 25% of all firms, in the old "Distress" zone. Since only a very small percentage of all firms fail each year and an average of about 3.5% of high-yield bond companies default each year, based on data over the last almost 50 years (see our default rate calculations, in Altman & Kuehne 2017 and Chapter 9 herein), the so-called Type II error (predicting default when the firm does not) has increased from about 5% in our original analysis to possibly 25–30% in recent periods. Hence, we do *not* recommend that users of our Z-Score model make their assessments of a firm's default likelihood based on a cutoff score of 1.8. Instead, we recommend using bond-rating-equivalents (BREs), based on the most recent median Z-Scores by bond-rating, such as the data listed in Figure 10.7. These BREs can then be converted into more granular PD estimates, as we now discuss.

PD Estimation Methods

Figure 10.8 lists two methods we have used over the years to estimate the probability of default and loss-given-default of a firm's bond issue at any point in time. The starting point in both methods, is a well-constructed and, if possible, intuitively understandable credit-scoring model. For example, the Z-Score on a new, or existing, debt issuer is then, in Method #1, assigned a Bond-Rating-Equivalent (BRE) on a representative sample of bond issues for each of the major rating categories

Method #1

- Credit scores on new or existing debt
- Bond rating equivalents on new issues (Mortality) or existing issues (Rating Agency Cumulative Defaults)
- Utilizing mortality rates to estimate marginal and cumulative defaults
- Estimating Default Recoveries and Probability of Loss

Or

Method #2

- Credit scores on new or existing debt
- Direct estimation of the probability of default based on logistic regressions
- Based on PDs, assign a rating

FIGURE 10.8 Estimating Probability of Default (PD) and Probability of Loss Given Defaults (LGD)

(see Figure 10.7) or, if available, more granular ratings with (+) or (–) "notches (S&P/Fitch) or 1, 2, 3 (Moody's). See Figure 10.15, at a later point, for the more granular categorization for another of the Altman Z-Score models, Z''-Scores.

In addition to the matching of Z-Scores by rating category, we also can assess the PD of an issue for various periods of time in the future. The more traditional time-dependent method is called "Cumulative Default Rates" (CDRs) as provided by all of the rating agencies and by several of the investment banks who provide continuous research on defaults, particularly for the speculative grade, or high-yield ("junk bond") market. This compilation, is an empirically derived PD estimate of bonds with a certain rating, for example, "B", at a point in time, and then the default incidence is observed 1, 2, ... 10 years, after that point in time. The estimate is for all B– rated bonds regardless of the *age* of the bond when it is first tracked. In our opinion, this PD estimate is more appropriate for *existing* bond issuer's debt, *not* for bonds when they are first issued. Almost all of the rating agencies, with the exception of FITCH, Inc., calculate the CDRs based on the number of issuers that default over time compared to the number of issuers with a certain rating at the starting point (regardless of the different ages of the bonds in the "basket" of, say, B-rated bonds). Therefore, on average, an S&P B-rated bond had about a 5% incidence of default within one year based on a sample of bonds from 1980–2016 (see Standard & Poor's, 2017).

Before the rating agencies first compiled their cumulative default rates, Altman (1989) created the "mortality rate" approach for estimating PDs for bonds of all ratings and specifically for newly issued bonds and based on dollar amounts of new issues-by-bond-rating, not by issuers. These mortality estimates are based on insurance-actuarial techniques to calculate the marginal and cumulative mortality rate, as shown in Figure 10.9. I felt, just like "people-mortality," there are certain characteristics of bonds, or loans, at birth that are critical in determining

$$MMR_{(r,t)} = \frac{\text{total value of defaulting debt from rating } (r) \text{ in year } (t)}{\text{total value of the population at the start of the year } (t)}$$

MMR = marginal mortality rate

One can measure the cumulative mortality rate (CMR) over a specific time period (1, 2, ..., T years) by subtracting the product of the surviving populations of each of the previous years from one (1.0), that is,

$$CMR_{(r,t)} = 1 - \Pi SR_{(r,t)}$$

$$t = 1 \rightarrow N$$

$$r = AAA \rightarrow CCC$$

here

$CMR_{(r,t)}$ = Cumulative Mortality Rate of (r) in$_{(t)}$,
$SR_{(r,t)}$ = Survival Rate in$_{(r,t)}$, $1 - MMR_{(r,t)}$

FIGURE 10.9 Marginal and Cumulative Mortality Rate Actuarial Approach

the likelihood of default over up to 10 years after issuance (the usual maturity of newly issued bonds). In addition, those characteristics can be summarized into an issue's (not an issuer's) bond rating at birth. Implicit in these PD estimates is the aging-effect of a bond issue, whereby the first year's mortality rate after issuance is relatively low compared to the second year; similarly, the second is usually lower than the third year's marginal rate, as shown in Figure 10.10. Note that the mortality rates in Figure 10.10 are based on the incidence of defaults for the 46-year period, 1971–2016. For example, the marginal default (or mortality) rate of a BB-rated issue for years 1, 2, and 3 after issuance is 0.92%, 2.04%, and 3.85% respectively. After three years, marginal rates seem to flatten out at between 1.5 and 2.5% per year.

Method #1's PD estimate is derived from Figure 10.9's equations and when adjusted for recoveries on the defaulted issue, we can derive estimates for Loss-Given-Default in Figure 10.11. This critical LGD estimate can be utilized to estimate expected losses in a bank's Basel II or III capital requirements, or for an investor's expected loss on a portfolio of bonds categorized by bond rating. (See our discussion of lender applications later from Figure 10.17). The earliest measures of LGD that I am aware of were from Altman (1977) and Altman Haldeman, and Narayanan (1977).

Method #2 utilizes a different approach to estimate PDs. Instead of using empirical estimates of defaults by bond rating, companies are analyzed with a logistic regression methodology, whereby the company is assigned a "0" or "1" dependent variable based on whether it defaulted or not at a specific point in time, and then a number of independent, explanatory variables are analyzed in the regression format to arrive at a PD estimate between 0 and 1.[3] The resulting

		Years After Issuance									
		1	2	3	4	5	6	7	8	9	10
AAA	Marginal	0.00%	0.00%	0.00%	0.00%	0.01%	0.02%	0.01%	0.00%	0.00%	0.00%
	Cumulative	0.00%	0.00%	0.00%	0.00%	0.01%	0.03%	0.04%	0.04%	0.04%	0.04%
AA	Marginal	0.00%	0.00%	0.19%	0.05%	0.02%	0.01%	0.01%	0.01%	0.01%	0.01%
	Cumulative	0.00%	0.00%	0.19%	0.24%	0.26%	0.27%	0.28%	0.29%	0.30%	0.31%
A	Marginal	0.01%	0.03%	0.10%	0.11%	0.08%	0.04%	0.02%	0.23%	0.06%	0.03%
	Cumulative	0.01%	0.04%	0.14%	0.25%	0.33%	0.37%	0.39%	0.62%	0.68%	0.71%
BBB	Marginal	0.31%	2.34%	1.23%	0.97%	0.48%	0.21%	0.24%	0.15%	0.16%	0.32%
	Cumulative	0.31%	2.64%	3.84%	4.77%	5.23%	5.43%	5.66%	5.80%	5.95%	6.25%
BB	Marginal	0.91%	2.03%	3.83%	1.96%	2.40%	1.54%	1.43%	1.08%	1.40%	3.09%
	Cumulative	0.91%	2.92%	6.64%	8.47%	10.67%	12.04%	13.30%	14.24%	15.44%	18.05%
B	Marginal	2.85%	7.65%	7.72%	7.74%	5.72%	4.45%	3.60%	2.04%	1.71%	0.73%
	Cumulative	2.85%	10.28%	17.21%	23.62%	27.99%	31.19%	33.67%	35.02%	36.13%	36.60%
CCC	Marginal	8.09%	12.40%	17.71%	16.22%	4.88%	11.60%	5.39%	4.73%	0.62%	4.23%
	Cumulative	8.09%	19.49%	33.75%	44.49%	47.20%	53.33%	55.84%	57.93%	58.19%	59.96%

*Rated by S&P at issuance.
Based on 3,359 issues.

FIGURE 10.10 Mortality Rates by Original Rating, All Rated Corporate Bonds,* 1971–2017
Source: Standard & Poor's (New York) and author's compilation.

		1	2	3	4	5	6	7	8	9	10
AAA	Marginal	0.00%	0.00%	0.00%	0.00%	0.01%	0.01%	0.01%	0.00%	0.00%	0.00%
	Cumulative	0.00%	0.00%	0.00%	0.00%	0.01%	0.02%	0.03%	0.03%	0.03%	0.03%
AA	Marginal	0.00%	0.00%	0.02%	0.02%	0.01%	0.01%	0.00%	0.01%	0.01%	0.01%
	Cumulative	0.00%	0.00%	0.02%	0.04%	0.05%	0.06%	0.06%	0.07%	0.08%	0.09%
A	Marginal	0.00%	0.01%	0.04%	0.04%	0.05%	0.04%	0.02%	0.01%	0.05%	0.02%
	Cumulative	0.00%	0.01%	0.05%	0.09%	0.14%	0.18%	0.20%	0.21%	0.26%	0.28%
BBB	Marginal	0.22%	1.51%	0.70%	0.57%	0.25%	0.15%	0.09%	0.08%	0.09%	0.17%
	Cumulative	0.22%	1.73%	2.41%	2.97%	3.21%	3.36%	3.45%	3.52%	3.61%	3.77%
BB	Marginal	0.54%	1.16%	2.28%	1.10%	1.37%	0.74%	0.77%	0.47%	0.72%	1.07%
	Cumulative	0.54%	1.69%	3.94%	4.99%	6.29%	6.99%	7.70%	8.14%	8.80%	9.77%
B	Marginal	1.90%	5.36%	5.30%	5.19%	3.77%	2.43%	2.33%	1.11%	0.90%	0.52%
	Cumulative	1.90%	7.16%	12.08%	16.64%	19.78%	21.73%	23.56%	24.41%	25.09%	25.48%
CCC	Marginal	5.35%	8.67%	12.48%	11.43%	3.40%	8.60%	2.30%	3.32%	0.38%	2.69%
	Cumulative	5.35%	13.56%	24.34%	32.99%	35.27%	40.84%	42.20%	44.12%	44.33%	45.83%

*Rated by S&P at issuance.
Based on 2,797 issues.

FIGURE 10.11 Mortality Losses by Original Rating, All Rated Corporate Bonds,* 1971–2017
Source: Standard & Poor's (New York) and author's compilation.

PDs are then assigned a rating-equivalent based on, for example, the percentage of bond issues that are AAA, AA, A ... CCC, in the "real world." This logistic structure is used widely in the academic literature and has been a standard technique ever since the early work of Ohlson (1980). We (e.g., Altman et al. [2010], Z-Metrics) have also used it for our hybrid-model estimations.

Which is the superior technique for estimating PD, Methods #1 or #2? I favor the BRE approach for newly issued debt (mortality rate approach) but for existing issues, the cumulative default rate method seems to be more appropriate. The reasons are that the mapping of PDs to BREs using mortality rates, or CDRs, are based on over one million issues and about 3,500 defaults over the past 45 years. Logistical regression models' PDs are a function of the sample characteristics used to build the model and the results are based on the logistic structure, which may not be representative of large sample properties. The beauty of logistical regression estimates, however, is that the analyst can access PDs directly from the results and avoid the mapping of scores as an intermediate step. Test of Types I and II accuracies are available for both methods, as well as statistical AUC (area under the curve) accuracy measures on both original and holdout samples. The latter is very important to help validate the empirical results from samples over time and from different industrial groups. From our experience, both methods have yielded very impressive Type I accuracies in numerous empirical tests.

Z-Score Model for Industrials and Private Firms

As noted earlier, the original 1968 Z-Score model was based on a sample of manufacturing, publicly held firms whose asset and liability size was no greater than $25 million. The fact that this model has retained its high Type I accuracy on subsequent samples of manufacturing firms (Figure 10.5) and is still to this date extensively used by analysts and scholars, even for nonmanufacturers, is quite surprising given that it was developed 50 years ago. It is evident, however, that nonmanufacturing firms, like retailers and service firms, have very different asset and liabilities structures and income statement relationships with asset levels, like the Sales/Total Assets ratio, which is considerably greater on average for retail companies than manufacturers, perhaps twice as high. And, given the 1.0 weighting for the variable (X_5) in the Z-model (Figure 10.3), most retail companies have a higher Z-Score than manufacturers. For example, even the beleaguered Sears Roebuck & Co. latest Z-Score (Figure 10.12) in 2016 was 1.3, a BRE of B−, compared to a "D" rating equivalent using the Z''-Score (discussed next); the latter model does not contain the Sales/Total Asset ratio and was developed for a broad cross-section of industrial sector firms, as well as for firms outside the United States (see Altman, Hartzell, and Peck 1995).

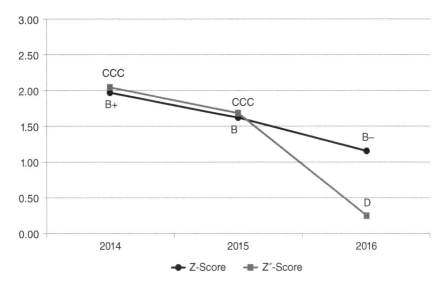

FIGURE 10.12 Z and Z″-Score Models Applied to Sears, Roebuck & Co.: Bond Rating Equivalents and Scores from 2014 to 2016
Source: S&P Capital IQ and NYU Salomon Center calculations.

To adjust for the industrial sector impact, we have built "second-generation" models for a more diverse industrial grouping, for example, the ZETA model (Altman, Haldeman, and Narayanan 1977) and for firms in emerging markets, (Altman, Hartzell, and Peck 1995). Additional Altman Z-Score models developed over the past 50 years for various organizational structured firms, for example, private firms (Z′), developed at the same time (1968) as the original Z-model; textile firms in France (Altman, Margaine, Schlosser, and Vernimmen 1974); industrials in the United States (ZETA, Altman et al. 1977); Brasil (Altman, Baidya, and Riberio-Dias 1979); Canada (Altman and Lavallee 1981); Australia (Altman and Izan 1982); China (Altman, Zhang, and Yen 2010); South Korea (Altman, Eom, and Kim 1995); non-US, Emerging Market (Z″) firms (Altman, Hartzell, and Peck 1995); SME models for the United States (Altman and Sabato 2007) and the UK (Altman, Sabato, and Wilson 2010); Italian SMEs and Minibonds (Altman, Esentato, and Sabato 2016); and for Sovereign Default Risk Assessment (Altman and Rijken 2011).

Private Firm Models

It has been most convenient to build credit-scoring models for publicly owned, listed companies in the United States and abroad due to data availability. Models for private firms can be built, indirectly, by using only those variables related to private firms, but based on publicly owned firm data, or by accessing databases which are populated by both publicly owned and private companies. The latter is especially available in several European countries via tax reporting and government Credit Bureau sources, for example, the UK, firm private databases, for example, from Bureau van Dijk (now owned by Moody's).

I used the indirect method in the Z'-Score models, discussed in (Altman 1983) and shown in Figure 10.13. The only difference from the original Z-Score model is the substitution of the book value of equity for the market value in X_4. Note that all of the coefficients are now different, but only slightly so, and the zones (Safe, Gray, and Distress) were slightly different, as well. There was some slight loss in accuracy by this model adjustment, but over the years this "private-firm-model" has retained its accuracy based on applications to individual private firm bankruptcies. These results have never been published, however.

I have also built numerous models for firms in non-U.S. countries generally following the pattern of first trying the original model on a sample of local firm bankrupts and nonbankrupts and then adding or subtracting variables thought to be helpful in those countries for more accurate prediction. In some cases, different criteria for the distressed-firm sample had to be used due to the lack of formal bankruptcies. An example was our China model (Altman, Zhang, and Yen 2010) which utilized firms classified as *ST (Special Treatment)* due to their consistent losses and book-equity dropping below par value. In others, such as in Australia (Altman and Izan 1982), the explanatory variables were all adjusted for industry-averages so that the model was thought to be more appropriate and accurate across a wide spectrum of industrial sectors. In the case of the sovereign risk assessment model (Altman and Rijken 2011), in addition to traditional financial ratios and market value levels and volatility measures, the authors added

$$Z' = .717X_1 + .847X_2 + 3.107X_3 + .420X_4 + .998X_5$$

$$X_1 = \frac{\text{Current Assets} - \text{Current Liabilities}}{\text{Total Assets}}$$

$$X_2 = \frac{\text{Retained Earnings}}{\text{Total Assets}}$$

$$X_3 = \frac{\text{Earnings Before Interest and Taxes}}{\text{Total Assets}}$$

$$X_4 = \frac{\text{Book Value Equity}}{\text{Total Liabilities}}$$

$$X_5 = \frac{\text{Sales}}{\text{Total Assets}}$$

FIGURE 10.13   Z-Score Private Firm Model
Source: Author's calculations.

macroeconomic variables, such as yield spreads and inflation indicators, in their *Z-Metrics* model applied to all nonfinancial, listed firms in order to assess the sovereign's private sector health. This modeling approach is applicable to any country in the world as long as data on listed or nonlisted private sector companies is available. See the discussion at a later point on the sovereign risk application, following Figure 10.17.

The Z″-score Model

As noted in Figure 10.14, we built a model (Z″-Score) for all industrial, manufacturing and nonmanufacturers in 1995 and first applied it to Mexican companies and then to other Latin American firms. It has since been successfully applied in the United States and in just about any other country, usually with superior accuracy compared to the original Z-Score model when the data includes nonmanufacturers. This Z″-Score model is also applicable to privately owned firms since X_4 is denominated in book equity to total liabilities, not market values. This substitution is particularly important for environments where the stock market is not considered a good valuation measure due to its size, scope, liquidity, or trading factors. In addition, note that the original fifth variable, Sales/Total Assets, is no longer in this model. We found that the X_5 variable was particularly sensitive to industrial sectors differences, for example retail or service firms versus manufacturing companies, and in countries whereby capital for investment in fixed assets was inadequate. Finally, this version of the Altman family of models that used Discriminant Analysis also has a constant term (3.25). The constant standardized the results such that scores slightly above or below zero are in the D-rated BRE (see Figure 10.15 for BREs more granular than the major rating categories). The Type I accuracy of the Z″-Score model over time is shown in Figure 10.16.

$$Z'' = 3.25 + 6.56X_1 + 3.26X_2 + 6.72X_3 + 1.05X_4$$

$$X_1 = \frac{\text{Current Assets} - \text{Current Liabilities}}{\text{Total Assets}}$$

$$X_2 = \frac{\text{Retained Earnings}}{\text{Total Assets}}$$

$$X_3 = \frac{\text{Earnings Before Interest and Taxes}}{\text{Total Assets}}$$

$$X_4 = \frac{\text{Book Value Equity}}{\text{Total Liabilities}}$$

FIGURE 10.14 Z″-Score Model for Manufacturers, Nonmanufacturer Industrials; Developed and Emerging Market Credits (1995)
Source: Author's calculations from Altman, Hartzell and Peck (1995).

$Z'' = 3.25 + 6.56X_1 + 3.26X_2 + 6.72X_3 + 1.05X_4$			
Rating	**Median 1996 Z″-Score**[a]	**Median 2006 Z″-Score**[a]	**Median 2013 Z″-Score**[a]
AAA/AA+	8.15 (8)	7.51 (14)	8.80 (15)
AA/AA−	7.16 (33)	7.78 (20)	8.40 (17)
A+	6.85 (24)	7.76 (26)	8.22 (23)
A	6.65 (42)	7.53 (61)	6.94 (48)
A−	6.40 (38)	7.10 (65)	6.12 (52)
BBB+	6.25 (38)	6.47 (74)	5.80 (70)
BBB	5.85 (59)	6.41 (99)	5.75 (127)
BBB−	5.65 (52)	6.36 (76)	5.70 (96)
BB+	5.25 (34)	6.25 (68)	5.65 (71)
BB	4.95 (25)	6.17 (114)	5.52 (100)
BB−	4.75 (65)	5.65 (173)	5.07 (121)
B+	4.50 (78)	5.05 (164)	4.81 (93)
B	4.15 (115)	4.29 (139)	4.03 (100)
B−	3.75 (95)	3.68 (62)	3.74 (37)
CCC+	3.20 (23)	2.98 (16)	2.84 (13)
CCC	2.50 (10)	2.20 (8)	2.57(3)
CCC−	1.75 (6)	1.62 (−)[b]	1.72 (−)[b]
CC/D	0 (14)	0.84 (120)	0.05 (94)[c]

[a] Sample size in parentheses.
[b] Interpolated between CCC and CC/D.
[c] Based on 94 Chapter 11 bankruptcy filings, 2010–2013.

FIGURE 10.15   U.S. Bond Rating Equivalents Based on Z″-Score Model
Source: Author calculations based on data from S&P Global.

SCHOLARLY IMPACT

Perhaps because of its simplicity, transparency and consistent accuracy over the years, the Z-Score models have been referenced and compared to in a large number of academic and practitioner studies in finance and accounting over the years. These references and comparisons have taken at least three forms. The first is to construct alternative models and frameworks to predict bankruptcy or defaults. The original model and its success, using a combination of financial and market valuation data with robust statistical analysis, made the task of default risk assessment more attractive for scientific work in many disciplines. It opened the door not only for finance and accounting scholars, but also statisticians and mathematicians, to find better and more efficient indexes and to examine new indicators and techniques, especially as expanded and more easily accessible databases became available.

New frameworks have involved seemingly more powerful statistical and mathematical techniques, such as logit, probit, or, quadratic nonlinear regressions, or artificial intelligence, neural networks, genetic algorithms, recursive partitioning, machine-learning, and structural, distance-to-default or hazard models, among others. Since the Z-score model was easily replicable, it usually was chosen by researchers to be compared in terms of accuracy of classification and prediction. These studies are too numerous to list individually, but probably number in the hundreds, including several by this author with numerous co-authors (see References).[4] The combination of simple, but theoretically well grounded, empirical analysis provided new and attractive avenues in bankruptcy research, laying the foundation for expanded modern understanding of bankruptcy prediction, see Scott (1981). For example, recent studies (Altman, Iwanicz-Drozdowska, Laitinen, and Suvas 2016; 2017) looked at several dimensions of bankruptcy prediction research for (a) long distance (10 years) time-series accuracy (Duffie, Saita, and Wang 2007), (b) a number (5) of different statistical techniques, and (c) numerous (34) different country databases and environments. Results covering 31 European countries and three others (China, Colombia, and the United States), showed that while models built specifically for individual countries usually outperformed the original Z-models, the added-value of new country-specific variables and data and numerous frameworks was not dramatic. Despite somewhat higher accuracies using the Z''-Score variables on data specific to each country, we found that the original weightings continued to show remarkable performance despite their determination more than two decades earlier.

Studies using accounting data, among other variables, potentially suffer when the data is either not very reliable, for example, from emerging markets, or is subject to earnings management manipulations. A recent study by Cho, Fu, and Yu (2012), reconstructed Z-Scores for this manipulation with the resulting accuracy improved.

The second dimension of Z-Score's scholarly impact is in its international "reach." Since the original model and its derivatives (e.g., Z''-Score) has stood the test of time, it has been widely applied in multiple settings, including application across all domains, with its sharp focus on a few key variables. Also important, is its robust empirical stability over long periods of time and its global applicability and understandability. We are familiar with Z-Score type models built and tested in at least 30 different countries, based on at least 70 individual articles, and even more in studies analyzing at least that many countries in a single study. Indeed, I helped assemble two special journal issues devoted to a large number of specific country models, see Altman (editor), 1984b and 1988. Those studies, and more, are also listed and described in Altman and Hotchkiss (2006).. More recent studies can also be found in our earlier discussion on scholarly impact and in Choi (1997).

The third impact dimension is related to corporate financial management, especially the important subject of optimal capital structure and the tradeoff between the tax advantage of debt financing and expected bankruptcy and other distress costs. My contribution to this question (Altman 1984a), discussed and measured empirically, for the first time, the so-called "indirect" bankruptcy costs (see Chapter 4 herein for more discussions).[5] In addition, since both tax benefits and bankruptcy costs are based on expected values, contingent upon the probability of bankruptcy, an important aspect of the tradeoff debate is that probability. We selected the Z-Score model's expected default probability algorithm, albeit an early version of the probability estimation technique, to complete the empirical measures for firms that went bankrupt in three different industrial sectors. Our findings were cited directly by an in-depth study from *The Economist* (Emmott, 1991), which highlighted Modigliani/Miller's "Irrelevance" theories compared to traditional optimal capital structure arguments. Perhaps the main difference between the two theories is the existence and magnitude of expected bankruptcy costs. These arguments are still one of the most important fundamental and debatable issues in modern corporate financial management and references to the bankruptcy cost measure can be found in countless corporate finance articles and in just about every basic and advanced relevant textbook.

FINANCIAL DISTRESS PREDICTION APPLICATIONS

Over the past 50 years, we have gleaned numerous insights and ideas from so many helpful, interested financial market practitioners and academic colleagues with respect to applications of the Z-Score models. For these insights, I will be forever grateful because it means so much to a researcher to see his/her scholarly contributions make its way into the "real world" to be applied in a constructive way.[6] Figure 10.16 provides lists of those applications whereby I, and others, have utilized the Altman Z-Score family of models for both external-to-the-firm (left column) and internal-to-the-firm (right column) and research (right column) analytics and application. There is not time or space in this chapter to discuss all of these applications. For this chapter, however, we will discuss just those listed in **bold** in Figure 10.16.

Lender Applications

Throughout this chapter, I have discussed a number of important applications of credit risk models, such as Z-Scores for lending institutions (see Chapter 11 herein

Number of Months Prior to Bankruptcy Filing	Original Sample (33)	Holdout Sample (25)	2011–2014 Predictive Sample (69)
6	94%	96%	93%
18	72%	80%	87%

FIGURE 10.16 Classification and Prediction Accuracy (Type I) Z-Score Bankruptcy Model*

*E. Altman and J. Hartzell, "Emerging Market Corporate Bonds: A Scoring System," Salomon Brothers Corporate Bond Research, May 15, 1995, Summarized in E. Altman and E. Hotchkiss, *Corporate Financial Distress and Bankruptcy*, third ed. (Hoboken, NJ: Wiley, 2006).

for more detailed discussions). These include the accept-reject decision (Altman 1970), estimates of the probability of default and loss-given-default (Altman 1989) and costs of errors in default-loss estimation (Altman et al. 1977, etc.). In addition to these generalized applications, the introduction of Basel II in 1999 drew upon Z-Scores and the structure proposed in CreditMetrics (Gupton, Finger, and Bhatia, 1977a). Later, Gordy (2000, 2003), among others, discussed the anatomy of credit-risk models and capital allocation under Basel II.

To File Chapter 11 or Not

One of the most interesting and rewarding applications of the Z-Score model, at least for Altman, was the essence of his testimony on December 5, 2008 before the U.S. House of Representatives Finance Committee's deliberation as to whether or not to continue to bail-out General Motors, Inc., and Chrysler Corporation or to "suggest" that these firms file for the "privilege" or "right" to reorganize under the protective confines of Chapter 11 of the U.S. Bankruptcy Code. This debate was available to very few firms in the history of the U.S. financial and legal system who had the opportunity to qualify for bailout with taxpayers monies. But, countless distress firms consider whether to file or not when they, or their creditors, face the prospect of the very survival of the company as a going concern.

The House invited panelists to discuss two issues; (1) CEOs of the big-three U.S. automakers in presenting their restructuring plan and strategies should they be granted an additional loan-subsidy (actually, Ford did not apply) from TARP funds, and (2) a panel of academics and practitioners to opine as to whether, or not, these large auto dealers should be bailed out or file for bankruptcy-reorganization like the vast majority of ailing companies. For Altman's testimony,[7] he presented arguments of the benefits of Chapter 11, such as the ability to borrow monies with D.I.P. (debtor-in-possession) financing and the "automatic-stay" on nonessential interest and principal existing loan obligations. In addition, we presented additional analysis on the then current, and historical, Z-Scores and their BREs. Data from Figure 10.18 was one of the primary determinants for my conclusion that GM was destined to go bankrupt, even with a temporary bailout, and should file for bankruptcy-reorganization as soon as feasible to do so.

Note that GM's Z-Score was in the CCC BRE, highly risky zone, for several years before the crisis in 2008, even when it was still rated investment-grade by all of the rating agencies, for example, in 2005, see Figure 10.17. In addition, at the time of my testimony in December 2008, GM's score was –0.63, deep into the "D" BRE zone. Hence, I strongly suggested "Chapter 11" filing and that GM should petition the Bankruptcy Court for a $50 billion D.I.P. loan – most likely from the Federal Government since none of the major banks at that time were in sufficient financial shape to offer that size loan. The House of Representatives, and particularly its Finance Committee members, voted to continue the bailout, despite my arguments. The US Senate, however, voted not to continue the bailout, but

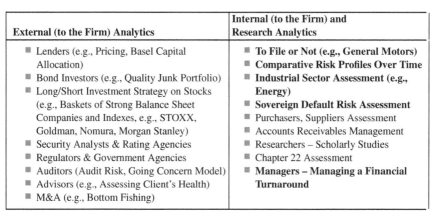

External (to the Firm) Analytics	Internal (to the Firm) and Research Analytics
▪ Lenders (e.g., Pricing, Basel Capital Allocation) ▪ Bond Investors (e.g., Quality Junk Portfolio) ▪ Long/Short Investment Strategy on Stocks (e.g., Baskets of Strong Balance Sheet Companies and Indexes, e.g., STOXX, Goldman, Nomura, Morgan Stanley) ▪ Security Analysts & Rating Agencies ▪ Regulators & Government Agencies ▪ Auditors (Audit Risk, Going Concern Model) ▪ Advisors (e.g., Assessing Client's Health) ▪ M&A (e.g., Bottom Fishing)	▪ **To File or Not (e.g., General Motors)** ▪ **Comparative Risk Profiles Over Time** ▪ **Industrial Sector Assessment (e.g., Energy)** ▪ **Sovereign Default Risk Assessment** ▪ Purchasers, Suppliers Assessment ▪ Accounts Receivables Management ▪ Researchers – Scholarly Studies ▪ Chapter 22 Assessment ▪ **Managers – Managing a Financial Turnaround**

FIGURE 10.17 Z-Score's Financial Distress Prediction Applications
Source: E. Altman, NYU Salomon Center.

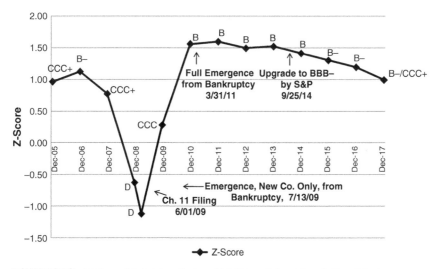

FIGURE 10.18 Z-Score Model Applied to GM (Consolidated Data): Bond Rating Equivalents and Scores, 2005–2017

President W. Bush, before leaving office, by Executive Order, provided the bailout to give GM and Chrysler more time to restructure. It was now "Obama's problem." GM's Z-score continued to crater in the early months of 2009 and despite management changes and the bailout, finally filed for bankruptcy under Chapter 11 on June 1, 2009. To assist the reorganization, Congress and the Bankruptcy Court provided a $50 billion DIP loan – that exact amount I suggested six months earlier!

In a remarkably short period, just 43 days, GM emerged from bankruptcy and was on its way toward once again being a going concern. Figure 10.17 shows the firm's improvement from deep into the D BRE to a B-rating BRE in about 12–18 months. The DIP loan was first exchanged for new equity and that equity was subsequently sold in the open market, whereby not only did the government not lose any of its "investment," it actually made a profit! GM today is a solid, thriving global auto competitor with, again, an investment-grade rating (BBB) achieved in 2014. Note that the Z-Score model, however, placed GM at the end of 2014 in the single-B BRE, not investment-grade; this low BRE continued through the end of 2018. So, while GM has indeed improved considerably since its bankruptcy, it still looked as noninvestment grade.

Comparative Risk Profile over Time

Students of history often ask the question of comparing a current situation with that of some past period(s). This query is particularly relevant in financial markets when the benchmark period in the past is related to some financial crisis and whether we can learn from the environment that existed then. Such is the case of the financial crisis of 2008/2009 and whether credit conditions today are similar, or not, to conditions just prior to that crisis. One metric that we have found useful in comparing credit market conditions over time is our Z-Score models. Was the average (or median) firm credit worthiness better, worse, or about the same in, say 2016, compared to 2007? One might have some priors based on related macro or micro observations, such as cash on the balance sheet, interest rates or GDP growth. A more holistic, objective measure, in my opinion, is one, that is based on default probabilities that consider multiple attributes, such as the Z-Score of a relevant sample of firms in the two periods.

Figure 10.19 shows the average and median Z and Z″ Scores for a large sample of high-yield bond issuers in 2007 and 2016, with some intermediate years between the two periods. Observing these comparisons, we find extremely helpful in clarifying certain conclusions based on the qualitative or quantitative opinions of some experts. We find that the average Z-Score was 1.95 (B+ BRE) in 2007 and 1.97 (also B+ BRE) in 2016 (3Q) and the median score was actually higher in 2007 (1.84) versus 1.70 in 2016. The average and median Z″-Scores, a measure probably more appropriate given that the high-yield firms in both periods came from many different industrial sector (see our discussion earlier about Z vs. Z″), were both higher in 2007 compared to 2016. Tests of the difference of means between 2007 and 2016 were found not to be significantly different, so our conclusion is that the average credit profile of risky-debt issuing firms was about the same in 2007 versus 2016. I leave it up to the reader to determine if this was good or bad news for default estimates in 2017 and beyond.

	Number of Firms	
	Z-Score	Z″-Score
2007	294	378
2012	396	486
2014	577	741
2016 (3Q)	581	742

Year	Average Z-Score/ (BRE)[*]	Median Z-Score/ (BRE)[*]	Average Z″-Score/ (BRE)[*]	Median Z″-Score/ (BRE)[*]
2007	1.95 (B+)	1.84 (B+)	4.68 (B+)	4.82 (B+)
2012	1.76 (B)	1.73 (B)	4.54 (B)	4.63 (B)
2014	2.03 (B+)	1.85 (B+)	4.66 (B+)	4.74 (B+)
2016 (3Q)	1.97 (B+)	1.70 (B)	4.44 (B)	4.63 (B)

[*]Bond rating equivalent.

FIGURE 10.19   Comparing Financial Strength of High-Yield Bond Issuers in 2007 and 2012/2014/3Q 2016
Source: Authors' calculations, data from Altman and Hotchkiss (2006), and S&P *Capital IQ/Compustat.*

Predicting Defaults in Specific Sectors

Over the years, default cycles usually produce particular carnage in one or more industrial sectors and, if persistent for several years, these sectors draw particular attention to researchers and practitioners. Hence, for example, railroads (Altman 1973), the textile industry model (Altman et al. 1974), U.S. airlines (Altman and Gritta 1984), broker-dealers (Altman 1976), and most recently, the energy and mining sectors in the United States, motivated specific analysis and tests.

A recent empirical test (Altman & Kuehne, 2017) of the Z-Score models analyzed its accuracy in the energy and mining sectors. Rather than build a model based on energy-firm data only, we decided to assess both the Z- and Z″-Score models on a sample of bankruptcies in 2015, 2016 and 2017, a period when energy related firms accounted for more than half of the total defaults in those years. Figure 10.20 shows the results of just the bankruptcies, that is, Type I accuracy, for two periods prior to the filing of the 31 firms with data available for a Z-Score test and even more firms (54) for the Z″-Score test. Our results were quite impressive, especially for the Z-Score model, which we built, as noted earlier, based only on manufacturing firm data. Indeed, 84% of the energy and mining companies had Z-Scores in the D rating (Defaulted BRE), based on data from one or two quarters prior to the filing, and the remaining five firms in the sample had a CCC or B– BRE. For data from 5 or 6 quarters prior CCC; that is, only 2 out of 31 had a B

BREs	Z-Score				Z″-Score			
	t – 1*		t – 2**		t – 1*		t – 2**	
	#	%	#	%	#	%	#	%
A								
BBB+								
BBB								
BBB–								
–BB+							1	2%
BB							0	0%
BB–							3	5%
B+					1	2%	1	2%
B			2	6%	3	5%	13	24%
B–					3	5%	6	11%
CCC+					1	2%	8	15%
CCC	5	16%	12	39%	2	4%	8	15%
CCC–					4	7%	9	16%
D	26	84%	17	55%	41	75%	6	11%
Total	31	100%	31	100%	55	100%	55	100%

*One or two quarters before filing.
**Five or six quarters before filing.

FIGURE 10.20 Applying the Z-Score Models to Recent Energy and Mining Company Bankruptcies, 2015–September 15, 2017
Source: S&P *Capital IQ.*

rating BRE. While the results for the Z''-Score model were not as accurate, 75% had a D BRE and the remaining firms had at least a B rating BRE, based on data from the last quarter prior to filing for bankruptcy, the results were still impressive and quite accurate.[8]

So, it appears that our original Z- and Z''-Score models retain their high accuracy level for distress prediction, even for some industries not included in our original tests. We are not able to generalize, however, as to all nonmanufacturers, especially service firms.

Sovereign Default Risk

An intriguing application of the Z-Score model is to use it to assess the default risk of sovereign nations' debt. We (Altman and Rijken 2011) were inspired by the World Bank study (Pomerleano 1998, 1999), which analyzed the causes of the financial crisis in Southeast and East Asia in 1997–1998 and found that the original Z-Score model clearly demonstrated that the most vulnerable country to private sector defaults *prior* to the crisis was South Korea. Indeed, Korea had the lowest average Z-Score for listed firms of all Asian countries, but was rated high investment-grade by all of the rating agencies in December 1996. Yet, it needed to be bailed out by the IMF shortly thereafter! This illustration inspired us, more than a decade later, to analyze sovereign default risk in a unique way.

The aggregation of Z-Scores, or in the case of Altman and Rijken's (2011) use of a more up-to-date version called Z-Metrics^T, proved to be exceptionally accurate in predicting the European countries with the most serious financial problems in the post-2008 crisis. Their "Bottom-Up" approach added a new microeconomic element to the arsenal of predictive measures for sovereign risk assessment, never before studied (See Chapter 13 herein for a detailed discussion).

Managing a Financial Turnaround

One of the most interesting and important applications of the Z-Score model, from an internal and active, rather than passive standpoint of the distressed firm, is to apply the model as a guide to a successful turnaround of the firm. I suggested this application and wrote up a case study on the GTI Corporation, Altman and LaFleur (1981), also found in Altman and Hotchkiss (2006) and in Chapter 12 of this volume. The idea is a simple one; if a model is effective in predicting bankruptcy, why can't it be helpful to the management of the distressed firm in identifying the strategies and their impact on performance metrics. In the case of the GTI Corp., the new CEO, James LaFleur, strategically simulated the impact of his management changes on the resulting Z-Scores, and only made those changes that resulted in an improved Z-Score. His strategy did result in a remarkably successful turnaround. Here again was an application of the Z-Score model that I (Altman) had never considered until a practitioner suggested its use.

CONCLUSION

The chapter has assessed the statistical and fundamental characteristics of Altman's 1968 Z-Score model over the 50 years since the model's introduction. In addition, we have listed a large number of proposed and experienced applications of the original Z-model and several subsequent ones, with a more detailed discussion of the specifics and importance of several of these applications. The 50-year old model has demonstrated an impressive resilience over the years and, notwithstanding the massive growth in the size and complexity of global debt markets and corporate balance sheets, has shown not only longevity as an accurate predictor of corporate distress, but also that it has been successfully modified for a number of applications beyond its original focus. The list, shown in Figure 10.17, is almost assuredly incomplete, especially in view of the large number of scholarly works that have cited the Z-Score models for a wide range of empirical research investigations. While we are surprised at the longevity of the Z-Score models' usefulness, we now cannot help but wonder what some analysts might conclude in the year 2068 about its 100-year track-record.

NOTES

1. See, for example, SP Global (2017).
2. See, for example, CreditRiskMonitor.com's revised FRISK^T Scoring system (2017).
3. Note that the 0/1 dependent variable is the regression equivalent to a discriminant analysis software program structure, but the resulting discriminant score cannot be interpreted as a PD.
4. Indeed, Bellovary, Giacomino, and Akers (2007) review 165 bankruptcy prediction models from 1965-2006, and a large number of similar articles. They conclude that MDA and neural networks are the most promising methods for bankruptcy prediction but that higher model accuracy is not guaranteed with a greater number of factors. According to the authors, "since Altman's study, the number and complexity of bankruptcy prediction models have dramatically increased: and Keasey and Watson (1991) attributed discriminant analysis as the main technique used in this knowledge field. Willer do Prado et al. (2016) found that logistic regression and neural networks also became popular after the 1990s, with logistic regression and discriminant analysis being the most used techniques up to the end date of their article's sample period in 2014. Willer do Prado et al. used bibliometric evaluation (see Pinto et al. 2014) to evaluate research about credit risk and bankruptcy using Reuters "Web of Science" database from 1968–2014. They found, through their exhaustive investigation, that the bankruptcy prediction field appeared to be multidisciplinary, spanning not only finance and accounting, but also operations research, management, mathematics, data processing, engineering and a broad range of statistical fields. Not surprisingly, they discovered an increased number of bankruptcy studies after the 2008 crisis. Finally, Willer do Prado et al. (2016) listed the 10 most-cited articles in the bankruptcy prediction field (Table 3 in their study), with Altman's (1968) work registering 1,483 Web of Science cites. The next most cited article being Huang, Chen, Hsu, Chen, and Wu (2004) with 250, and down to Hillegeist, Keating, Kram, and Lundstedt's(2004) 165 cites.
5. We apologize to the many authors whose published studies are not specifically cited and, also, we appreciate immensely their interest and attention to our models' extensions and tests, over time.
6. Indeed, the Z-Score model has even made its way into a novel written by a bestselling author, see Thomas Pynchon's (2013) *Bleeding Edge*
7. The complete testimony can be found on YouTube, "Altman testimony before the U.S. House of Representatives Finance Committee," December 5, 2008.
8. We have also tested our models for the Type II error. Results show a reasonably high Type II error, although the overall accuracy is still impressive. Note that the sample sizes are different for the comparison of Z and Z" models.

CHAPTER 11

Applications of Distress Prediction Models

By External Analysts

In Chapter 10, we discussed the evolution of the Z-Score model, and other distress prediction techniques, and included a list of applications of the models. Figure 11.1 shows that list again. In the next three chapters, we revisit this applications subject, elaborating on many of suggested, or already experienced, applications, including several that were not discussed earlier. This first chapter concentrates on those applications performed by analysts who are external to the distressed debtor (left-hand column of Figure 11.1) in order to improve their position or to exploit profitable opportunities presented by distressed firms and their securities. Chapter 12 explores in greater depth, via a case study, the unique application related to managerial efforts for formulating and implementing a turnaround strategy. And in Chapter 13, we apply our updated distress prediction model, called Z-Metrics™, in a "bottom-up" analysis of the default assessment of sovereign debt.

Lenders and Loan Pricing

Perhaps the most obvious application of distress prediction/credit scoring models is in the lending function. Banks and other credit institutions are continuously involved in the assessment of credit risk of corporate, consumer, sovereign, and structured counterparties. The importance of credit-scoring models for specifying the probability of default (PD) has been heightened and motivated immensely by the requirements of Basel II and Basel III and the necessity for banks to develop and implement internal rating based (IRB) models. We hope that the decision of U.S. regulators (e.g., the Federal Reserve Board) *not* to require most banks in the United States to conform to Basel II will not serve to demotivate banks from

External (to the Firm) Analytics	Internal (to the Firm) and Research Analytics

▪ Lenders (e.g., Pricing, Basel Capital Allocation)
    * Bond Investors (e.g., Quality Junk Portfolio)
▪ Long/Short Investment Strategy on Stocks (e.g., Baskets of Strong Balance Sheet Companies and Indexes) e.g., STOXX, Goldman, Sachs, Morgan Stanley
▪ Security Analysts and Rating Agencies
▪ Regulators and Government Agencies
▪ Auditors (Audit Risk, Going Concern Model)
▪ Advisors (e.g., Assessing Client's Health)
▪ M&A (e.g., Bottom Fishing)
	▪ To File or Not (e.g., General Motors)
	    * Comparative Risk Profiles Over Time
	▪ Industrial Sector Assessment (e.g., Energy)
	▪ Sovereign Default Risk Assessment
	▪ Purchasers, Suppliers Assessment
	▪ Accounts Receivables Management
	▪ Chapter 22 Assessment
	▪ Researchers – Scholarly Studies
	▪ Managers – Managing a Financial Turnaround (see Chapter 13)

FIGURE 11.1   Z-Score's Financial Distress Prediction Applications
Source: E. Altman, NYU Salomon Center.

developing these models – but we fear that it will, in many cases. On the other hand, banks in many other parts of the world, particularly Europe, have been encouraged, with great success, to modernize their credit risk systems by the requirements under Basel II.

One of the important dimensions of the lending function is to specify the "price" of credit (i.e., the appropriate interest rate). The use of credit scoring models permits the specification of the PD in the determination of LGD in the pricing algorithm. For example, in Figure 11.2, we use a corporate loan rated as BBB by an internal scoring model to begin the LGD process (see our discussion in the prior Z-Score chapter) and pricing decision. The PD and recovery rate assumptions are given (0.3% per year and 70.0%, respectively) and the *expected* loss of 0.09% per year is quantified.

The next step is to add the required amount to the loan price based on the *unexpected* loss. We can do this in two ways – based on either *economic capital* criteria or *regulatory capital* requirements. Economic capital requires an additional cost in making the loan for unanticipated losses based on the degree of conservatism of the lending institution (i.e., its own risk preference). For example, the bank with a high-risk avoidance preference – that is, one that wants to attain a high credit rating for itself – will require a very high confidence interval for not exceeding a particular loss. In our example, we utilize a six-standard-deviation requirement, sometimes referred to as the required average return on capital (RAROC) approach of an AA bank. The estimated standard deviation of 50 basis points per year (given) is then multiplied by 6 to arrive at the required amount of capital (300bp) for this lending capital requirement, which is then multiplied by the bank's net opportunity cost (15%) for not investing and this product is added to the expected loss and

Given: Five-Year Senior Unsecured Loan
Risk Rating = BBB
Expected Default Rate = 0.3% per year (30 b.p.)
Expected Recovery Rate = 70%
Unexpected Loss (O') 50 b.p. (0.5%) per year
BIS Capital Allocation = 8%
Cost of Equity Capital = 15%
Overhead + Operations Risk Charge = 40 b.p. (0.4%) per year
Cost of Funds = 6%
Loan Price[1] = 6.0% + (0.3% × [1 − .7]) + (6[0.5%] × 15%) + 0.4% = 6.94%
Or
Loan Price[2] = 6.0% + (0.3% × [1 − .7]) + (8.0% × 15%) + 0.4% = 7.69%

[1] Internal Model for Capital Allocation
[2] BIS Capital Allocation Method

FIGURE 11.2   Risk-Based Pricing: An Example

other costs to arrive at the required economic capital (6.94%). We suggest using the net cost of equity (cost of equity minus the risk-free rate) for the calculation of the opportunity cost (15%). We also factor in other operating costs by adding an estimated 40 basis points per year to cover such noncredit items as overhead and operating risk charges, (the latter is required under Basel III).

So, for the economic capital computation, in our example, the result is a required price, or interest rate, of 6.94%. This compares to the old Basel I regulatory capital requirement calculation, based on a flat 8% instead of the 3% (6 × 0.5%) economic capital, and the regulatory capital interest rate is higher by 0.75%, at 7.69%. One can now see why most banks will prefer the Basel II framework. Again, accurate scoring models are critical to the modern pricing structure. Even if a bank, or nonregulated institution, does not use or cannot use economic capital pricing criteria due to competitive conditions, the economic pricing analytics based on risk rating criteria will be helpful to ascertain how far from the actual price charged is the one based on economic pricing. Note that our example does not incorporate correlation and concentration issues in the pricing function. Such factors add to the complexity of credit decisions and should be considered by the portfolio management group of the financial institution.

Bond Investors

Beyond the financial institution lender of commercial loans, other financial institutions (e.g., shadow banks) and individuals can profit from a well-tested and appropriate credit scoring system in their fixed income strategies. Perhaps the most obvious application is to determine whether to invest in a debt instrument selling at, or near, par value. The determination of PDs is important for investment-grade as well as junk bonds. Indeed, about 22% of all defaulting issues from 1971–2017 and 19% from 2007–2017 were originally rated as investment grade by the professional rating agencies![1] So the professional manager should include default risk analysis, as well as yield and concentration considerations, in his or her deliberations of investment grade decisions. If the investment-grade company has a financial profile of a lower-rated entity, the required rate of return should reflect that. The calculated financial requirement can be determined by both counterparties' bond rating equivalent (BRE), discussed earlier in Chapter 10.

For bonds selling in a distressed condition (i.e., 1,000 basis points or more above Treasuries), the key question to ask is whether the company will continue to migrate to an even lower credit quality (or, in fact, default) or whether the PD is sufficiently low to assess that its price will return to par (assuming it had already migrated down, perhaps even to a "fallen angel" downgraded condition). The upside potential, from distressed to par, provides equity-type returns that are far greater than returns expected from a typical debt portfolio. Indeed, the average return on a portfolio of non-investment-grade bonds in 2009 was about 60%, as a large proportion of those corporate bonds that were distressed as of the end of 2008 returned to par or above-par value in about one year. Probably the onset of the benign (forgiving) credit cycle had a lot to do with this incredible run, but one's conviction to select securities that are selling at "deep junk" or distressed-yield levels can be heightened when the credit model ascribes a higher BRE to the security than the one derived from the market. This "Quality-Junk" investment strategy can be shown in Figure 11.3, whereby higher-yield bonds from companies in quadrants A and B, with high BREs, are particularly attractive candidates.

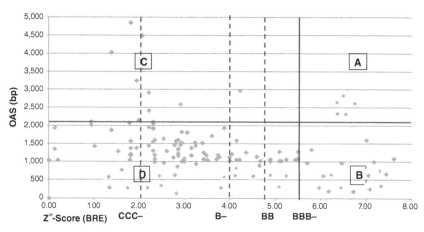

$$Z'' = 3.25 + 6.56X1 + 3.26X2 + 6.72X3 + 1.05X4$$
$$X1 = CA - CL / TA; X2 = RE / TA; X3 = EBIT / TA; X4 = BVE / TL$$

A = Very High Return / Low Risk
B = High Return / Low Risk
C = Very High Return / High Risk
D = High Return / High Risk

FIGURE 11.3 Return/Risk Tradeoffs – Distressed and High-Yield Bonds, as of December 31, 2012

The Type II error – that of selling, or not buying, the distressed security when in fact its price returns to par – is always possible but the cost involved is not an important one for most traditional debt investors. Where it does matter, however, is in the case of a distressed debt or highly leveraged hedge fund investor. We strongly believe that a disciplined investor will find credit scoring and default risk models of considerable benefit in the investment process. And, combining rigorous credit risk default probability models with the selection of low-volatility distressed securities (see Altman, Gonzalez-Heres, Chen, and Shin 2014) for a discussion of this yield-spread criteria) will, in our opinion, provide a potentially very attractive fixed-income investment strategy. That is, to select distressed or attractively high-yield spread companies with acceptable default probability attributes.

Common Stock Investing

We were surprised to find that in our discussions with data vendors, like S&P Global's Capital IQ (part of S&P Global's Global Market Intelligence) and Bloomberg, LP, that there were more "hits" from equity-market investors/analysts on the Altman Z-Score "page" than from Bond and Fixed Income investors/analysts. So it seems that equity investors are interested in preservation of capital as well as in achieving "alpha" returns or matching some index of performance. One way of preserving capital is to avoid investing in securities of companies that eventually go bankrupt. In most cases, defaulting companies' equity securities are completely or mostly wiped out.

In addition to avoiding financial distress of corporate issuers, investors take comfort with strategies that provide a "strong balance sheet," whether for value investing or in the selection of companies that have desirable growth prospects *and* low probabilities of default. Such a strategy might fit into a long-short equity approach. And, we have found several investment management "packagers" who offer investors this long-short strategy through equity "baskets," whereby the selection of "long" purchases are firms' equities with high Z-Scores and the "shorts" are selected from those firms with low Z-Scores. A variation on the long-short strategy approach is to just concentrate on going long or going short, but not both, based on the Z-Score criterion.

Examples of equity selection "baskets" are Goldman Sachs' "Strategy Baskets," which combined the preservation of capital goal with dividend yields and growth prospects.[2] The main idea was that companies with strong balance sheets, based on Z-Scores above a certain high level, for example, 4.0, will outperform firms with weak balance sheets, for example, below 2.0, especially in tight or stressed credit conditions. So, a long-short strategy based on these criteria will provide excess returns, especially when combined with high dividend growth prospects, if both are available. The Goldman Long GSTHWBAL/Short SPX strategy returned 12.9% for the period February 2008 to December 2008 compared to an S&P 500 performance of –31.2%, as reported in Kostin, Fox, Maasry, and Sneider (2008).

A more recent example of a common stock strategy based on Z-Scores is the "STOXX Strong Balance Sheet Indices," a strategy that selects the financially fittest companies on the European, Japanese, and U.S. Stock Exchanges.[3] Launched in 2014, their indices are based on relatively high Z-Scores' companies with a three-year track record of scores of 3.5 and higher. Financial firms are excluded from the STOXX indices since the Z-Score model did not include this sector. Indeed, it is the case that the original Z-Score model's data contained only manufacturing firms (see discussion in Chapter 10) but it has been liberally applied across all industrial sectors in some cases.

Many of the vendors who reference the Z-Score selection strategy discuss the approach with particular reference to distressed market conditions, see for example, Gutzeit and Yozzo (2011) and the aforementioned Goldman Sachs 2008 publication. We believe that the strategy has both equity and fixed-income securities potential in any market condition, but in most cases will trail market indices' performance in strong bull markets. For one thing, the equity's past performance is encompassed in the Z-Score via the X_4 variable, market value of equity/total liabilities, and if the market has been strong, high Z-Score firms may have peaked, or at least performed very well already. And many times, low Z-Score firms, if they do not default, will outperform market indices in strong bull markets.

Other Investment Applications

A number of other investment banks and their brokerage operations have used the Z-Score for a variety of analytical purposes to serve their clients' specific needs. For example, Morgan Stanley Dean Witter utilized a "Z-Score" methodology to quantify retail REIT portfolios credit risk exposure based on their tenants' financial health, see Ostrowe & Calderon (2001). JPMorgan's Asia Pacific Equity Research, Smith & Dion (2008), applied Z-Scores to information technology and health care sectors in Asia. And Merrill Lynch, Bernstein (1990), analyzed the financial health of S&P industrials during the entire decade of the 1980s and found that despite record levels of earnings and cash flows, the overall financial strength based on Z-Scores had deteriorated. We addressed this seeming anomaly in Chapter 10 on the Z-Score's 50-year retrospective. The financial health of firms in 2016 were compared to 2007. Bernstein's concern was prescient as the market had serious problems with a big spike in defaults in late 1990 and early 1991. Stronger Z-Score firms, however, made a remarkable comeback in their stock prices and also with high-yield bond returns later in 1991.

Finally, Altman and Brenner (1981) utilized the Z-Score to analyze a short-sale strategy, whereby a portfolio of "shorts" was accumulated based on the first time a Company's Z-Score dropped below a cutoff-score. After adjusting for market returns, this short-sale strategy demonstrated significant excess returns for at least 12 months after the Z-Scores' deterioration was first indicated. Their analysis of a two-factor model's findings indicated market inefficiency since the

Z-Score's variables and the weightings were public information at the time of our tests. A single-factor market model's results, however, were mainly consistent with market efficiency tests in that the "new" Z-Scores seemed to be digested in market prices instantaneously.

Security Analysts

One of the fundamental axioms in finance is that the debt analyst is, and should be, far more focused on the downside possible movement of securities than is the common equity analyst. So, an obvious tool for the debt analyst is a default prediction model. Both traditional ratio analysis, and one or more distressed prediction models, would seem to be a prudent addition to the security analyst process. As noted earlier, we were surprised to learn that equity analysts log-into our Z-Score model from data providers more than do debt analysts.

The analyst should also consider a type of pro-forma distressed prediction treatment of the entity, especially where a heavily leveraged condition is likely to change through a series of steps (e.g., asset sales and debt paydowns). We have seen in Chapter 6 of this volume that highly risky capital structures can lead either to a healthy and high return scenario or to a default, depending on management's success in reducing debt and improving its rating equivalent status. The analyst earns his or her analytical status, in our opinion, by providing realistic expectations and forecasts about the likely success of management to achieve its target capital structure and cash flow goals. It is clear that highly leveraged companies, especially just after a major restructuring transaction, will be evaluated by most credit scoring techniques as a distressed situation. We advocate realistic pro-forma scenario analytics, as well as current financial statement criteria, in the analysis process.

Regulators

Regulatory institutions, particularly bank examination department personnel, should be aware of and comfortable in evaluating credit risk systems used by constituent banks and other institutions. One of the challenges of the new Basel III framework is the role of the regulators, especially when evaluating the Pillar I capital requirement specifications of banks and in the Pillar II regulatory oversight function. Bank examiners need to be trained to evaluate the various credit scoring and probability of default/loss given default (PD/LGD) estimates used by banks. And, the various possible PD assessments on the same counterparty from numerous banks will serve as useful inputs in this process.

In addition, it is quite common for a nation's central bank, like the Banque de France, Banca Italia (now the ECB), or the U.S. Federal Reserve, to utilize its own credit scoring evaluation system to assess the credit quality of bank portfolios and to monitor the assessments made by their individual bank customers.

Auditors

In a very early application of the original Z-Score model, we discussed (Altman and McGough 1974) the potential use of credit scoring models to assist the accounting firm audit function to assess the *going-concern* qualification condition of customer accounts. We concluded, at that time, that while a reasonable proportion (about 40%) of bankrupt companies did receive a going-concern qualification in the year just prior to filing, the Z-Score approach predicted well over 80% of those entities as in the "distress-zone". Note that we now advocate using the bond rating equivalent (BRE) benchmark (see Chapter 10). Indeed, the accounting profession is very sensitive to its responsibility and has argued that it should not be liable for not qualifying a firm's financials just because it is in a highly risky, low rating-equivalent condition.

Our experience with accounting going-concern models involved building one for the now defunct Arthur Andersen (AA) franchise in their desire to establish criteria for going-concern qualifications and to use in assessing whether to take-on a new auditing client. We called that model "A-Score," which could be interpreted as with Altman, Arthur, or Andersen. Instead of using publicly available bankrupt firm data, like we did in building the Z-Score model (1968), we utilized a combination of public firm bankruptcies and firms which defaulted from Arthur Andersen's historic database, as well as those that did not. Utilizing a logistical regression technique applied to this database, resulted in a highly accurate audit-risk model, which was used for many years at the discretion of their local offices. Unfortunately, it was apparently not used in the infamous Enron case, which led to AA's downfall. Tests using our Z''-Score model (Chapter 10) did show that Enron's BRE fell to single-B where the agencies' rating was triple – B!

Despite the obvious potential conflict between a very rigorous professional auditor's posture with respect to going-concern qualifications, and its desire to not cause problems for its customers and perhaps lose an account that it gives a going-concern qualification, it would seem that auditors should be aware of and, indeed, use credit scoring models and other failure assessment techniques. These approaches can add objective feedback in their own assessments as well as in their discussions with the clients about their financial condition and plans going forward.

Bankruptcy Lawyers

One of the most prominent players in the bankruptcy and distressed firm arenas is the bankruptcy lawyer. The decision to take a firm into bankruptcy is a momentous one. While most bankruptcies are involuntary and decided on only as a last resort, whether to file and the timing of the filing are critical decisions. The longer a firm puts off the decision, the less likely, in most cases, it is that the reorganization will be successful. At the same time, if a successful turnaround can be managed out of court, then most likely the costs of financial distress will be lower than if the turnaround is achieved in bankruptcy (see Gilson, John, and Lang 1990). Usually, the bankruptcy lawyer is the prime adviser to the management of the firm as to whether to file and when to do so. Typically, the bankruptcy law firm will recommend to the ailing company potential consultants and paths of actions to take surrounding the bankruptcy or out-of-court restructuring decision. Such consultants, also discussed later in this chapter, might be turnaround and other restructuring specialists, bankers for bailout financing or, DIP financing, and so forth.

While there may be unmistakable signals of financial distress, lawyers can also productively use financial distress prediction models in their advisory work for clients. Whether the firm is mildly or deeply distressed is an important determinant. As we will explore in Chapter 12 of this book, management itself can use such models in its determination. Lawyers can benefit from the implications of a failure prediction model in such areas as the failing company doctrine (see below) and the once-important issue of deepening insolvency. We now turn to these topics.

Legal Applications

Failure prediction models have had a number of direct applications in the legal arena over the years. These include such areas as (1) the failing company doctrine, (2) avoidance of pension obligations, and increasingly lately (3) the fiduciary responsibility of owners, managers, directors, and other corporate insiders like professional advisers. The last area relates to a concept and condition known as "deepening insolvency."

Failing Company Doctrine

One defense against an antitrust violation by firms attempting to merge is the argument that an otherwise illegal merger should be permitted to occur if one or both of the merging entities would have failed anyway and its market share would likely have been absorbed by the other entity. We wrote in detail about this so-called *failing company doctrine* in the first edition of this volume (Altman 1983) and in Altman and Goodman (1981, 2002), but space limitations in this edition preclude an exhaustive treatment.

Essentially, the failing company doctrine can be invoked if it can be shown that, while competitors that are trying to merge are unquestionably linked, either geographically or by market segment, at least one party was on the verge of bankruptcy and extinction. Examples might include two newspapers competing in a standard metropolitan area and one paper's demise would almost certainly result in its market share going to its closest competitor. This occurred in the antitrust dispute involving the *Detroit News* and the *Detroit Free Press* in the early 1980s. Using several of the Z-Score models, one of the authors argued, in a deposition, that at least one of these entities would likely fail within a short period of time – both were in very bad shape. The court's solution was to permit the merger but to require the independence of the editorial staffs. Both entities remained in existence but the ownership with respect to revenues, costs, and profits was combined. Concerns about newspaper costs, labor relations, and other negative antitrust results were outweighed by the likely scenario of a single major newspaper for the city if the merger was not permitted.

Another example was the potential combination of two low-priced beer companies in the Northeast of the United States in the early 1070s – Schaefer Beer and Schmidt's of Philadelphia. Both firms competed for the low-end price customers of the beer market. We argued that Schaefer was a likely failing company and that while Schmidt's would surely absorb Schaefer's market share if the merger were permitted, it would happen anyway if it were not. The plaintiff, in this case Schaefer itself, which did not want to be taken over by Schmidt's, argued that it was not failing since it was not receiving a going-concern qualification from its auditors (see earlier discussion) and its major creditor was not calling in its now long overdue loans, which had been nonperforming for almost two years. The judge agreed and essentially ruled that Schaefer was not failing because it had not failed – yet! Another beer, Stroh's of Milwaukee, which did not compete directly with either company, soon purchased Schaefer. A related issue to this case is whether the firm was in a so-called zone of insolvency or not. We now turn to that issue.

Deepening Insolvency[4]

The theory of deepening insolvency, as discussed in Kurth (2004), originated with two federal cases in the early 1980s, *In re Investors Funding Corporation* and *Schacht v. Brown*. The simple argument was made that a corporation is not a biological entity for which it can be assumed that any act that extends its existence is beneficial to it. This argument is in stark contrast to the fundamental premise of the turnaround management industry and the principles underlying the Bankruptcy Code that the estate, involving creditors, shareholders, and employees, typically benefits if a distressed company can be reorganized successfully. A deepening insolvency argument, on the other hand, argues that the efforts to save an obviously dying entity can benefit some at the considerable expense of others. For example, the managers, advisers, and others trying to save the entity receive payments during the failed turnaround period, which results in lower recoveries after bankruptcy to others, such as creditors.

Deepening insolvency was increasingly being recognized, Kurth observes, as an independent course of action. This action could argue that a bankrupt company, or its representatives, may recover damages caused by professionals, such as advisers, accountants, investment bankers, and attorneys, who have either facilitated the company's mismanagement or misrepresented its financial condition in such a way as to conceal its further deterioration from an insolvent condition into deepened insolvency and bankruptcy. One recognized method of calculating damages is to measure the extent of the company's deepening insolvency.

Several questions emerged around this legal argument. How do you know that a firm is in the zone of insolvency, and how do you measure its deepening condition? Is it enough to simply say that as long as a firm has not gone bankrupt, defaulted on its debt obligations, or participated in a distressed restructuring (e.g., an equity-for-debt swap), it is not in the zone of insolvency? We do not believe so! A firm may be in an insolvent condition, but still not be defaulted.

The courts seemed to rely on a comparison between the fair market value of the firm's assets and the market value of its liabilities to determine whether it was insolvent. This criteria is essentially the basis for distressed prediction from the KMV model, first commercialized in the 1990s and sold to Moody's in 2002. See our references to KMV in Chapter 10 of this book and a more in-depth discussion in Caouette, Altman, Narayanan, and Nimmo (2008). Incidentally, we would argue that the appropriate comparison benchmark for assets is not the market value of debt, but its book value, since the latter is what needs to be repaid to creditors. This is especially true if you measure asset value as the sum of the market value of debt plus equity – as most financial economists do.

In any event, we would also argue that a reasonable test of whether a firm is in the insolvency zone is to calculate its bond rating equivalent or failure score using *several* statistical measures. In particular, we advocate using the Z-Score models and other techniques, such as the Moody's/KMV expected default frequency (EDF) approach. If both classify the firm as "in default" (e.g., a Z-Score below zero and a KMV EDF of 20 or more), then its likely survival as a non-bankrupt or nondefaulted entity is seriously in doubt. Deteriorating scores will indicate a deepening condition, although we cannot argue that the deterioration is linear with respect to the change in score. Certainly, a firm with a Z-Score of –2.2 is in worse condition than one with a score of –0.2. The use of the Z-Score model in deepening insolvency cases and analyses was discussed by Appenzeller and Parker (2005).

A final note about the deepening insolvency legal claim. Most legal analysts point out that the original purpose of the argument was based on the bankruptcy that resulted after the insiders had perpetrated a type of Ponzi scheme or some other action that resulted in fraud or embezzlement, enriching its operators or advisers at the expense of unsuspecting creditors. And, it is argued that the guilty parties knew, or should have known, that the firm's chance of survival was unlikely. So, while we can help to specify whether or not the firm has a failing company profile, we cannot say for sure whether some turnaround strategy could not be successful.

Certainly, if the firm's true condition was known, but not revealed by those who could profit from continuing existence, then a legal cause for damages would seem to be valid. If, however, everything was revealed and best efforts were made to protect the remaining interests of owners, we would be reluctant to say that a firm's Z-Score in the distressed zone, or a rating equivalent of D, means that it could not be saved. What we are arguing for is a clear and unambiguous metric of a firm's financial condition rather than relying only upon an expert's fair valuation of the firm's assets.

The next chapter of this book shows how a manager, with the assistance of the Z-Score model, successfully used his business acumen and judgment to manage a financial turnaround. The essence of this case study is that all the indicators of financial distress were transparent and the strategy of simulating corrective actions with respect to a likely outcome on the firm's "health index" was not only appropriate, but also prudent. Even if these actions had failed, we do not believe that deepening insolvency would have been a legitimate argument in this case.

Bond Raters

While bond-rating agencies do not use failure prediction models to reach their rating conclusions, we would argue that the results of one or more well-tested and successful models could assist in the process. Obviously, Moody's saw great benefits in the output from the KMV EDF model since it paid a handsome sum to acquire KMV. Yet raters legitimately argue that such items as industry analysis, interviews with management, and a longer-perspective "through the cycle" approach will determine their rating designations and that a model's point-in-time perspective should not be the basis for rating decisions. See Loeffler (2004) and Altman and Rijken (2004 and 2006) for discussions on rating stability, accuracy, and comparisons between through-the-cycle versus point-in-time models.

Risk Management and Strategy Advisory Consultants

Basel II has also been an important catalyst for the growth in risk-management consulting firms. Entities such as Oliver-Wyman (purchased by Mercer in 2002), Algorithmics (purchased by the Fitch Group in 2005 and sold to IBM in 2011), RiskMetrics (purchased by MSCI in 2010), Kamakura, CreditSights, and the risk-management divisions of the other major rating agencies (e.g., KMV/Moody's, SPGlobal's Risk Solutions, Fitch Information Services) have prospered as the appetite for modern credit risk systems has grown. Most of these firms have developed credit risk tools that include scoring, structural, or hybrid combinations of these two credit-scoring models. In addition, consulting and credit risk assessment entities, providing services related to valuation and portfolio management, might find that objective credit risk tools are helpful in their assessment of clients, for example, CreditRiskMonitor's FRISK system.

A related area of management consulting advice can be in mergers and acquisitions (M&A). A distressed firm could be encouraged to solicit a purchasing/strategic partner when its condition, assessed objectively, indicates going-concern problems. This would especially be helpful if potential acquirers do not share the same internal assessment. Obviously, the price of the acquisition will be reduced if it is generally known that the firm is in a highly distressed condition. Accurate early-warning models, however, can give a competitive advantage to users – whether they are the target or the acquiring firm.

Restructuring Advisers: Bankers, Turnaround Crisis Managers, and Accounting Firms

Four of the most prominent types of professionals that have emerged as important players in the distressed firm industry are (1) restructuring specialists, usually from boutique investment banks, (2) corporate turnaround or crisis managers from operations management consultants, (3) restructuring consultants from traditional accounting firms and (4) asset-backed lenders. Figure 11.4 illustrates their relationships, overlaps and primary functions in the distressed firm industry.

The competitive landscape in the turnaround consulting industry has evolved over the years. In the mid to late 1980s, the market was led by accounting firms and the efforts by the larger investment banks that had placed large amounts of debt in the highly leveraged restructuring boom (primarily ill-fated leveraged buyouts and leveraged recapitalizations). For example, Drexel Burnham Lambert was a leading proponent of the out-of-court restructuring in the 1980s. In the early 1990s,

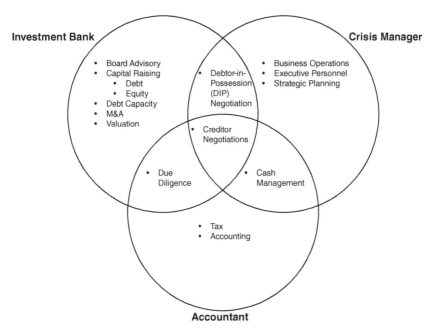

FIGURE 11.4   Comparison of Financial Advisers' Roles
Source: Lazard Freres, 2005.

with the huge increase in large Chapter 11 bankruptcies, some investment banks and new boutique divisions of smaller banks sprung up to fill the void caused by conflicts of interest from the larger investment bank under-writing firms. More of these boutiques, sometimes as offshoots from the large accounting firms or invest-ment banks, emerged in the 2001–2005 period, and the larger underwriters put resources into distressed refinancing, especially since M&A activity slackened in that period.

In addition to these advisers and consultants, a fairly active banking market emerged to provide funding at the bankrupt stage, such as debtor-in-possession (DIP) financing, and exit-financing at the emergence stage. Many of the major commercial banks and also firms like GE Capital, Congress Finance, and CIT Finance became major players in this sector in the period from 2000–2007. After the financial crisis of 2008/2009, however, distressed firm lenders abandoned the market due to their own capital constraints, for example, GE and CIT.

Although definitely not mutually exclusive, the turnaround consultants play a role as advisers for the restructuring of the firms' assets (operations consultants known as crisis managers) and liabilities (restructuring bankers and accountants). The number of specialists in these fields swelled to over 20,000 globally in 2017, and many are members of the increasingly prominent Turnaround Management Association (TMA; website: www.turnaround.org). This professional, educa-tional, and networking organization, based in Chicago, Illinois, had 55 chapters globally (33 in the United States and 22 abroad) with over 9,000 members in 2017.[5]

Turnaround managers can be hired to assist a firm to avoid filing for bankruptcy or, once in a legal bankrupt condition, to assist in the reorganization of the firm's assets and liabilities. Probably the three largest turnaround management consulting organizations are Alvarez and Marsal, AlixPartners, and FTI. These firms also typically provide financial and capital structure advice. While these three firms are relatively large, each with more than 1,000 full-time employees, most turnaround management companies are quite small, with fewer than 10 full-time consultants. There are also a number of relatively large turnaround management specialists in the 20- to 100-employee range that can provide a full spectrum of advisory services. Often, these professional consultants have had prior experience as full-time corporate managers in such areas as finance, mar-keting, operations, human resources, and information systems and have chosen to work with ailing companies to assist in their corporate renewal. In addition to their primary operation functions, these firms often assist in such areas as creditor negotiations, cash, and even strategic management. For example, McKinsey & Co. recently started a sizeable restructuring group. In the past 20 years, a new position has been created to deal with the many complex reorganization issues of companies in crisis – the chief restructuring officer (CRO) – not to be confused with another CRO acronym, the chief risk officer.

Corporate restructuring advisers from investment banks tend to specialize on the "right-hand side" of the balance sheet, with particular emphasis in assisting the management of distressed companies (or the operations turnaround specialist) to acquire needed capital during the restructuring period and to sell unprofitable or other assets to acquire cash. One type of financing that has proven to be crucial to the early phase of the Chapter 11 process is debtor-in-possession (DIP) financing, whereby the new creditors typically have a super priority status over all existing creditors. We discussed this financing mechanism earlier in this book. At the other end of the restructuring period is the so-called exit financing, whereby the firm needs to emerge from Chapter 11 with sufficient capital to conduct business on a going-concern basis. Again, the boutique investment bank is a key player in these exit-financings. Indeed, many times the provider of DIP financing is the same lender of exit financing and/or the distressed investor in the pre-emergence phase becomes the new owner of the emerged firms' equity securities.

Some of the larger restructuring advisers that specialize in advising debtors are the boutique investment bankers like Lazard Freres, PJT Partners (formerly the Blackstone Group), Miller Buckfire, N. M. Rothschild & Sons, Evercore, and Greenhill, although there are also several smaller successful operations. On the creditor advisory side, the largest advisers are Houlihan Lokey Howard & Zukin, Jefferies, Chanin, FTI, and Giuliani Partners. The latter two are carve-outs or sales of divisions from accounting firms.

Bankruptcy Reform Bill Impact on Advisory Firms

Leading up to the Bankruptcy Abuse Prevention and Consumer Protection Act of 2005 (BAPCPA), restrictions in the 1978 Bankruptcy Code had historically limited participation of the larger investment banks in Chapter 11 restructuring work, especially if a bank was the underwriter of the debtor's securities within three years before the bankruptcy. While there were exceptions to this restriction and sometimes a formal appeal from one of the stakeholders groups was necessary to eliminate a major investment banking firm (e.g., the switch from UBS to Lazard as the debtor's adviser in the 2005 Trump Casinos & Entertainment Chapter 11), large underwriters were generally restricted under the premise that they were not disinterested professional.

With the passage of BAPCPA, the competitive landscape for distressed firm advisory work changed somewhat. Signed into law on April 17, 2005, and implemented October 17, 2005, this law impacted corporate bankruptcies, as well as consumer filings. It now permits heretofore excluded investment banks to be advisers as long as there is no clear conflict with the financing or other advisory work that helped cause the bankruptcy. Disinterestedness still requires that an entity not have an interest materially adverse to the interests of the estate or any class of creditors as equity security holders, by reason of any relationship or connection or interest in the debtor. So, while a current or former underwriter may be deemed not to still be disinterested, the prospect that other disinterested bulge-bracket firms will be eligible is very real. In fact, the large investment banks have not entered the financial advisory market in any material way, but continue to provide asset-backed financing, for example, DIP loans, in the distressed firm market.

We advocate that both the turnaround managers (who are trying to save the business either before filing for Chapter 11 or while in bankruptcy-reorganization) and restructuring debtor advisers can effectively use distress-prediction credit scoring models as an early warning tool to assess the financial health of an enterprise or as a type of post restructuring barometer of the health profile of an entity as it emerges. If the firm still looks like a distressed, failing entity upon emergence, then its chances for subsequent distress, indeed the "Chapter 22" situation, would likely be higher than the renewal process should provide. Unfortunately, based on the frequent occurrence of "Chapter 22" or other forms of continued distress (like distressed sales), it appears that the restructuring process is not always successful. Indeed, Gilson (1997), in a study of the success of Chapter 11s, found that too often firms emerge with excessive leverage or operating problems, and Hotchkiss (1995) found that emerging firms often do not perform as well as their industry counterparts. On the other hand, Eberhart, Altman and Aggarwal (1999) found that emerging equities do extremely well in the post-Chapter 11 one-year trading period. And recent (2003–2004) evidence (e.g., J.P. Morgan 2004) supports that conclusion.

In conclusion, we advocate that firms be advised to emerge looking like going concerns and that rights offerings as well as equitylike securities, including options and warrants for junior creditors and old equity holders, be used wherever possible so as not to burden the "new" firm with too much fixed cost debt in the early years after emergence.

Government Agencies and Other Purchasers

Many of the larger U.S. government agencies have a policy to screen their vendors as to their staying power and independence from government support should they become distressed. A related issue is whether the vendor will be able to deliver the goods and services that are contracted for. We have learned, over the years, that one of the screens used by government agencies and their auditing counterparts is the Z-Score model(s). For example, the U.S. Department of Defense had this policy, as did the government's Accounting Audit Agency. If an entity fares poorly by the Z-Score screen, then it will be screened even more closely and/or passed over as a possible vendor. As such, both the government and those firms seeking to become, or remain, suppliers to the federal or state government should understand the pros and cons of using an automated financial early-warning model, such as Z-Score. We would especially suggest that any agency using such an approach do so in conjunction with other screening tools – especially qualitative methods like interviews with existing customers of the vendor.

The use of financial screening models by purchasing agents should not be restricted to public agencies. Indeed private enterprises should also be concerned about the health of their suppliers. This is especially true if the purchaser practices something like a just-in-time (JIT) inventory approach to its production process. For example, computer manufacturers, like IBM or Dell, wanted to be assured that the keyboards in their PC fabrications be available at the precise time that the rest of the computer is about ready to be shipped. Another industry where the health of vendors, and of the manufacturers themselves, was of vital concern is the U.S. auto industry in 2005–2007. As its fragile condition became more obvious, going-concern and staying power probabilities were a pervasive issue. Indeed, many medium and large auto-part suppliers succumbed even before the major auto manufacturers bankruptcies in 2009 (i.e., Chrysler and GM); some as early as 2004–2005. A related issue to the manufacturer is the cost of supporting a vendor if the latter is sustaining continuing losses and is in jeopardy of failing and having to be bailed out. This was a common occurrence in Japan, and may still be.

M & A Applications

A final left-hand column of Figure 11.1 application of our Z-Score model relates to both highly leveraged finance (e.g., LBOs) takeovers and ordinary merger and acquisition selections. As noted earlier, distress predictions are fundamental to any merger transaction that involves massive debt financings. The high profile risk of default involving LBO takeovers was perhaps the major cause of the financial crisis of the late 1980s to early 1990s. Analysis of the acquired firm's existing default vulnerability, as well as post-LBO scenarios, should be a prominent focus for the private-equity fund.

With respect to ordinary M&A activity, acquirers are sometimes interested in "bottom-fishing" to pay a very small price to a firm with good restructuring potential. A reasonable rule of thumb is to purchase a firm in distress for no more than three-times EBITDA. Of course, many distressed firms have negative EBITDA, so a prospective estimate of cash flow is relevant. We discuss distressed firm valuations earlier in this volume (Chapter 5).

NOTES

1. Our estimate of fallen-angel defaults is based on an analysis of about 3,800 defaulting issues from the Altman/NYU Salomon Center's Master Default database.
2. See Kostin et al., Goldman Sachs (2008) for a more detailed discussion of this strategy-basket. The Goldman strategy has been offered to investors now for about a decade.

3. See Eibl, STOXX (2014) for more detail on this Strategy-Index and various internet browsers for commentary and portfolio results.
4. We refer the reader to Chapter 6 herein for current discussion of governance issues and fiduciary responsibilities based on the insolvency of the company.
5. Dr. Altman serves the TMA as Chairman of its Academic Advisory Council and contributes, along with market practitioners, to the organization's monthly publication, the *Journal of Corporate Renewal*, as well as at periodic conferences.

CHAPTER 12

Distress Prediction Models

Catalysts for Constructive Change
-Managing a Financial Turnaround*

We have frequently been asked by managers and analysts the difficult question, "Now that your model has classified the entity as having a high probability of failure, what should be done to avoid this dismal fate?" Not being an operating manager or turnaround consultant, we had to throw up our hands and reluctantly reply, "Get yourself some new management or specialists in crisis management," or, even less satisfying, "That is your problem!" Needless to say, these answers were not accepted with applause, nor did the response capture the spirit of a true early-warning system. Such a system usually connotes prescribed rehabilitative action when the warnings are in other areas, such as medicine, weather, or military science. Unfortunately, management science applications of early warning systems are typically unique to the entity, and it is difficult to generalize rehabilitative prescriptions.

Our attitude toward this important and inevitable outgrowth of distress prediction has changed. One important incident has taught us a valuable lesson, one which is, we believe, transferable to other crisis situations. The lesson emanated not from a conceptual, academic analysis of the problem but from the application

---

*This chapter has been derived and updated from an article by E. Altman and J. Lafleur, "Managing a Return to Financial Health," *Journal of Business Strategy* (Summer 1981). The story of GTI's turnaround was written for the popular press by Michael Ball (*Inc.*, December 1980).

of the Z-Score model (Chapter 10 herein) to a real-world problem by a remarkably perceptive chief executive officer (CEO). Let us review the case of GTI Corporation, a manufacturer of parts, subsystems, and processing equipment for the computer, automotive, and electronic industries. GTI Corporation was listed on the American Stock Exchange. Although this situation took place over 40 years ago, it is still as relevant in 2018 as it was in 1975.

Active versus Passive Use of Financial Models

Statistically verified predictive models have long been used in the study of business. Generally, these models are developed by scientists and tested by observers who do not interact with or influence the measurements of the model. Consequently, the models, when valid, have predicted events with satisfactory accuracy, and business analysts regard them with a reasonable degree of confidence. As discussed briefly in Chapter 10 of this book, this passive use of predictive models for credit analysis, investor analysis, and so on overlooks the possibility of using them actively. In the active use of a predictive model, the role of the observer is shifted to that of a participant. For example, a manager may use a predictive model that relates to business affairs of a company by deliberately attempting to influence the model's measurements. Hence the manager can make decisions suggested by the parameters of the model in order to control its prediction.

In a specific case, we will discuss how the Z-Score bankruptcy predictor model was used *actively* to manage the financial turnaround of a company that was on the verge of bankruptcy. A series of management decisions was made to "foil" the model's prediction of bankruptcy. These decisions, many of which were specifically motivated by considering their effect on the financial ratios in the model, led directly to the recovery of the company and the establishment of a firm financial base.

Earlier in this book, we indicated that management could declare bankruptcy once the indication was that a firm was headed toward bankruptcy – in other words, that its overall financial profile was consistent with that of other firms that had gone bankrupt in the past. It took GTI Corporation, and specifically the management strategy formulated and implemented by its CEO, Jim LaFleur, to turn the model inside out and show it ability to help shape business strategy to *avert* bankruptcy.

What the Z-Score Told GTI

Jim LaFleur, a Cal Tech graduate and successful entrepreneur and business executive, had recently retired but remained a director of several companies, one of which was the GTI Corporation. During the first six months of 1975, GTI had suffered the following financial results:

Working capital decreased by $6 million.

Retained earnings decreased by $2 million.

A $2 million loss was incurred.

Net worth decreased from $6.207 million to $4.370 million.

Market value of equity decreased by 50%.

Sales decreased by 50%.

Earlier in LaFleur's career, he had noticed an article in *Boardroom Reports* about the Z-Score. LaFleur immediately saw the potential application of the bankruptcy predictor to the problem at GTI. As we showed in Chapter 10, the original Z-Score model is of the following form:

Factor	Definition	Weighting Factor
X_1	Working capital/Total assets	1.200
X_2	Retained earnings/Total assets	1.400
X_3	Earnings before interest and taxes/Total assets	3.300
X_4	Market value of equity/Book value of liabilities	0.600
X_5	Sales/Total assets	0.999

LaFleur, a member of the audit committee, was asked to replace the existing CEO. Plugging in the preliminary numbers for the five ratios, LaFleur put the Z-Score predictor to work for GTI; the resulting Z-Score was 0.7. At that level, the predictor indicates a condition of financial distress with a high probability of bankruptcy. When more accurate numbers were inserted into the Z-Score formula, it fell even lower, to 0.38, about half the earlier calculation. The prognosis was grave since GTI's bond rating equivalent (BRE) was at best CCC–, but actually very close to the median Z-Score of a bankrupt company of about zero (see Figure 10.7).

A Tool for Recovery

Despite its portent of doom, the Z-Score was also seen as a management tool for recovery. The predictor's five financial ratios were the key to the Z-Score movement, either up or down. While the previous management had inadvertently followed a strategy that had decreased the ratios and caused the Z-Score to decline (see Figure 12.1), GTI's new management decided to reverse the plunge by deliberate management actions. Before each decision, LaFleur and his team simulated the decision's impact on the model. Inherent in the Z-Score predictor was the message that *underutilized assets* could be a major contributor to the deterioration of a company's financial condition. Such deterioration had taken place at GTI over several years. The company's total assets had grown out of proportion to other financial factors. We have found this to be the case in many business failures, particularly larger ones.

FIGURE 12.1 Z-score Distressed Firm Predictor: Application to GTI Corporation, 1972–1975

By using retrospective analysis. LaFleur concluded that the Z-Score could have predicted GTI's deterioration toward financial distress. For example, historical data in 1975 showed that GTI's Z-score started to dive precipitously at least two years prior to the spring of 1975.

Taking Quick Action

At year-end 1974, GTI's Z-Score BRE fell to about BBB (Figure 12.1)[1] and its earnings per share (EPS) had fallen to $0.19. GTI's Z-Score had been falling for several years, even during periods when the company's profits were rising. That was further evidence of the predictor's validity and suggested its ability to help set strategy to guide the company's recovery.

The Effects of Growth Fever

For more than two years, LaFleur had cautioned against what appeared to be overaggressive policies of debt and excessive expansion by GTI's operations. The warnings, unfortunately, had little effect. Along with most of the industry, GTI had succumbed through the 1960s and into the 1970s to a highly competitive growth fever. During those years, many managers focused almost entirely on their profit and loss (P&L) statements. They were willing to borrow what was necessary to increase sales and profits. With stock values rising, they expected to obtain very favorable equity funding in the future to pay off the accumulated debt. Does this sound like *déjà vu* in the late 1990s and again in 2007? That strategy served them well until the economic downturn of 1972. Then, with profits falling, many companies had trouble servicing the debt that had looked so easy to handle a few years earlier. GTI started losing money in early 1975. Before that profit slide could be stopped, GTI's 1975 first six-month net loss accumulated to over $2.6 million on sales of $12 million, a loss of $1.27 per share.

Also, during the month of May, a member of the audit committee discovered information indicating that the figures for the first quarter of 1975 were reported incorrectly. As the evidence developed during the ensuing audit meetings, it was obvious that the company's problems were serious. GTI's auditing firm began a thorough reexamination of the company's first-quarter activities. The auditors quickly confirmed that there was, indeed, a material discrepancy in the figures and set to work revising first-quarter figures. As chairman of the audit committee, LaFleur contacted the Securities and Exchange Commission (SEC), disclosing the discrepancy and promising to define and correct it. He also asked the American Stock Exchange to halt trading of the company's stock. By finding and reporting the errors quickly, GTI had the stock back trading in less than 10 days. No delisting of the stock ever occurred, and the company even received compliments from some observers on its rapid self-policing action.

At that point, GTI's board of directors chose a new executive team, asking LaFleur to become part of management and take over as chairman and chief executive officer. Having observed GTI going into debt to finance its operations over several prior years, even with record sales and profits on paper, LaFleur was determined to find the underlying problems. It didn't take long. Inventory, out of control, revealed itself as a major contributor to the company's ballooning assets. In many instances, returned goods had been set aside and not properly accounted for. Adding to that difficulty, work-in-process was grossly out of proportion to sales. Again, these symptoms seem to be common among corporations in crisis.

Genesis of Strategy

From this new evidence of excess assets, a recovery strategy began to emerge. It was to find ways to decrease GTI's total assets without seriously reducing the other factors in the numerators of the Z-Score's X ratios: working capital, retained earnings, earnings before interest and taxes, market value of equity, and sales. GTI started looking for assets that were not being employed effectively – that is, not making money. When identified, such assets were sold and proceeds used to reduce the company's debt. Having conceived the strategy, LaFleur began to implement the actions to eliminate GTI's excess assets. Excess inventory was sold as quickly as possible, even at scrap value in some cases. The effect was a decrease in the denominators of all five X ratios simultaneously. It is not enough simply to sell assets – the proceeds must be utilized as soon as possible. GTI's Z-Score rose accordingly.

While the bankruptcy predictor was originally designed for an observer's analysis of a company's condition, GTI used it as an aid to managing company affairs. The predictor actually became an element of active strategy to avoid GTI's impending bankruptcy.

	Pennsylvania	Indiana	New York	California	West Germany	
Operations	$1	$1	$1	$1	$1	$5
Marketing	$1	$1	$1	$1	$1	$5
Engineering	$1	$1	$1	$1	$1	$5
Finance	$1	$1	$1	$1	$1	$5
	$4	$4	$4	$4	$4	$20

FIGURE 12.2   Function/Location Matrix

Stopping the Cash Bleed

In quick order, GTI's cash bleed was stanched. The staffs at two unprofitable West Coast plants were sliced to a skeleton crew within 10 days, and the corporate staff at headquarters was pared from 32 to 6. A year earlier, with company's profits at $1.5 million, the corporate staff expense had been over $1 million! All capital programs were frozen. Only the most critical production needs, repair, and maintenance were authorized. GTI asked its creditors for additional short-term credit, then pushed strenuously ahead on its collections. Inventories were placed under strict control. Taking effect, these measures got cash and expenses under control and improved debt service capability. Reducing costs further took more analysis. A management function/location matrix, a "job versus cost" grid, was constructed for each of GTI's plants. The grid showed each executive's job, what work he or she performed, and how much that job cost the company. When overlaps or duplications were found, jobs were consolidated. The grid is illustrated in Figure 12.2 (actual dollars in each function/location not indicated). Where the revenues from different locations did not cover the identifiable costs, it was clear that a problem existed.

Finding Lost Profits with Employee Assistance

Employees were also involved in the turnaround. A simple questionnaire was handed out to the 250 employees of GTI's largest plant in Saegertown, Pennsylvania, asking their opinions on why the plant was no longer profitable. The implied question was about the underutilized assets that had depressed GTI's Z-Score. The employees knew what was wrong! They were specific about how to improve the use of their machines. Many of the suggestions were implemented, and productivity improved. Eventually, however, this plant was sold and product lines moved elsewhere. Several weeks later, similar questions were asked at GTI's plant in Hadley, Pennsylvania. The employee responses resulted in changing the plant's organization from functional to product line, another move that more effectively employed the company's assets. Because they participated in the changes, the plant's employees really worked to make the organization succeed. After a few

Z-Score

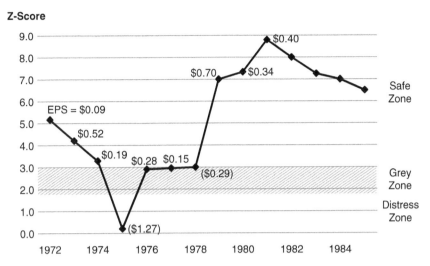

FIGURE 12.3 Z-Score Distressed Firm Predictor: Application to GTI Corporation, 1972–1984

weeks, the plant began to return to profitability. In fact, profitable product lines were moved from Saegertown to Hadley.

Those profits were the forerunners of profits that would be produced in other parts of the company as time went on. The Z-Score, while it did not jump much as a result of those profits, did begin to react. By mid-1976, after slanting down for three years, the Z-Score bottomed out and started up (Figure 12.3). GTI began turning the corner.

Selling Off a Product Line

Though cost reduction and increased profits had eased the problems, GTI needed stronger recovery actions. The function/location matrix analysis was extended to include products and was used to rate product profitability throughout the company. Plans were made to eliminate the losers and strengthen the winners. As a result, late in 1976, GTI sold one of its major underutilized assets. GTI's crystal base product line had appeared fairly strong, but the product matrix analysis presented a different view. Crystal bases were not complementary to GTI's other products, and though the line had been marginally profitable in the past, demand for its products was likely to decrease. The line also appeared to need a great deal of capital to be competitive in the future. The cash generated by the sale of the crystal-base product line was used to reduce debt. The consequent simultaneous decrease of both total assets and debt produced a dramatic effect. The Z-Score leaped from under 1.0 to 2.95. In one transaction, GTI zoomed almost all the way into the Z-Score predictor's "safe" zone. Although to outside observers the company did not appear to turn around for another year and a half, LaFleur felt the firm was on the road to recovery with the sale of the crystal base product line. The company had come from almost certain bankruptcy to the stage where it could begin contemplating new products. In less than 18 months, the Z-Score had climbed from 0.38, in the near-death bankrupt zone, almost all the way to the Z-Score's safe zone (Figure 12.3). With heightened confidence in the model, GTI started working to put the Z-Score firmly in the safe zone. Since the company's improving stability and profitability were corroborating the Z-Score approach, GTI's headquarters staff began figuring how a proposed new product or financial transaction would affect the rising Z-Score. Further, GTI extended the product evaluation matrix from simple profit and loss to multiyear projections of return on assets. This involved taking a hard look at projected working capital and capital expenditure requirements, product by product. The analysis established what costs would be if the company attempted to expand within its current markets.

Progress in Operations

While doing this planning, GTI continued to make progress on the operations side, finishing 1976 with $0.28 earnings per share and an increasing Z-Score as well. In 1977, earnings sagged to $0.15 per share and – $0.29 in 1978; but with an improving overall financial condition, GTI's Z-Score continue gradually to rise. The company even bought out a competitor's glass-seal product line with notes secured by the acquired assets and with negligible adverse impact on the Z-Score.

Then in 1978, GTI boosted its Z-Score again by shutting down an entire division that made ceramic capacitors and selling its assets. That transaction, again based on the strategy of selling underutilized assets to pay off debt, occurred later than it should have. This was a case of emotion interfering with a rational, proven strategy. LaFleur had been swayed toward saving this technically interesting product line, though the Z-Score strategy consistently suggested disposal. Though delayed, the difficult disposal decision was made. As a result of the closing of the capacitor division and the sale of its assets, GTI's 1978 bottom line sustained a $0.29 per share loss, but the Z-Score increased as the company paid off more debt. As anticipated, operating profits continued to gain throughout the year, paving the way for a strong 1979. Once again, the asset-reduction strategy had worked.

Into the Safe Zone

After 1978, GTI's Z-Score continued to climb, rising through the safe zone, and 1979 pretax profits reached $1.9 million and $0.70 per share, on sales of $21 million. From a balance sheet viewpoint, in five years GTI's strategy had decreased the debt-to-equity ratio from 128% to 30% and increased stockholders' equity from $3.5 million to $4.7 million. The debt-to-equity (market equity) ratio improved even more in 1981, to just under 10%! During 1980 and 1981, GTI further consolidated its financial position and increased stockholders' equity. The company continued its policy of conservative financial management augmented by close attention to the Z-Score in business decision making. That spirit was reflected in actions taken in August 1980, when the company entered into an agreement with a research and development (R&D) limited partnership to investigate several new projects for the partnership. Further, to raise return on investment and to provide more funds for new growth in electronics, GTI disposed of a metal and plastics product line in early 1981. In July 1981, GTI negotiated a 10-year loan for $1 million at a fixed 16% to provide increased working capital. Yes, interest rates were that high in the early 1980s!

As a result of careful financial planning and continued profitability, GTI attained positive flow from the net interest of its investment and loan portfolio. Essentially, the company had internalized control of the Z-Score, because it has sufficient funds available to pay off debt. In terms of the model, the firm could actively impact and control the financial ratio (X_4), the relation between its equity value and outstanding debt, GTI's Z-Score zoomed to 7.0 in 1979 and continued to rise to 8.8 as of year-end 1981, mainly as a result of improvement in X_4. This rise is all the more impressive in view of the recessionary period in the early 1980s, including a drop in 13 earnings in 1980. GTI, before being purchased by a Scandinavian firm in 1992, was listed on the American Stock Exchange. GTI continued as a financially sound division pursuing new avenues with controlled growth. In major part, that success came about from implementing a financial strategy suggested by the Z-Score bankruptcy predictor model.

Conclusion to the GTI Story

We believe that certain predictive models offer opportunities to be used as management tools. Supporting that view, GTI's employment of the Z-Score bankruptcy predictor has been described as a specific illustration of how a seemingly passive model can be used actively with substantial success. With emphasis made on prudent selection and use, managers are encouraged to search out and review predictive models that relate to their companies' activities. Improved business strategies could well result. It is quite conceivable that large number of firms in a distressed situation can learn from and perhaps be put on the road to recovery by the strategies used by GTI Corporation.

In addition to the prescriptive use of a financial model in a turnaround strategy, we advocate the Z-Score model's use as a type of barometer to any restructuring. We often hear how turnaround managers use the model to identify distressed firms.

An additional use is to test the viability of the restructuring *before* sending the firm back into action. Not only is it important to avoid Chapter 11, it is also important to reduce the chance of a Chapter 22!

NOTE

1. Note that we had not established BREs as early as 1975, but the score in 1974 was close to the so-called Gray Zone (see Chapter 10).

CHAPTER 13

A Bottom-Up Approach to Assessing Sovereign Default Risk*

In our list of Applications of Distressed Prediction models in Chapter 10 and 11, for example, Figure 11.1, we noted extending micro-oriented corporate models to the sovereign risk arena. This chapter now goes into greater depth on that theme. In the past decade, bank executives, government officials, and many others have been sharply criticized for failing to anticipate the global financial crisis. The speed and depth of the market declines shocked the public. And no one seemed more surprised than the credit rating agencies that assess the default risk of sovereign governments, as well as corporate issuers operating within their borders.

Although the developed world had suffered numerous recessions in the past 150 years, this most recent international crisis in the wake of the mortgage-backed security downfall in 2008–2009, raised grave doubts about the ability of major banks and even sovereign governments to honor their obligations. Several large financial institutions in the United States and Europe required massive state assistance to remain solvent, and venerable banks like Lehman Brothers even went bankrupt. The cost to the United States and other sovereign governments of rescuing financial institutions believed to pose "systemic" risk was so great as to result in a dramatic increase in their own borrowings. The general public in the United States and Europe found these events particularly troubling because they had assumed that elected officials and regulators were well-informed about financial risks and capable of limiting serious threats to their investments, savings, and pensions.

---

*This chapter is an updated and expanded version of Altman and H. Rijken (2011).

High-ranking officials, central bankers, financial regulators, ratings agencies, and senior bank executives all seemed to fail to sense the looming financial danger. This failure seemed even more puzzling because it occurred years after the widespread adoption of advanced risk management tools. Banks and portfolio managers had long been using quantitative risk management tools such as value at risk (VaR). And they should also have benefited from the additional information about credit risk made publicly available by the market for credit default swaps (CDSs). Instead, these mechanisms were cited by many as adding to the depths of the crisis.

But, as financial market observers have pointed out, VaR calculations are no more reliable than the assumptions underlying them. Although such assumptions tend to be informed by statistical histories, critical variables such as price volatilities and correlations are far from constant and thus difficult to capture in a model. The market prices of options – or of CDS contracts, which have options "embedded" within them – can provide useful market estimates of volatility and risk. And economists have found that CDS prices on certain kinds of debt securities increase substantially before financial crises become full-blown. But because there is so little time between the sharp increase in CDS prices and the subsequent crisis, policy makers and financial managers typically have little opportunity to change course.[1]

Most popular tools for assessing sovereign risk are effectively forms of top-down analysis. For example, in evaluating particular sovereigns, most academic and professional analysts use macroeconomic indicators such as GDP growth, national debt-to-GDP ratios, deficits to GDP, and trade deficits, and so on, as gauges of a country's economic strength and well-being. But, as the recent Euro debt crisis has made clear, such macro approaches, although useful in some settings and circumstances, have clear limitations.

In this chapter, we present a totally new method for assessing sovereign risk, a type of bottom-up approach that focuses on the financial condition of an economy's private sector. The assumption underlying this approach is that the fundamental source of national wealth, and of the financial health of sovereigns, is the economic output and productivity of their corporate entities. To the extent we are correct, such an approach could provide financial professionals and policy makers with a more effective means of anticipating financial trouble, thereby enabling them to understand the sources of problems before they become unmanageable.

In the pages that follow, we present Z-Metrics™, as a practical and effective tool for estimating sovereign risk. Developed in collaboration with the Risk Metrics Group, now a subsidiary of MSCI, Inc., Z-Metrics is a logical extension of the Altman Z-Score technique that was introduced in 1968 and has since achieved considerable scholarly and commercial success (see Chapter 10 herein). Of course, no method is infallible, or represents the best fit for all circumstances. But by focusing on the financial health of private enterprises in different countries, our system promises at the very least to provide a valuable complement to, or reality check on, standard macro approaches. Before we delve into the details of Z-Metrics, we start by briefly reviewing the record of financial crises to provide some historical perspective. Next we attempt to summarize the main findings of the extensive academic and practitioner literature on sovereign risk, particularly those studies designed to test the predictability of sovereign defaults and crises.

With that as background, we then present our Z-Metrics system for estimating the probability of default for individual (nonfinancial) companies and show how that system might have been used to anticipate many developments during the recent EU debt crisis. In so doing, analyze the publicly available corporate data from 2008–2015 for nine European countries, both to illustrate our model's promise for assessing sovereign risk and to identify the scope of reforms that troubled governments must consider not only to qualify for bailouts and subsidies from other countries and international bodies, but to stimulate growth in their economies.

More specifically, we examine the effectiveness of calculating the *median company five-year probability of default of the sovereign's nonfinancial corporate sector*, both as an absolute measure of corporate risk vulnerability and a relative health index comparison among a number of European sovereigns, and including the United States as well. Our analysis shows that this health index, measured at periods prior to the explicit recognition of the crisis by market professionals, not only gave a distinct early warning of impending sovereign default in some cases, but also provided a sensible hierarchy of relative sovereign risk. We also show that, during the recent European crisis, our measures not only compared favorably to standard sovereign risk measures, notably rating agency credit ratings, but performed well even when compared to the implied default rates built into market pricing indicators such as CDS spreads, while avoiding the well-known volatility of the latter.

Our aim here is not to present a "beauty contest" of different methods for assessing sovereign risk in which one method emerges as the clear winner. What we are suggesting is that a novel, bottom-up approach that emphasizes the financial condition and profitability of a nation's private sector can be effectively combined with standard analytical techniques and market pricing to better understand and predict sovereign health. And our analysis has one clear implication for policy makers: that the reforms for assisting ailing nations should be designed, as far as possible, to preserve the efficiency and value of a nation's private enterprises.

Modern History Sovereign Crises

When thinking about the most recent financial crisis, it is important to keep in mind how common sovereign debt crises have been during the past 150 years – and how frequently such debacles have afflicted developed economies as well as emerging market countries. Figure 13.1 shows a partial list of financial crises (identified by the first year of the crisis) that have occurred in "advanced" countries. Overall, Latin America seems to have had more recent bond and loan defaults than any

Austria			1893,	1989					
Brazil			1898,	1902,	1914,	1931,	1939		
Canada			1873,	1906,	1923,	1983			
Czechoslovakia	1870,	1910,	1931,	2008					
China			1921,	1939					
Denmark		1877,	1885,	1902,	1907,	1921,	1931,	1987	
DEU			1880,	1891,	1901,	1931,	2008		
GBR			1890,	1974,	1984,	1991,	2007		
Greece			1870,	1894,	1932,	2009			
Italy			1887,	1891,	1907,	1931,	1930,	1935,	1990
Japan			1942						
Netherlands		1897,	1921,	1939					
Norway		1899,	1921,	1931,	1988				
Russia			1918,	1998					
Spain			1920,	1924,	1931,	1978,	2008		
Sweden		1876,	1897,	1907,	1922,	1931,	1991		
United States			1873,	1884,	1893,	1907,	1929,	1984,	2008

FIGURE 13.1   Financial Crises, Advanced Countries 1870–2010 Crisis Events
(First Year)
Source: IMF Global Financial Stability Reports (www.imf.org), Reinhart and Rogoff
(2010), and various other sources, such as S&P's economic reports.

other region of the world. But if we had included a number of now developed Asian countries among the "advanced" countries, the period 1997–1999 would be much more prominent.

The clear lesson is that sovereign economic conditions appear to spiral out of control with almost predictable regularity and then require massive debt restructurings and/or bailouts accompanied by painful austerity programs. Recent examples include several Latin American countries in the 1980s, Southeast and East Asian nations in the late 1990s, Russia in 1998, and Argentina in 2000. In most of those cases, major problems originating in individual countries not only imposed hardships on their own people and markets, but had major financial consequences well beyond their borders. We saw such effects recently as financial problems in Greece and other southern European countries not only affected their neighbors, but threatened the very existence of the European Union.

Such financial crises have generally come as a surprise to most people, including even those specialists charged with rating the default risk of sovereigns and the enterprises operating in these suddenly threatened nations. For example, it was not long ago that Greek debt was investment grade, and Spain was rated Aaa as recently as June 2010.[2] And this pattern has been seen many times before. To cite just one more case, South Korea was viewed in 1996 as an "Asian Tiger" with a decade-long record of remarkable growth and an AA– rating. Within a year however, the country was downgraded to BB–, a "junk" rating, and the county's government avoided default only through a $50 billion bailout by the IMF. And it was not just the rating agencies that were fooled; most of the economists at the brokerage houses and bond insurance companies also failed to see the problems looming in Korea.

What Do We Know about Predicting Sovereign Defaults?

There is a large and growing body of studies on the default probability of sovereigns, by practitioners as well as by academics.[3] A large number of studies, starting with Frank and Cline's 1971 classic, have attempted to predict sovereign defaults or rescheduling using statistical classification and predicting methods like discriminant analysis, as well as similar econometric techniques.[4] And in a more recent development, some credit analysts have begun using the "contingent claim" approach[5] to measure, analyze, and manage sovereign risk based on Robert Merton's classic "structural" approach (1974). But because of its heavy reliance on market indicators, this approach to predicting sovereign risk and credit spreads has the drawback of producing large – and potentially self-fulfilling – swings in assessed risk that are attributable solely to market volatility.

A number of recent studies have sought to identify global or regional common risk factors that largely determine the level of sovereign risk in the world, or in a region such as Europe. Some studies have shown that changes in both the risk factor of individual sovereigns and in a common time-varying global factor affect the market's repricing of sovereign risk.[6] Other studies, however, suggest that sovereign credit spreads are more related to global aggregate market indexes, including U.S. stock and high-yield bond market indexes, and global capital flows than to their own local economic measures.[7] Such evidence has been used to justify an approach to quantifying sovereign risk that uses the local stock market index as a proxy for the equity value of the country.[8] Finally, several very recent papers focus on the importance of macro variables such as debt service relative to tax receipts and the volatility of trade deficits in explaining sovereign risk premiums and spreads.[9]

A number of studies have also attempted to evaluate the effectiveness of published credit ratings in predicting defaults and expected losses, with most concluding that sovereign ratings, especially in emerging markets, provide an improved understanding of country risks for investment analytics.[10] Nevertheless, the recent EU debt crisis would appear to contradict such findings by taking place at a time when all the rating agencies and, it would seem, all available models for estimating sovereign risk indicated that Greece and Spain were still classified as investment grade.[11] What's more, although most all of the studies cited above have been fairly optimistic about the ability of their concepts to provide early warnings of major financial problems, their findings have either been ignored or have proven ineffective in forecasting most economic and financial crises.

In addition to these studies, a handful or researchers have taken a somewhat different bottom-up approach by emphasizing the health of the private sectors supporting the sovereigns. For example, a 1998 World Bank study of the 1997 East Asian crisis by Pomerleano (1998)[12] used our average Z-Score of listed (non-financial) companies to assess the "financial fragility" of eight Asian countries and, for comparison purposes, three developed countries and Latin America. Surprising many observers, the average Z-Score for South Korea at the end of 1996 suggested that it was the most financially vulnerable Asian country, followed by Thailand, Japan, and Indonesia. A traditional macroeconomic measure like GDP growth would not have predicted such trouble since, at the end of 1996, South Korea had been growing at double-digit rates for nearly a decade.[13]

The Z-Metrics™ Approach[14]

In 2009, we partnered with RiskMetrics Group with the aim, at least initially, of creating a new and better way of assessing the credit risk of *companies*. The result was our new Z-Metrics approach. This methodology might be called a new generation of the original Z-Score model of 1968. Our objective was to develop up-to-date credit scoring and probability of default metrics for both large and small, public and private, enterprises on a global basis.

In building our models, we used multivariate logistic regressions and data from a large sample of both public and private U.S. and Canadian nonfinancial sector companies during the 20-year period 1989–2008.[15] We analyzed over 50 fundamental financial statement variables, including measures (with trends as well as point estimates) of solvency, leverage, size, profitability, interest coverage, liquidity, asset quality, investment, dividend payout, and financing results. In addition to such operating (or "fundamental") variables, we also included equity market price and return variables and their patterns of volatility. Such market variables have typically been used in the "structural distance-to-default measures" that are at the core of the KMV model[16] now owned by Moody's (see Crosbie and Bohn 2002 for Moody's documentation).

In addition to these firm-specific variables, we also tested a number of macroeconomic variables that are often used to estimate sovereign default probabilities, including GDP growth, unemployment, credit spreads, and inflation. Since most companies have a higher probability of default during periods of economic stress – for example, at the end of 2008 – we wanted to use such macro variables to capture the heightened or lower probabilities associated with general economic conditions.[17]

The final model, which consists of 10 fundamental, market value, and macroeconomic variables, is used to produce a credit score for each public company. Although our primary emphasis was on applying Z-Metrics to publicly traded companies, we also created a private firm model by using data from public companies and replacing market value with book value of equity. The next step was to use a logit specification of the model (described in the Appendix) that we used to convert the credit scores into probabilities of default (PDs) over both one-year and five-year horizons. The one-year model is based on data from financial statements and market data approximately one year prior to the credit event, and the five-year model includes up to five annual financial statements prior to the event.

To test the predictive power of the model and the resulting PDs, we segregated all the companies in our sample into "cohorts" according to whether they experience "credit events" that include either formal default or bankruptcy (whichever comes first). All companies that experienced a credit event within either one year or five years were assigned to the "distressed" or "credit event" group (with all others assigned to the nondistressed group).

Our test results show considerable success in predicting defaults across the entire credit spectrum from the lowest to the highest default risk categories. Where possible, we compared our output with that of publicly available credit ratings and existing models. The so-called "accuracy ratio" measures how well our model predicts which companies do or do not go bankrupt on the basis of data available before bankruptcy. The objective can be framed in two ways: (1) maximizing correct predictions of defaulting and nondefaulting companies and (2) minimizing wrong predictions (Type II accuracy).

As can be seen in Figure 13.2, our results, which include tests on actual defaults during the period 1989–2009, show much higher Type I accuracy levels for the Z-Metrics model than for either the bond rating agencies or established

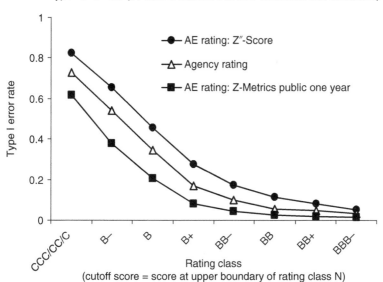

FIGURE 13.2 Type I Error for Agency Ratings, Z″-score, and Z-Metrics Agency Equivalent (AE Ratings (1989–2008): One-Year Prediction Horizon for Publicly Owned Firms

models (including an older version of Z-Scores). At the same time, our tests show equivalent Type II accuracies at all cutoff levels of scores.[18]

Perhaps the most reliable test of credit scoring models is how well they predict critical events based on samples of companies that were not used to build the model, particularly if the events took place after the period during which the model was built (after 2008, in this case). With that in mind, we tested the model against actual bankruptcies occurring in 2009, or what we refer to as our "out-of-sample" data. As with the full test sample results shown in Figure 13.2, our Z-Metrics results for the "out of sample" bankruptcies of 2009 outperformed the agency ratings and the 1968 Z-Score and 1995 Z″-Score models using both one-year and five-year horizons.

A "Bottom-Up" Approach for Sovereign Risk Assessment

Having established the predictive power of our updated Z-Score methodology, our next step was to use that model (which, again, was created using large publicly traded U.S. companies) to evaluate the default risk of European companies. And after assuring ourselves that the model was transferable in that sense, we then attempted to assess the overall creditworthiness of sovereign governments by aggregating our Z-Metrics default probabilities for individual companies and then estimating both a median default probability and credit rating for different countries.

In conducting this experiment, we examined nine key European countries over three time periods, end of 2008, 2009, and 2010 (Figure 13.3) and again in 2015 (Figure 13.4), when the crisis was well known. People clearly recognized the crisis and concern for the viability of the European Union in June 2010, when Greece's debt was downgraded to noninvestment grade and both Spain and Portugal were also downgraded. Credit markets, particularly CDS markets, had already recognized the Greek and Irish problems before June 2010. Market prices during the first half of 2010 reflected high implied probabilities of default for Greece and Ireland, but were considerably less pessimistic in 2009. By contrast, we display the Z-Metric median PD estimates alongside sovereign CDS spreads over both periods.[19] Our PD estimates were uniformly higher (more risky) in 2009 than early in 2010, even if the world was more focused on Europe's problems in the latter year. In this sense, our Z-metrics PD might be viewed as providing a leading indicator of possible distress. It should be noted that the statistics in Figures 13.3 and 13.4 report only on the nonfinancial private sector, while those in Figure 13.6, to be discussed, include results from our banking credit risk model, as well.

As of year-end 2009, our Z-Metrics' five-year PDs for European corporate default risk placed Greece (10.60%) and Portugal (9.36%) in the highest risk categories (ZC-ratings), followed by Italy (7.99%), Ireland (6.45%), and Spain (6.44%), all in the ZC category. Then came Germany and France (both about 5.5% – ZC+), with the UK (3.62%) and the Netherlands (3.33%) at the lowest risk levels (ZB– and ZB). The United States looked comparatively strong, at 3.93% (ZB–).

For the most part, these results are consistent with how traditional analysts ranked sovereign risks. Nevertheless, there were a few surprises. The UK had a fairly healthy private sector, and Germany and France were perhaps not as healthy as one might have thought. The UK's relatively strong showing might have resulted from the fact that our risk measure at this time did not include financial sector firms, which comprised about 35% of the market values of listed UK corporates and were in poor financial condition. And several very large, healthy multinational entities in the UK index might have skewed results a bit. The CDS/five-year market's assessment of UK risk was harsher than that of our Z-Metrics index in 2010, with the median of the daily CDS spreads during the first four months implying a 6.52% probability of default, about double our Z-Metrics median level. Greece also had a much higher CDS implied PD at 24.10%, as compared to 10.60% for Z-Metrics. (And, of course, our choice of the *median* Z-Metrics PD is arbitrary, implying as it does that fully 50% of the listed companies have PDs higher than 10.60%.)

We also observed that several countries had relatively high standard deviations of Z-Metrics PDs, indicating a longer tail of very risky companies. These countries included Ireland, Greece and, surprisingly, Germany, based on 2010 data. So, while almost everyone considers Germany to be the benchmark-low risk country in Europe – for example, its five-year CDS spread was just 2.67% in 2010, even lower than the Netherlands (2.83%), we are more cautious based on our broad measure of private sector corporate health.

2010 Results and the 75th Percentile Firm as a Measure of Sovereign Default Risk

Figure 13.3 shows the weighted-average median PDs for 11 (including now Sweden and Belgium) European countries and the United States as of the end of 2010. The results show the large difference between Greece (15.28%) and all the rest, but also that the "big-five PIIGS" stand out as the clear higher risk domains. Indeed, we felt that Italy could have been considered the "fulcrum" country to decide the ultimate fate of the Euro (see Altman's "Insight" article in the Financial Times, June 21, 2011).

Figure 13.4 shows our results through 2016 for the 75th percentile firm in each country. We find that the 75th percentile firm is our *best estimate* of the sovereign's risk of default and can be compared to the implied PD from the market, via the CDS premium (Figure 13.5).

CDS Implied PDs

Figure 13.5 shows the implied PDs for the "Big-Five" European high-risk countries from the start of 2009 to the end of 2017. Note that while the PDs, based on

	Z-Metrics PD Estimates: Five-Year Public Model			Five-Year Implied PD from CDS Spread*		
Country	Listed Companies	Y/E 2010 Median PD	Y/E 2009 Median PD	Y/E 2008 Median PD	2010	2009	2008
Netherlands	85	3.56%	3.33%	5.62%	2.03%	2.83%	6.06%
United States	2226	3.65%	3.93%	6.97%	3.79%	3.28%	4.47%
Sweden	245	3.71%	5.31%	6.74%	2.25%	4.60%	6.33%
Ireland	29	3.72%	6.45%	7.46%	41.44%	12.20%	17.00%
Belgium	69	3.85%	5.90%	5.89%	11.12%	4.58%	5.53%
UK	507	4.28%	3.62%	5.75%	4.73%	6.52%	8.13%
France	351	4.36%	5.51%	7.22%	4.51%	3.75%	4.05%
Germany	348	4.63%	5.54%	7.34%	2.50%	2.67%	3.66%
Italy	174	7.29%	7.99%	10.51%	9.16%	8.69%	11.20%
Spain	91	7.39%	6.44%	7.39%	14.80%	9.39%	8.07%
Portugal	33	10.67%	9.36%	12.07%	41.00%	10.90%	7.39%
Greece	93	15.28%	10.60%	11.57%	70.66%	24.10%	13.22%

*Assuming a 40% recovery rate (R); based on the median CDS spread (s). PD computed as $1 - e^{(-5*s/(1-R))}$.

FIGURE 13.3 Financial Health of the Corporate, Nonfinancial Sector: Selected European Countries and United States in 2008–2010
Sources: RiskMetrics Group (MSCI), Markit, Compustat.

	Z-Metrics PD Estimates*: Five-Year Public Model								
Country	Listed Companies (2Q16)**	75th Percentile PD							
		2Q16	2015	2014	2013	2012	2011	2010	2009
Ireland	26	3.0%	3.4%	3.0%	3.2%	3.0%	8.1%	8.6%	11.0%
Sweden	181	3.8%	5.1%	6.1%	4.1%	5.6%	8.3%	6.8%	8.0%
UK	499	5.1%	5.3%	5.6%	4.9%	6.3%	10.4%	5.7%	9.3%
Netherlands	75	7.3%	7.7%	8.9%	6.8%	5.8%	9.1%	5.7%	6.7%
France	348	7.9%	8.8%	9.4%	9.2%	10.2%	13.0%	8.5%	10.3%
Germany	317	8.2%	9.7%	9.9%	10.4%	8.0%	10.6%	9.7%	11.9%
Poland	171	12.5%	10.4%	10.9%	17.5%	25.5%	28.5%	15.2%	17.1%
Italy	169	12.6%	15.0%	15.2%	18.4%	23.3%	26.4%	14.1%	18.1%
Spain	85	12.9%	17.0%	13.4%	20.8%	21.2%	22.6%	13.2%	12.7%
Portugal	36	28.5%	39.4%	22.5%	27.5%	26.3%	42.4%	22.2%	22.1%
Greece	78	57.4%	38.1%	46.7%	60.0%	60.8%	59.2%	46.9%	31.4%
Australia	361	8.7%	7.3%	9.8%	7.7%	8.7%	10.3%	7.4%	7.8%
United States	2,450	3.2%	3.5%	3.6%	3.7%	4.6%	11.7%	8.0%	11.5%

*Since the Z-Metrics Model is not practically available for most analysts, we could substitute the Z″-Score method (available from <altmanZscoreplus.com>).
**Sales > € 50mm.

FIGURE 13.4 Financial Health of the Corporate, Nonfinancial Sector: Selected European Countries and Australia/United States in 2008–2Q16
Sources: RiskMetrics Group (MSCI), Markit, Compustat Global.

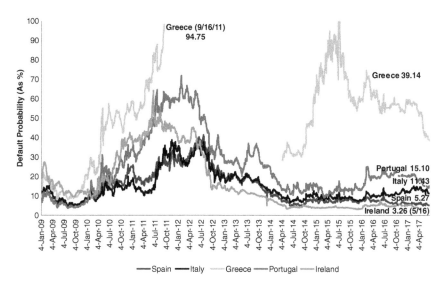

FIGURE 13.5   Five-Year Implied Probabilities of Default (PD) from Capital Market CDS Spreads,* January 2009–June 2017
*Assuming a 40% recovery rate (R); based on the median CDS spread (s). PD computed as $1 - e^{(-5*s/(1-R))}$.
Source: Author's calculations.

CDS spreads and assuming a 40% recovery rate, all came down from their highs, all still implied a considerable default risk until 2016. Indeed, as of mid-January 2012, the Greek CDS implied probability of default increased to almost 95%, and Italy, the subject of our fulcrum risk country "Insight" piece, increased from 19% in July 2011 to 35% in January 2012. As noted, Greece actually did default later in 2012.

2010 versus 2009

As noted earlier from Figure 13.4, our PD estimates for early 2009 were uniformly higher (more risky) than those for early 2010. One important reason for the higher PDs in 2009 is the significant impact of the stock market, which is a powerful variable in the Z-Metrics model – and in many other default probability models (notably, Moody's KMV, see Crosbie and Bohn, 2002). Recall that the stock markets were at very low levels at the end of 2008 and into the early months of 2009, while there was a major recovery later in 2009 and in early 2010.

A Model That Includes the Financial Sector

A valid question about a "micro-bottom-up" approach to sovereign risk assessment is the importance of the financial sector. We are hesitant to include early-warning

Country	Nonfinancial Firms PD (%)	Nonfinancial Firms Weight	Banking Firms PD (%)	Banking Firms Weight	Weighted Average (%)	Rank	CDS Spread PD (%)***	Rank
Netherlands	3.56	0.977	11.1	0.023	3.73	1	2.03	1
Sweden	3.71	0.984	17.3	0.016	3.93	2	2.25	2
Belgium	3.85	0.972	12.4	0.028	4.21	3	11.12	8
France	4.36	0.986	14.0	0.014	4.49	4	4.51	5
UK	4.28	0.977	15.5	0.023	4.54	5	4.73	6
Germany	4.63	0.983	13.1	0.017	4.77	6	2.50	3
United States	3.65	0.837	13.8	0.163	5.30	7	3.79	4
Spain	7.39	0.948	10.9	0.052	7.57	8	14.80	9
Italy	7.29	0.906	20.0	0.094	8.48	9	9.16	7
Ireland	3.72	0.906	77.6	0.094	10.65	10	41.44	11
Portugal	10.67	0.971	12.1	0.029	10.71	11	41.00	10
Greece	15.28	0.921	30.1	0.079	16.45	12	70.66	12

*Based on the Z-Metrics Probability Model.
**Based on Altman-Cziel-Rijken Model (Preliminary).
***PD based on the CDS Spread as of April 26, 2011.

FIGURE 13.6   Weighted Average Median Five-Year (PD) for Listed Nonfinancial* and Banking Firms** (Europe and United States), 2010

indicators of banks, for example, due to the relatively poor track record (in our opinion) of banking predictive models. Still, it is worth trying, and in Figure 13.6, we introduce the results of a preliminary model (Altman, Cziel, and Rijken 2016) based on a number of fundamental banking variables for defaulting and nondefaulting U.S. banks, then applied to European banking entities. This model is, admittedly, a work-in-process.

Based on 2010 results, the weighted-average median default probability for Greece increased to 16.45, from 15.28%, based only on the nonfinancial sector. At that time, the CDS spread indicated a 70.66% five-year default probability.

Comparing PD Results Based on Privately Owned versus Publicly Owned Firm Models

As shown in Figures 13.3 and 13.4, the improvement (reduction) in Z-Metrics PDs for most countries in 2010 – a period in which most EU sovereigns appeared to be getting riskier – looks attributable in large part to the stock market increases in almost all countries. But to the extent such increases could conceal a deterioration of a sovereign's credit condition, some credit analysts might prefer to have PD estimates that do not make use of stock market data.

With this in mind, we applied our private firm Z-Metrics model to evaluate the same nine European countries and the United States. The private and public firm models are the same except for the substitution of equity book values (and volatility of book values) for market values. Note, we did the same adjustment for our original Z-Score model. This adjustment is expected to remove the capital market influence from our credit risk measure.

Correlation of Sovereign PDs: Recent Evidence on Z-Metrics versus Implied CDS PDs

As a final test of the predictive power of our approach, we compared our Z-Metrics five-year median PDs for our sample of nine European countries (both on a contemporary basis and for 2009) with the PDs implied by CDS spreads in 2010. The contemporary PD correlation during the first third of 2010 was remarkably high, with an R^2 of 0.82. This was a period when it was becoming quite evident that certain European countries were in serious financial trouble and the likelihood of default was not trivial. But if we go back to the first half of 2009, the correlation drops to an R^2 of 0.36 (although it would be considerably higher, at 0.62, if we excluded the case of Ireland). Ireland's CDS implied PD was considerably higher in 2009 than 2010 (17.0% versus 12.0%), while the Z-Metrics PD was relatively stable in the two years (7.5% and 6.5% respectively).[20] In 2010, whether we calculate the correlation with or without Ireland, the results are essentially the same (0.82 and 0.83). Incidentally, it is remarkable that the PDs for Ireland from our bottom-up approach showed impressive strength, that is, low PDs, throughout the entire time series from 2009 to 2017, despite its need for a bailout solely due to the financial sector. Indeed, our results are consistent with a "buy" strategy once the financial sector's rescue was in place.

Given the predictive success of Z-metrics in the tests already described, we were curious to find out whether it could be used to predict capital market (i.e., CDS) prices. So, we regressed our public firm model's 2008 Z-Metrics median, nonfinancial sector PDs against implied CDS PDs one year later in 2009. Admittedly, this sample was quite small (10 countries) and the analysis is for only a single time-series comparison (2008 vs. 2009). Nevertheless, these two years spanned a crucial and highly visible sovereign debt crisis, whereas the PDs implied by prior years' Z-Metrics and CDS showed remarkably little volatility.[21]

Figure 13.7 summarizes the results of our public versus private firm Z-Metrics models comparative PD (delta) results for 2010 and 2009. For eight of the 10 countries, use of the private firm model showed smaller reductions in PDs when moving from 2009 to 2010 than use of the public model. Whereas the overall average improvement in PDs for the public firm model was a drop of 1.91 percentage points, the drop was 0.79% for our private firm model. These results are largely the effect of the positive stock market performance in late 2009 and into 2010. But improvements in general macro conditions, along with their effects on traditional corporate performance measures, also helped improve (reduce) the PDs. Moreover, in two of these eight countries – the UK and France – not only did the public firm model show an improved (lower) PD, but the private firm model's PD

Country	Number Listed Companies		Public-Firm Z-Metrics Model			Private-Firm Z-Metrics Model		
	2010	2009	PDs 2010	PDs 2009	Delta*	PDs 2010	PDs 2009	Delta*
Netherlands	61	60	3.33%	5.62%	−2.29%	5.25%	6.00%	−0.75%
UK	442	433	3.62%	5.75%	−2.13%	6.48%	5.97%	+0.49%
United States	2226	2171	3.93%	6.97%	−3.04%	4.28%	4.80%	−0.52%
France	297	294	5.51%	7.22%	−1.71%	7.33%	7.19%	+0.14%
Germany	289	286	5.54%	7.34%	−1.80%	6.29%	7.56%	−1.27%
Spain	82	78	6.44%	7.39%	−0.95%	8.06%	9.32%	−1.26%
Ireland	28	26	6.45%	7.46%	−1.01%	6.31%	6.36%	−0.05%
Italy	155	154	7.99%	10.51%	−2.52%	8.14%	9.07%	−0.89%
Portugal	30	30	9.36%	12.07%	−2.71%	8.73%	9.62%	−0.89%
Greece	79	77	10.60%	11.57%	−0.97%	11.03%	13.93%	−2.90%
Average			6.28%	8.19%	−1.91%	7.19%	7.98%	−0.79%

*Negative sign means improved credit risk.

FIGURE 13.7 Private versus Public Firm Model PDs in 2010 and 2099

actually got worse (increased) in 2010 (as indicated by the positive delta in the last column of Figure 13.7).

The correlation between our Z-Metrics PDs and those implied by the CDS one year later proved to be remarkably strong, with an r of 0.69 and an R^2 of 0.48 (Figure 13.8). In sum, the corporate health index for our European countries (plus the United States) in 2008 explained roughly half of the variation in the CDS results one year later.[22] More tests need to be run for us to be even more confident of the results.

A potential shortcoming of our approach is that we are limited in our private sector corporate health assessments to data from listed, publicly held firms. This is especially true for relatively small countries like Ireland (with just 28 listed companies), Portugal (with 30), Greece (79), Netherlands (61), and Spain (82). Since the private, nonlisted segment is much larger in all of the countries, we are not clearly assessing the health of the vast majority of its firms and our sovereign health index measure is incomplete.[23]

But if the size of the listed firm population is clearly a limitation in our calculations, there does not seem to be a systematic bias in our results. To be sure, the very small listings in Ireland, Portugal, and Greece appear heavily correlated with their high PDs, but the country with the lowest PD (the Netherlands) also has a very small listed population. Another potentially important factor is that the listed population in countries like the UK and the Netherlands is represented quite heavily by multinational corporations that derive most of their income from outside their borders.[24]

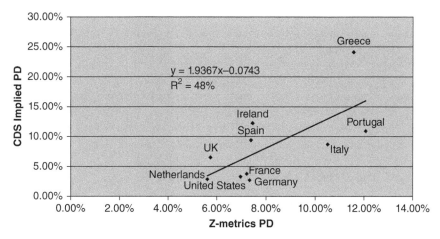

FIGURE 13.8 2008 Z-Metrics PD vs 2009 CDS Implied PD
Source: Data from Figures 13.3 and 13.5.

Conclusion and Implications

As the price for bailing out distressed sovereigns, foreign creditors, especially the stronger European nations, were demanding a heavy dose of austerity. Several governments, including those of Greece, Ireland, Spain, Portugal, Italy, and the UK, did enact some painful measures. Others, such as France and Hungary, either resisted austerity measures or faced significant social unrest when austerity measures were proposed. These measures typically involve substantial cuts in cash benefits paid to public workers, increases in retirement age, and other reduced infrastructure costs, as well as increased taxes for companies and individuals. The objective was to reduce deficits relative to GDP and enhance the sovereigns' ability to repay their foreign debt and balance their budgets.

While recognizing the necessity of requiring difficult changes for governments to qualify for bailouts and subsidies, we caution that such measures should be designed to inflict as little damage as possible on the health and productivity of the private enterprises that ultimately fund the sovereign. The goal should be to enable all private enterprises with clear going concern value to pay their bills, expand (or at least maintain) their workforces, and return value to their shareholders and creditors (while those businesses that show no promise of ever making a profit should be either reorganized or liquidated). For this reason, raising taxes and imposing other burdens on corporate entities is likely to weaken the long-run financial condition of sovereigns.

To better estimate sovereigns' risk of default, we propose that traditional measures of macroeconomic performance be combined with more modern techniques.

Along with the intuitive appeal of such an approach and our encouraging empirical results, the probabilities of sovereign default provided by aggregating our Z-Scores across a national economy can be seen, at the very least, as a useful complement to existing methods and market indicators – one that is not subject to government manipulation of publicly released statistics. Using our approach, the credit and regulatory communities could track the performance of publicly held companies and the economies in which they reside – and by making some adjustments, unlisted entities as well. And if sovereigns were also willing to provide independently audited statistics on a regular basis, so much the better.

Appendix: Logit Model Estimation of Default Probabilities

We estimated our credit scoring model based on a standard logit-regression functional form whereby:

$$CS_{i,t} = \alpha + \Sigma B_j X_{i,t} + \varepsilon_{i,t} (1)$$

$CS_{i,t} = Z\text{–}Metrics\ credit\ score\ of\ company\ i\ at\ time\ t$

$B_j = variable\ parameters\ (or\ weights)$

$X_{i,t} = set\ of\ fundamental,\ market\ based\ and\ macroeconomic\ variables\ for$

   $firm\ i\ quarter\ t\ observations$

$\varepsilon_{i,t} = error\ terms\ (assumed\ to\ be\ identically\ and\ independently\ distributed)$

$CS_{i,t}\ is\ transformed\ in\ to\ a\ probability\ of\ default\ by\ PD_{i,t} = \dfrac{1}{1 + \exp\ (CS_{i,t})}$

- We compare Z-Metrics results with issuer ratings. To ensure a fair comparison, credit scores are converted to agency equivalent (AE) ratings by ranking credit scores and by matching exactly the actual agency rating distribution with the AE rating distribution at any point in time.
- We also compare our Z-Metrics results to the well-established Altman Z''-score (1995) model.[25]

NOTES

1. See, for example, Hekran Neziri's "Can Credit Default Swaps Predict Financial Crises?" in the Spring 2009 *Journal of Applied Economic Sciences,* Volume IV/Issue 1(7). Neziri found that CDS prices had real predictive power for equity markets, but that the lead time was generally on the order of just one month.
2. On April 27, 2010, Standard & Poor's Ratings Services lowered its long- and short-term credit ratings on the Hellenic Republic (Greece) to noninvestment grade BB+; and on June 14, 2010, Moody's downgraded Greece debt to Ba1 from A2 (four notches), while Spain was still Aaa and Portugal was A1. S&P gave similar ratings.

3. One excellent primer on sovereign risk is Babbel's (1996) study, which includes an annotated bibliography by S. Bertozzi on external debt capacity that describes many of these studies. Babbel lists 69 potentially helpful explanatory factors for assessing sovereign risk, all dealing with either economic, financial, political, or social variables. Except for the political and social variables, all others are macroeconomic data and this has been the standard until the past few years. Other work worth citing include two practitioner reports – Chambers (1997) and Beers et al. (2002) – and two academic studies – Smith and Walter (2003), and Frenkel, Karmann, and Scholtens (2004). Full citations of all studies can be found in the References section at the end of this book.

4. Including Grinols (1976), Sargen (1977), Feder and Just (1977), Feder, Just, and Ross (1981), Cline (1983), and Schmidt (1984).

5. Gray, Merton, and Bodie (2006, 2007).

6. See Baek, Bandopadhyaya, and Du (2005). Gerlach, Schulz, and Wolff (2010) observe that aggregate risk factors drive banking and sovereign market risk spreads in the Euro area; and in a related finding, Sgherri and Zoli (2009) suggest that Euro area sovereign risk premium differentials tend to move together over time and are driven mainly by a common time-varying factor.

7. See Longstaff, Pan, Pedersen, and Singleton (2007).

8. See Oshiro and Saruwatari (2005).

9. These include Haugh, Ollivaud, and Turner's (2009) discussion of debt service relative to tax receipts in the Euro area; Hilscher and Nosbusch (2010) emphasis on the volatility of terms of trade; and Segoviano, Caceres, and Guzzo's (2010) analysis of debt sustainability and the management of a sovereign's balance sheet.

10. For example, Remolona, Scatigna, and Wu (2008) reach this conclusion after using sovereign credit ratings and historical default rates provided by rating agencies to construct a measure of ratings implied expected loss.

11. To be fair, S&P in a *Reuter's* article dated January 14, 2009, warned Greece, Spain, and Ireland that their ratings could be downgraded further as economic conditions deteriorated. At that time, Greece was rated A1 by Moody's and A– by S&P. Interestingly, it was almost a full year later on December 22, 2009, that Greece was actually downgraded by Moody's to A2 (still highly rated), followed by further downgrades on April 23, 2010 (to A3), and finally to junk status (Ba1) on June 14, 2010. As noted earlier, S&P downgraded Greece to junk status about three months earlier. Greece eventually defaulted in 2012.

12. This study, which is based on a longer article by the author (1997). Taking a somewhat similar approach, many policy makers and theorists have recently focused on the so-called *shadow banking system*. For example, Gennaioli, Martin, and Rossi (2010) argued that the financial strength of governments depends on private financial markets and its ability to attract foreign capital. They concluded that strong financial institutions not only attract more capital but their presence also helps encourage their governments to repay their debt. Chambers of S&P (1997) also mentions the idea of a "bottom-up" approach but not to the assessment of sovereign risk, but to a corporate issuer located in a particular country. He advocates first an evaluation of an issuer's underlying creditworthiness to arrive at its credit rating and then considers the economic, business and social environment in which the entity operates. These latter factors, such as the size and growth and the volatility of the economy, exchange rates, inflation, regulatory

environment, taxation, infrastructure, and labor market conditions are factored in on top of the micro variables to arrive at a final rating of the issuer.

13. Afterwards, the World Bank and other economists, such as Paul Krugman, concluded that that crony capitalism and the associated implicit public guarantees for politically influential enterprises coupled with poor banking regulation were responsible for the crisis. The excesses of corporate leverage and permissive banking were addressed successfully in the case of Korea and its economy was effectively restructured after the bailout.

14. For more details, see Altman et al. 2010 "The Z-Metrics™ Methodology for Estimating Company Credit Ratings and Default Risk Probabilities," RiskMetrics Group, continuously updated, available at the MSCI website and Professor Altman's website.

15. Our model's original sample consisted of over 1,000 U.S. or Canadian nonfinancial firms that suffered a credit event and a control sample of thousands of firms that did not suffer a credit event, roughly a ratio of 1:15. After removing those firms with insufficient data, the credit event sample was reduced to 638 firms for our public firm sample and 802 observations for our private firm sample.

16. Adapted for sovereigns by Gray, Merton, and Bodie (2007).

17. In all cases, we carefully examined the complete distribution of variable values, especially in the credit-event sample. This enabled us to devise transformations on the variables to either capture the nature of their distributions or to reduce the influence of outliers. These transformations included logarithmic functions, first differences and dummy variables if the trends or levels of the absolute measures were positive/negative.

18. We assessed the stability of the Z-Metrics models by observing the accuracy ratios for our tests in the in-sample and out-of-sample periods and also by observing the size, signs, and significance of the coefficients for individual variables. The accuracy ratios were very similar between the two sample periods and the coefficients and significance tests were extremely close.

19. The median CDS spread is based on the daily observations in the six/four-month periods. The median Z-Metrics PD is based on the median company PDs each day and then we calculated the median for the period. The results are very similar to simply averaging the median PDs as of the beginning and ending of each sample period.

20. No doubt the CDS market was reacting quite strongly to the severe problems in the Irish banking sector in 2009, while Z-Metrics PDs were not impacted by the banks. This implies a potential strength of the CDS measure, although the lower CDS implied PD in early 2010 was not impressive in predicting the renewed problems of Irish banks and its economy in the fall of 2010.

21. The last time an entire region and its many countries had a sovereign debt crisis was in Asia in 1997-1998. Unfortunately, CDS prices were not prominent and the CDS market was illiquid at that time.

22. Several other nonlinear structures (i.e., power and exponential functions) for our 2009 Z-Metrics versus 2010 CDS implied PDs showed similar results. In all cases, we are assuming a recovery rate of 40% on defaults in calculation of implied sovereign PDs.

23. We suggest that complete firm financial statement repositories, such as those that usually are available in the sovereign's central bank be used to monitor the performance of the entire private sector.

24. Results showing the percentage of "home-grown" revenues for listed firms across our European country sample were inclusive, however, as to their influence on relative PDs.

25. As noted in Chapter 10, our original Z-Score model (1968) is well known to practitioners and scholars alike. It was built, however, over 40 years ago and is primarily applicable to publicly-held manufacturing firms. A more generally applicable Z''-Score variation was popularized later (Altman, Hartzell and Peck, 1995) as a means to assess the default risk of nonmanufacturers as well as manufacturers, and was first applied to emerging market credits. Further, the Altman Z-Score models do not translate directly (unless BREs are used) into a probability of default rating system, as does the Z-Metrics system. Of course, entities that do not have access to the newer Z-Metrics system can still use the classic Z-Score frameworks.

CHAPTER 14

The Anatomy of Distressed Debt Markets

In the 1993 and again in the 2006 editions of this book, we wrote that the market for investing in distressed securities, mainly debt obligations (the so-called *vulture markets*), had captured the interest of increasing numbers of investors and analysts. These investors, sometimes categorized as "alternative asset" institutions, mainly hedge funds, now can convincingly argue that the market has matured into a genuine asset class, with a reasonably long history of data on return and risk attributes. And, we have been there every step of the way, researching its growth and performance, documenting its dynamics and nurturing the asset class growth with statistics and analytics.[1]

Our fascination with distressed firms and their outstanding securities began when the chairman of a lender and investor enterprise, the Foothill Group (now part of Wells Fargo), in debt of firms that were either already bankrupt or deeply distressed with the likely prospect of defaulting on its obligations, came to me (Altman) with an assignment to provide a descriptive and analytical white paper on what was generally known as "distressed" debt. This resulted in two monographs, one on Distressed Bonds (Altman 1990) and a second on distressed loans (Altman 1992). Our first task was to carefully define this market and after getting several interesting, but not sufficient, definitions from practitioners, such as bonds selling for less than 80% of par value, we established essentially two precise categories: (1) bonds or loans whose yield to maturity (later amended to option-adjusted yield) was equal to or greater than 10% (1,000 bps) above the 10-year U.S. government bond rate (later amended to be the comparable duration in the U.S. government securities) and (2) those bonds or loans of firms who have defaulted on their debt obligations and were in their restructuring, usually, Chapter 11, phase. The former was categorized as "Distressed" and the latter as "Defaulted." We also included the equity securities of those firms, but did not attempt any documentation on distressed equity at that time.

In addition to an increasing number of articles and reports (discussed later in Chapter 15) on the distressed securities market and to the two aforementioned studies, several books related to the subject have been published including Ramaswami and Moeller (1990), Altman (1991), Rosenberg (1992, 2000), Branch and Ray (1992, 2007), Moyer (2005), Whitman and Diz (2009), and Gilson (2010). Many of these sources will also be reviewed in the next chapter.

The purpose of this chapter is to document the descriptive anatomy of the distressed debt market's size, growth, major strategies and characteristics, and participants. Chapter 15 explores its performance attributes, reviewing the relevant 30-year period from 1987–2017, with particular emphasis on the most recent years and new empirical results.

SIZE OF THE DISTRESSED AND DEFAULTED DEBT MARKET

When we first defined and investigated the overall distressed and defaulted debt market in 1990, we estimated that the publicly traded and privately issued market was about $300 billion (face value) and $200 billion (market value) – see Figure 14.1. The huge increase in distressed firms in the 1989–1991 period, from the fallout of the highly leveraged restructuring movement in the United States in the 1980s, was the primary contributor to these totals. Coupled with huge, excess returns to risk alternative asset investors of over 40% in 1991, the market's growth gave birth to a professional asset class that was heretofore populated

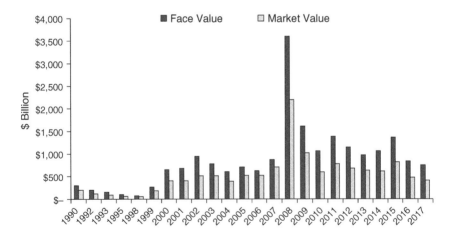

FIGURE 14.1 Value of Distressed and Default Debt
See definitions of the calculation estimates in Figure 14.2.
Not all years estimated in this mostly annual survey.
Source: Professor Edward I. Altman and Brenda Keuhue estimates, NYU Salomon Center (2018).

only by investors who specialized in unique situations (event driven) and sporadic, anecdotal references to those events, like railroad bankruptcies following the great depression and continuing into the 1960s and 1970s, or LBO flame-outs in the 1980s (see some excellent "stories" by Rosenberg (1992 and 2000). The next great growth catalyst was the massive defaults in 2000–2002, when the distressed and defaulted debt market's size again surged, this time to $940 billion (face value) and $500 billion (market value) in 2002.

The growth followed the benign credit cycle's impact on the reduced market size from 1993 to 1998. The number of distressed and defaulted companies grew to unprecedented levels following the combined factors of Russia's meltdown and Long-Term Capital Management's demise in 1998; massive fraud among large, heavily leveraged companies (like Enron, WorldCom, Adelphia, and Global Crossing, among others); the burst of the telecom bubble; and a host of bankruptcies in such industries as airlines, steel, health care, and retailers.

The face value size of the market then remained at between $700 and $1,000 billion until the next major spike in defaults and bankruptcies during and immediately after the great financial crisis (GFC) of 2008–2009, when the face value of distressed and defaulted debt zoomed to about $3.6 trillion and the market value increased to over $2.0 trillion. These totals notably included Lehman Brothers' over $600 billion in liabilities. Indeed, the number of billion-dollar liability bankruptcies in 2009 surged to 49 filings.[2] The median number of these

	Face Value ($)			Market Value ($)			
	31 Dec 2015	31 Dec 2016	31 Dec 2017	31 Dec 2015	31 Dec 2016	31 Dec 2017	Market/ Face Ratio[d]
Public Debt							
Defaulted	287.97	298.03	270.91[a]	86.39	119.21	94.82	0.35
Distressed	395.26	119.78	102.49[b]	256.92	77.86	66.62	0.65
Total Public	683.23	417.81	373.40	343.31	197.07	161.44	
Private Debt							
Defaulted	287.97	298.03	270.91[c]	172.78	193.72	176.09	0.65
Distressed	395.26	119.78	102.49[c]	296.44	89.83	76.87	0.75
Total Private	683.23	417.81	373.40	469.23	283.56	252.96	
Total Public and Private	1,366.46	835.63	746.80	812.54	480.63	414.40	

[a]Calculated using: (2016 defaulted population) + (2017 defaults) − (2017 emergences) − (2017 distressed restructurings).
[b]Based on 6.11% of the size of the high-yield market ($1.677 trillion).
[c]Based on a private/public ratio of 1:1.
[d]The market/face value ratio was 0.40 for public defaulted debt, 0.65 for public distressed debt, 0.65 for private defaulted debt and 0.75 for private distressed debt in 2016.

FIGURE 14.2   Estimated Face and Market Values of Defaulted and Distressed Debt, 2015–2017 (Dollars in Billions)
Source: NYU Salomon Center and estimates by Professor Edward I. Altman.

mega-billion-dollar filings has been 21 per year over the past 30 years, with about that number even in benign credit years like 2017 and likely in 2018. Indeed, as of the first half of 2018, already 14 firms filed for Chapter 11 reorganization with more than $1 billion in liabilities. Since the GFC, the U.S. high-yield and distressed debt market has been in a benign credit cycle with much below average default rates and high recovery rates, low yield-spreads and abundant liquidity, even for the seriously distressed firms. Hence, the 2017 total of distressed and defaulted public and private debt has decreased to an estimated $747 billion (face) and $414 billion (market value) – see Figure 14.2 for these totals and the methodology for our estimates. This is still a substantial market size poised to grow considerably when the next credit crisis hits!

DISTRESSED SECURITIES INVESTORS

We observed that the number of distressed debt investors grew impressively following the massive defaults of 1990–1991, and our 1992 estimate was that perhaps as much as $100 billion was under dedicated management in this sector. Most of these investors were very focused and specialized only on distressed debt and, in some cases, the equities of companies emerging from Chapter 11. In the early 1990s, there were about 60 of these alternative investment companies; see Altman (1991). These institutions, mainly hedge funds, ranged from assets of a few hundred million dollars to mega-firms with scores of portfolio managers and analysts and as much as $10 billion to $15 billion under management. We estimate that the number of so-called vulture investors has grown to perhaps 200 in the United States and 100 operating internationally in 2018. Usually, they are now called "credit" or "event-driven" strategic investors.

There is no definitive estimate of the total funds under distressed securities management, but an educated guess is about $400–$500 billion at the start of 2018. Periodically, large investors and private equity firms may become interested in these securities and possibly the control of distressed or bankrupt companies. Our estimate does include hedge funds and private equity groups with significant assets in distressed debt and a few mutual funds,[3] as part of a number of different investment strategy portfolios.

So, from a supply and demand aspect, the market has matured in 2017 whereby the huge disequilibrium of the early 1990s and again in 2002 has narrowed considerably, as demand catches up to supply. And with so many sharp penciled investment analysts, pricing has become more efficient and anomalies more difficult to detect.

Investment Strategies

In conjunction with the impressive growth in the distressed debt market's size and diversity, another source of interest in these debt and emerging equity securities has been the stories about some spectacular vulture-investor successes. Despite these unique episodes, usually involving large bets on corporate turnarounds, and the recent increase in new investors and capital, the formula for successful investing will continue to be a complex set of skills involving fundamental valuation of debt and equity assets and technical, legal, and fixed income knowledge, complemented by a patient, disciplined, and, at times, highly proactive approach to asset management. We always tell our students that in order to be successful in distressed investing, one should not consider this field solely as a fixed income credit or an equity play, but rather as a combination of both with a number of credit-related substrategies that provide a more modern, rigorous risk-return framework. In addition, the attraction of this asset class is not only in its stand-alone individual security performance but also, very importantly, in its extremely low return correlation with other asset classes.

Figure 14.3 shows the correlation between the S&P 500 stock index monthly returns versus both high-yield and defaulted debt indexes. The latter are based on our Altman-Kuehne Defaulted Bond and the Combined Defaulted Bond and Bank Loan Indexes. The periods covered are the last three stressed credit cycles: 1990/1991, 2001/2002, and, the most recent, 2008/2009 (through March). We also observe the correlations for the recent benign credit cycle since January 2010, as well as the entire sample period 1987–2018. The results are quite revealing.

Typically during stressed credit cycles (and the subsequent recovery), correlations between the stock market and risky debt markets are quite low – 12% in 1990/1991, 23% in 2001/2002, and, not shown, –16% and 43% in their subsequent recoveries. Over the entire sample period since we have been tracking defaulted debt as an asset class (1987–present), the correlation between the S&P and defaulted bond returns is only 39%, and a moderate 59% for the high-yield market and stock market returns. However, since the most recent economic and financial collapse of 2008–early 2009, the latter's correlation spiked enormously to over 70%. In the most recent cycle (January 2010–June 2018), the correlation between defaulted bonds and bank loans and the S&P 500 Stock Index was 37%, but 72% between the S&P 500 and FTSE's High-Yield Bond Index! On any given day, it is likely that if there is bad news about financial or default related uncertainties, both risky bond and stock markets will decline, and the opposite is true if the news is positive. Since the news, of late, has been "choppy," both stock market and high-yield bond prices have also been quite volatile but trending lower. It is interesting to note that the same correlation of 0.72 was found between the European high-yield bond market and the STOXX 600 stock market in the post-2009 period. Of further interest is the correlation between distressed and defaulted debt and government bond returns that is negative in all time periods.

With respect to investment strategies, Figure 14.3 illustrates three major types, and several additional substrategies, as well as target returns for distressed debt investors. The portfolios of these investors typically consist of public defaulted bonds, private loans, high yield bonds and leveraged loans that are distressed, and

		FTSE HY Index	S&P 500 Stock Index
Stressed Cycle I[a] 01/1990–12/1991 (24 obs.)	Defaulted Bond Index	68%	12%
	S&P 500 Stock Index	48%	
Stressed Cycle II[b] 01/2001–12/2002 (24 obs.)	Defaulted Bond and Bank Loan Index	76%	23%
	S&P 500 Stock Index	54%	
Stressed Cycle III 01/2008–03/2009 (15 obs.)	Defaulted Bond and Bank Loan Index	80%	73%
	S&P 500 Stock Index	73%	
Recovery Cycle 04/2009–04/2011 (25 obs.)	Defaulted Bond and Bank Loan Index	71%	65%
	S&P 500 Stock Index	67%	
Full Sample Period 01/1987–06/2018 (378 obs.)	Defaulted Bond and Bank Loan Index[c]	62%	39%
	S&P 500 Stock Index	59%	
Most Recent Cycle 01/2010–06/2018 (102 obs.)	Defaulted Bond and Bank Loan Index	52%	37%
	S&P 500 Stock Index	72%	

[a]Correlation between Defaulted Bond Index and S&P 500 during recovery cycle was –16%.
[b]Correlation between Defaulted Bond and Bank Loan Index and S&P 500 during recovery cycle was 43%, and the Defaulted Bond Index and the S&P was 49%.
[c]For the period from 1987–1995 only Defaulted Bond Index Returns were used in the calculations.

FIGURE 14.3   Total Monthly Return Correlations on Various Asset Class Indexes during Stressed and Recovery Credit Cycles
Source: E. Altman and B. Kuehne.

residual cash and its equivalent. In addition, investors may hold other instruments for hedging purposes, such as credit derivatives or short-sale positions.

Active Control Investors

The active control strategy involves mainly the "big boys and girls" of the distressed debt buy-side industry. The strategy requires a significant capital investment in specific company securities so that the distressed investor can possibly get control of the entire entity. In a sense, this is basically a private-equity strategy, except the initial vehicle for getting involved, and eventually gaining control, is usually bank loans and/or public bonds. In addition, control often requires a subsequent injection of equity capital to help ensure the successful rehabilitation and turnaround of the firm. There have been some spectacular successes of distressed firms control in such industries as movie cinemas, steel, and retailing.

A related strategy to active control investing involves the purchase of several companies in the same industrial segment, leading to a combined "roll-up strategy" and eventual running of, or sale of, the combined company. An example of this is W.L. Ross's roll-up of the U.S. steel industry in the early 2000s, which took over two years, involving such major firms as LTV Corporation, ACME Steel, Bethlehem Steel, Weirton, and Georgetown Steel, and the eventual sale of the roll-up, International Steel Group (ISG), to another entity, the Mittal Steel Group. Other examples are Philip Anschutz's efforts in the cinema industry and Eddie Lampert's purchase of Kmart and then adding Sears. The objectives of these bankruptcy acquisitions were to unlock dormant earning power, restructure the firms' liabilities, and reduce costs dramatically. The resulting success, if any, usually shows up in the aftermarket of the emerging equity, which was exchanged for the debt purchased earlier at a significant discount to par value.

One of the primary motivations of these distressed asset purchases is the concept that focused new management, with sufficient equity capital and capital market credibility, can turn around the struggling entity. In the case of the steel industry, a critical factor was the shedding of some or all of the huge legacy costs involving retired employee pension and health care benefits – almost impossible to do except under the more flexible negotiating environment of the bankruptcy courts.

As indicated in Figure 14.4, the active control strategy through distressed security acquisition requires significant ownership of at least one of the major impaired liability classes (e.g., unsecured bonds and/or loans), and usually requires

Active Control	Active/Noncontrol	Passive
Requires one-third minimum to block and one-half to control; may require partner(s)	Senior secured, senior unsecured	Invest in undervalued securities trading at distressed levels
Take control of company through debt/equity swap	Active participation in restructuring process; influence process	Substrategies: trading/buy-hold/senior or senior secured/sub debt/
Restructure or even purchase related businesses; roll-up; equity infusion; run company	Exit via debt or equity (post-Chapter 11) markets; generally do not control	"busted converts"/capital structure arbitrage/long-short, value
Exit two to three years	Holding period of one to two years	Trading oriented; sometimes restricted
Larger or mid-cap focus	Larger or mid-cap focus	Holding period of six months to one year generally, sometimes longer
Target return: 15–25% per year	Target return: 12–20%	Target return: 10–15%

FIGURE 14.4 Investment Styles and Target Returns in Distressed Debt Investing
Source: Authors' compilation.

an equity infusion by the distressed debt investor, or a strategic partner, assuming control of the company via an equity-for-debt swap or exchange after the firm emerges from Chapter 11. Then, either managing the company for an indefinite period of time or selling the, hopefully, now rehabilitated company in two to three years, with a target return on investment of at least 15–25% per year. In most cases, the focus is on larger or mid-cap companies. Indeed, a number of prominent private-equity firms entered the distressed debt market after the huge growth years of 2001 and 2002 and again after 2009. Typically, either a new fund is organized within the investment firm or a new unit is first organized to gain experience. See Hotchkiss & Mooradian (1997) for a rigorous analysis (288 firms) of the role of activist distressed investors and their success compared to restructured firms that do not have such investors. The activist investors outperform based on postrestructuring operations results. Jiang, Li, and Wang (2012), followed-up with their study on hedge funds and private equity funds' (474 cases) presence in the Chapter 11 process. Their effect on the debtor results in more likely a loss of exclusive rights of the reorganization plan, more CEO turnover and higher payoff to junior claims.

Active Noncontrol Investors

A second strategy also involves active involvement by the distressed investor but does not require controlling the entity after the reorganization period. The investor will actively participate in the restructuring process by being a member of the creditors' committee and/or by arranging for post filing, for example, DIP financing. The investor will often retain the equity for a period of perhaps six months to two years after the firm emerges and even place a member of its firm on the board of directors of the emerged company.[4] Since the capital requirement is less than the control strategy, the target return is lower, perhaps 12–20% per year. Again, the focus is typically on larger or mid-cap companies – at least in terms of sales and liabilities – --and usually involves several active/noncontrol investors.

Passive Investors

A fairly common type of strategy followed by distressed investors is to purchase a distressed company's bonds or loans, expecting that the firm will turn around and not go bankrupt. The upside potential is to purchase the debt from a heavily discounted price, say 50–60% of par, and increase either to par value, or to at least a significant increase in price, before the firm may accomplish a distressed restructuring or acquire sufficient capital to engineer a tender offer at a smaller discount from par than the original purchase price paid by the distressed investor. Such was the case of *Level 3 Communications* in 2004 when it tendered for some of its outstanding public bonds at about a 15% discount. On the news of the tender offer, the bonds immediately responded favorably, rising to the mid 80% of par value.

Related to the distressed, but not defaulted, debt strategy is the ability *to forecast* whether the firm will go bankrupt. Such techniques as our Z-Score models, KMV's EDF model, CreditSights' BondScore model, or perhaps other failure prediction methods could be used to assess default probability (these models were discussed in Chapter 10 herein). The prospect of an increase of value from a distressed state to par value in a relatively short period of time (e.g., 6 to 12 months) was achieved in many cases by both active and passive investors as well as outsized returns and securities of firms already in bankruptcy in such years as 2003 and 2009. In 2009, the rate of return on Defaulted Bonds was an incredible 96.4% and 32.8% on Defaulted Loans, a combined Bond/Bank Loan return of 56%.

Passive investors also invest in the securities of defaulted and bankrupt companies, usually waiting until the condition of the company is so drastic that a formal reorganization structure has commenced. The prospect of an increase in value after reaching some price nadir is the motivating factor. In the bond timeline, the formal bankruptcy petition is filed several months after the firm has missed an interest payment and defaulted. Indeed, in about 50% of over 2,000 cases that we have studied, the Chapter 11 petition follows the default date rather than both occurring simultaneously.

After bankruptcy, the firm's prospects become clearer, many times assisted by debtor-in-possession (DIP) financing, and the prices of debt securities start to rise based on the anticipated valuation of the reorganized firm. After a plan is submitted and confirmed by the court, based either on the affirmation of the creditors or, in some cases, by a "cram-down" by the bankruptcy judge, the firm emerges, usually one to three months after confirmation. New securities are exchanged for the old debt and, in some cases, the old equity, which may also participate in the newly emerged entity. Upon emergence, those still holding the debt securities will usually receive new debt and/or new equity. A study by Eberhart, Aggarwal, and Altman (1999) found that these new equities, a type of initial public offering (IPO), did extremely well in the post emergence 240-day period. The performance of these equities, however, was mediocre at best during the benign credit cycle years of 1993–1999, but they performed well in the post-2002 period. Indeed, in 2003 and 2004 emerging equities had amazing returns, probably averaging well over 100%.

The passive investor is basically a trader in specific distressed securities with an investment time horizon of less than one year in many cases. Hedging techniques are often utilized to protect against positions moving dramatically against the investor. These may include a short sale of the underlying equity of the company following the purchase of a distressed bond or loan that is thought to have a good chance for improvement, but could default instead. The hedge can also be achieved by purchase of credit insurance, usually via the credit default swap market, whereby the distressed investor will receive par plus interest if the firm defaults on one or more companies' securities. Of course, all of these hedging instruments are costly and may not need to be exercised. We will soon analyze one of the hedging strategies called capital structure arbitrage.

Valuation of Distressed Securities

As in most investment strategies, the key to successful results is a careful and comprehensive valuation process, whether the security be debt or equity. And, distressed investing is no exception. Only, as noted earlier, the process is more complex, since it usually considers both bond and equity issues, as well as capital structure nuances. For a detailed description of these factors, see our discussion in Chapter 5 of this volume, as well as comprehensive treatments in Altman (1991), Jeffries & Company (2003), Branch and Ray (2007), Moyer (2005), and Whitman and Diz (2009).

CAPITAL STRUCTURE ARBITRAGE[5]

Capital structure arbitrage can loosely be defined as having simultaneous positions in two or more of a particular issuer's debt or equity securities. Arbitrage implies that an investor attempts to capture price anomalies that may exist temporarily between different securities. In a broader sense, capital structure arbitrage may be used to tailor the risk-reward parameters of a particular trade to desired levels. The hedging nature of this strategy is particularly attractive in the hedge fund industry.

Consider the simplest possible two-tiered capital structure of a company that has one issue of zero coupon debt outstanding and common equity below it. Structural bond pricing models, for example, KMV, would say that the equity could be modeled as a call option on the firm's assets. Put-call parity would imply that the debt could be modeled as a risk-free zero coupon bond plus a put option with a strike price equal to the amount of debt outstanding. In this simple case, if we know the value of the firm, then the implied volatilities of both the put option and the call option should be equal. To the extent that they are different, an arbitrage profit opportunity exists.

Unfortunately, the application of pure structural models to identify capital structure opportunities in practice is complicated by the fact that capital structures are much more complex than the simple two-tiered model mentioned earlier, and certain variables are not clear. First, firm value is not readily observable. Many companies often segment their capital flexibility and liquidity against the objective of minimizing the weighted average cost of this capital. Thus a company may have a number of outstanding securities, including secured floating-rate bank debt, fixed rate senior debt, subordinated debt, convertible bonds, and preferred stock, in addition to common equity. Since there are likely to be various debt maturities, cash-pay instruments, imbedded options, and guarantees, modeling of the capital structure and applying arbitrage techniques in practice is a challenge.

The reader should recognize that binary outcomes at specific points in time, while useful for illustrative purposes, do not often reflect reality. Defaults can occur continuously over time. As such, cumulative default probability functions are necessary to model the outcomes accurately. Similarly, recovery rates, as a function of time, might be influenced by future operating cash flows, asset depreciation, potential future changes to the capital structure of the company, and supply-demand conditions in the distressed debt market (see Chapter 16 herein on discussion on recovery rates and Altman et al. [2002, 2005], among a myriad of other reasons). From today's price, future states of nature would form a continuum along a number of different dimensions, rather than just two discrete outcomes.

Capital structure arbitrage strategies based on seniority differences can similarly be arranged between secured bank debt and unsecured bonds, as well as between bonds and common stock. These trades can be structured in a bearish or bullish fashion. Hedge ratios can be determined by optimizing a variety of parameters. For example, a trader might determine a hedge ratio so that expected return is maximized (on a defined amount of capital) subject to constraints on the maximum loss he or she would incur under different future economic scenarios.

Capital structure arbitrage trades can also be structured with two *pari passu* seniority securities that have different maturities. In this case, one knows that the recovery rates in a default scenario will be equal. That the prices and yields of the two securities are different today implies something about the perceived cumulative default probability curve of the issuer and, in some cases, the securities' relative coupon rates.

For an issuer that has various *pari passu* bonds outstanding (with many different maturities), one can construct a term structure of yields—a yield curve. For a nondistressed credit, this yield curve will usually be normal, that is, monotonically increasing as a function of maturity. However, as the issuer's credit quality deteriorates into distress, the yield curve will usually invert, that is, shorter-dated bonds will have higher yields (to maturity). Longer-dated bonds, while trading with lower yields than their shorter-maturity brethren, will also have lower dollar prices and be trading closer to expected ultimate default recovery rates. One can understand this phenomenon if one realizes that the shorter-dated paper is subject to a lower aggregate default probability than the longer-dated paper. The shorter-dated paper may have some reasonable chance of getting paid off at par in the near term, while the longer-dated bonds will just have to stick it out and take whatever comes their way. While the one-year bonds come with a greater yield (higher reward), they also come with a higher dollar price and have further to fall in the event of default.

CONCLUSION

The next chapter will assess the development of the distressed debt market by first reviewing several relevant empirical studies on the risk/return attributes of distressed debt portfolios. In addition, we will document the returns on defaulted debt based on the Altman/Keuhne Indexes of monthly total return performance for bonds (1987–2018) and loans (1995–2018) and a combined index of bonds and loans. Finally, we present some recent empirical work on the performance of defaulted bonds for the period from default to emergence from bankruptcy and for the six-month period prior to default. This, we hope, will provide a fairly complete profile of distressed debt investing.

NOTES

1. See the Altman/Kuehne (annually since 1987) Reports, now published by the NYU Stern Salomon Center, for example, "The Investment Performance & Market Dynamics of Defaulted & Distressed Bonds & Bank Loans: 2017 Review and 2018 Outlook" March 1, 2018), and our latest Companion Annual Report, "Defaults and Returns in the High-Yield Bond Market: 2017 Review and 2018 Outlook," February 2, 2018.
2. The NYU Salomon Center compiles statistics on Chapter 11 bankruptcies of more than $100 million in liabilities from 1971 to the present.
3. For example, Franklin Templeton's Mutual Series Funds (Short Hills, NJ) have several funds with distressed debt as one of their several value investment strategies.
4. Li and Wang (2016) find that more than half of the firms emerging from bankruptcy in their sample have at least one director from distressed hedge funds or private equity funds sitting on their boards.
5. This discussion in part was provided by Allan Brown, an ex-portfolio manager for the distressed debt funds at Concordia Advisors.

CHAPTER 15

Investing in Distressed Firm Securities

In the prior chapter, we began our discussion of distressed debt securities by observing its anatomy from its inception; its genesis, growth, investment strategies and some initial exploration of its valuation attributes. We now proceed to examining the detailed aspects of investing in distressed and defaulted debt securities of firms, which are either close to or already in bankruptcy, and in some cases, the equity of the firms that succeed in emerging from the bankruptcy-reorganization process. In order to rigorously assess whether a class of securities can legitimately be labeled as an "asset class," we need a relatively long history of returns to investors. In this case, we now have a 30-year history, from the late 1980s through 2017, to assess the return/risk tradeoff and compare it to other asset classes.

Our concentration will primarily be on the bonds and loans of firms either in "distressed" state, namely where the option-adjusted-spread over comparable duration in U.S. government bonds is at least 1,000 basis points (bp), or the securities are traded during the bankruptcy-reorganization Chapter 11 process. And, in some cases, we assess the returns to debt investors after their securities are exchanged to the equity of the emerged firm. Of particular interest will be performance data of bonds by their seniority. In most cases, we will compare these returns with time-dependent returns of two other asset classes – the S&P 500 stock index and an index of returns on high-yield bonds. But first we examine a detailed literature review of distressed investing.

LITERATURE REVIEW

The profit-making potential of securities selling at discount prices makes distressed securities very attractive to the educated and aggressive investor. Altman (1991) published the first definitive guide to this market. Materials for this guide were derived from his first practitioner monograph-white paper (Altman 1990), which documented and analyzed the unique class of distressed bonds.

An early book by Ramaswami and Moeller (1990) undertook a study to examine the impact of the 1978 Bankruptcy Reform Act on investors who own shares or bonds in distressed corporations. Demonstrating that high average returns often accompany wise investment choices, the authors explain how to spot potential investment targets, assess investment risk, and profit from investing in firms undergoing reorganization following a bankruptcy filing. Branch and Ray (1992) added to the early works on the risky, but lucrative, opportunities to invest in the securities of troubled companies. As companies in distress go through an informal or formal workout of problems, the investment implications for the securities of firms in each of these stages are considered. An expanded book by the same authors (2007) presented their unique perspective on the valuation of distressed securities. Moyer's (2005) book also concentrated on valuation issues of distressed firms' securities, as did Jefferies (2003), Whitman and Diz (2009), and Chapter 5 herein. Finally, Stark, Siegel, and Weisfelner (2011) edited a compendium of articles on contested valuation issues in corporate bankruptcy.

Altman (2014) explored the scope and importance of the distressed debt market and its market participants, and presents new and potentially important data on recent trends in the outcomes of Chapter 11 bankruptcy reorganization filings. This work will possibly contribute to the current investigation by the "bankruptcy industry" on the possible revision of the U.S. Bankruptcy Code. Such questions as to the relative success, or not, of the Chapter 11 process, the time in bankruptcy for various outcomes of the process, the impact of prepackaged restructuring on the outcomes, and the recovery rate to various creditor classes were examined.

Rosenberg (1992 and again in 2000) described the big-profit-world of distressed investing, where active vulture investors cast their sights on distressed concerns. Vulture investing became more common during the period involving major bankruptcies in the 1980s and 1990s when declaring bankruptcy became commonplace among debt-heavy companies. Schultze and Lewis (2012) also share their insights and experience as "vulture" investors, advocating this type of investing as critical to the economic ecosystem. Altman and Eberhart (1994) showed that despite reasonably good average return performance for defaulted bonds, there are some stark differences between bonds with different seniorities. Only the two most senior tranches of defaulted bonds, that is, senior secured and senior unsecured, did well. The junior defaulted bonds, that is, subordinated and discounted bonds, however, did not do well at all. Fridson and Gao (2002) pursued this same theme by analyzing data on defaulted bonds during the period January 1980–July 1992. Specifically, they found that less risky senior issues produced higher returns than the riskier subordinated issues. However, they later updated their analysis and showed that this anomaly became less pronounced. Also, as the prior results became more publicized, the demand for the most senior debt securities increased, making excess returns less likely.

Hotchkiss and Mooradian (1997) provided a rigorous analysis of 288 bankrupt firms and separated their sample by analyzing the role of activist distressed investors and their post-emergence operational performance vs. those filings that did not have an active-control investor. The activist investors firms significantly outperformed those firms that did not have any activist investors. Jiang, Li, and Wang (2012) followed up on this theme with their study on hedge funds presence in the Chapter 11 process. Analysis of 474 Chapter 11 cases revealed that the debtor was (1) more likely to lose their exclusivity to file a reorganization plan, (2) have their CEOs removed during the process, (3) result in a higher probability of emergence, and (4) higher payoffs to junior claimants, when hedge funds were actively involved. At the same time, they argued that these outcomes do not, usually, come at the expense of other claimholders. Lim (2015) also assessed the role of activist hedge funds in distressed firms by analyzing 469 firms with out-of-court or in-court restructurings. She concluded that activist hedge fund involvement was associated with a higher probability of prepackaged Chapter 11s and these restructurings were faster than those without an active-control investor. Her conclusion was that activist hedge funds can create value by enabling efficient contracting.

Wang (2011) also studied bond returns in bankruptcy. Collecting prices on the 1st and 2nd month after Chapter 11 filing and in the month of plan confirmation, his study showed that senior bonds had a 20% annualized average gain while junior bonds had more than 20% annualized average *loss* during the restructuring period. Altman and Benhenni (2017) undertook an up-to-date empirical performance analysis of defaulted bonds from default to emergence covering the time period 1987–2016 and also after the 2005 Bankruptcy Code's extensive revision. Results showed excess return performance over the period from the default date to emergence, especially in the period after the bankruptcy code was revised. We will expand on these results at a later point.

Some recent research has concentrated on the influence that the ownership structure and the type of claims have on the outcome of bankruptcy plans. Ivashina, Iverson, and Smith (2016) analyzed a sample of 136 filings, covering 71,000 investor-claimants. The authors conclude that trading claims during the reorganization leads to (1) higher concentration of ownership, particularly of those claims which are eligible to vote on the bankruptcy plan; (2) that active investors, are the largest net buyers of these claims; (3) that while the initial concentration of investor claims is most important for the coordination of a prepackaged or prearranged plan, it is the concentration of claims during the process that has the most influence on the speed of the restructuring, the probability of a successful emergence or (unsuccessful) liquidation and for the size of the recovery rates for the different claim classes. Ivashina and Iverson (2018) focus on the role of trade claimants in the bankruptcy process. (For a primer on the overall role of trade claims in bankruptcy, see Altman's (1991) chapter in his early book.) The authors find that large trade creditors' decision to sell their receivables of bankrupt firms are predictive of lower recovery rates and oftentimes sell ahead of less-informed other creditors.

Eberhart, Aggarwal, and Altman (1999) were the first, and one of the few studies, to analyze the post-emergence performance of equity securities from 131 cases. These equities, similar in many respects to equity IPOs, did extremely well in the 240-day post-emergence period and demonstrated excess returns compared to the overall stock market using a two-factor asset-pricing model. Although the post-emergence performance was not superior in some benign credit cycle periods, for example, 1993–1999 and 2009–2010 (see Barron's 2010), emerging equities had amazingly positive returns in other years. These returns encouraged one market participant, Jefferies & Co., to start an index of post-emergence equities called "The Jefferies Re-Org Index (SM)" (2006). This firm's interest in this niche asset class cooled considerably in subsequent benign credit periods and the index was discontinued. New Generation Research (Boston, MA) also has an index of post-Chapter 11 common stock performance. Li and Zhong (2013) analyzed the trading pattern and performance of bankrupt firm equities during the Chapter 11 reorganization period and found that, in general, equities incurred significant excess losses, especially those stocks with higher levels of uncertainty and more binding short-sale constraints. The authors assessed performance of 602 filings from 1998–2006, the former year when daily data was first available from the stock market "pink sheets." They found active trading in many stocks and returns were positively related to asset values, asset volatility, the risk-free rate and the expected duration of the reorganization period and negatively related to the amount of liabilities.

Using credit spread volatility, Altman, Gonzalez-Heres, Chen, and Shin (2014) showed that, for the analysis period from July 1997 to December 2012, the mean annualized return value of a lower-volatility distressed bond portfolio outperformed that of a higher-volatility distressed portfolio by 12.22%. Additionally, the lower-volatility portfolio also significantly outperformed the High-Yield Index and Distressed Bond Index. Controlling for market volatility by normalizing each individual bond's spread volatility by the market's overall volatility, the lower-volatility portfolio outperformed the higher-volatility portfolio by 11.09%.

Gande, Altman, and Saunders (2010) found that the loan market is informationally more efficient than the bond market around events such as loan default, bond default, and bankruptcy dates. Specifically, risk-adjusted loan prices were found to fall more than risk-adjusted bond prices of the same borrower *prior to* an event date, but to fall less in the periods *surrounding* an event date. Controlling for security-specific characteristics, such as maturity, size, seniority, collateral, covenants, and for multiple measures of cumulative abnormal returns, they find that the results are highly robust, based on different empirical methodologies. Das and Kim (2014) discussed how to restructure a portfolio of distressed debt, and attribute portfolio gains to restructuring and portfolio effects. They develop a model for the pricing and optimal restructuring of distressed debt portfolios and show that even under moderate deadweight costs of bankruptcy, restructured debt return distributions are very attractive to fixed-income investors. They also discuss how investing in portfolios of distressed debt involves divergence from standard portfolio construction paradigms in two ways. First, the return distribution of distressed debt itself depends on restructuring efforts by the investor, and is therefore no longer exogenous. The gains to constructing portfolios of distressed debt come from: (1) the adjustments and restructuring of individual loans by the investor, and (2) the diversification and optimal portfolio construction across all loans. Second, the distribution of returns is highly non-Gaussian, and the Markowitz mean-variance paradigm was found to be not applicable.

MEASURING AND INVESTING IN DEFAULTED AND DISTRESSED DEBT SECURITIES

The main focus of academic and professional investment research in the risky corporate credit market has typically been on the performance of high-yield bonds, including, importantly, the default and recovery rates of those bonds that default. See our discussions in Chapters 2 and 9 on high-yield bonds and an in-depth discussion on recovery rates on defaulted debt in Chapter 16 herein. There is very little similar work on defaulted bonds and loans. This is understandable, as the high-yield bond market is significantly larger and has a 40-year track record, with coverage and data from many sources. To address this issue, we have accumulated data on the monthly performance of defaulted bonds and loans. The performance data is based on the Altman/Kuehne indexes of monthly returns for securities from default to emergence (or liquidation).[1] These indexes, originally constructed in 1987, and first used in research in Altman (1990), are used extensively by distressed debt investment managers as an important benchmark of their performance, as well as by analysts and researchers.

Our performance statistics on bonds goes back to 1987, and on defaulted loans to 1996. As of December 31, 2017, the number of *issues* in the defaulted bond index was 55, only slightly more than half the number at year-end 2016 (101), and about one-quarter of its previous highs in the early 1990s and 2001 (see Figure 15.1). Over the 31-year period of 1987–2017, the average number of bond issues ranged from a high of 231 in 1992 to a low of 36 in 1998. It should be noted that the number and amount of defaulted issues is considerably greater than those aggregated in Altman and Kuehne (2018b), since the index totals are limited to any one issuer comprising no more than 10% of the index's total market value, and we include only issues for which we find consistent monthly quotes.[2] Figure 15.2 shows the number and amount of Bank Loan facilities in our second Defaulted Debt Index.

Year-End	Number of Issues	Number of Firms	Face Value ($ Billions)	Market Value ($ Billions)	Market/ Face Ratio
1987	53	18	5.7	4.2	0.74
1988	91	34	5.2	2.7	0.52
1989	111	35	8.7	3.4	0.39
1990	173	68	18.7	5.1	0.27
1991	207	80	19.6	6.1	0.31
1992	231	90	21.7	11.1	0.51
1993	151	77	11.8	5.8	0.49
1994	93	35	6.3	3.3	0.52
1995	50	27	5.0	2.3	0.46
1996	39	28	5.3	2.4	0.45
1997	37	26	5.9	2.7	0.46
1998	36	30	5.5	1.4	0.25
1999	83	60	16.3	4.1	0.25
2000	129	72	27.8	4.3	0.15
2001	202	86	56.2	11.8	0.21
2002	166	113	61.6	10.4	0.17
2003	128	63	36.9	17.7	0.48
2004	104	54	32.1	16.9	0.53
2005	98	35	29.9	17.5	0.59
2006	85	36	31.2	23.3	0.75
2007	48	17	13.8	6.3	0.46
2008	77	28	29.6	4.5	0.15
2009	91	34	45.5	15.1	0.33
2010	53	16	26.4	8.3	0.31
2011	57	19	18.0	6.1	0.34
2012	62	21	14.6	5.2	0.36
2013	45	20	12.1	4.3	0.36
2014	49	20	11.8	3.0	0.25
2015	88	32	23.4	2.8	0.12
2016	101	35	39.7	16.2	0.41
2017	**55**	**26**	**19.1**	**5.5**	**0.29**
Average Annual	**97**	**43**	**21.5**	**7.5**	**0.38**

FIGURE 15.1 Size of the Altman-Kuehne Defaulted Bond Index, 1987–2017
Source: NYU Salomon Center.

Market-to-Face Value Ratios

Figure 15.3 shows the time series trend in the market-to-face value ratios of defaulted bonds and bank loans. As of year-end 2017, the market-to-face value ratio for defaulted bonds was 29%, and was approximately 10 percentage points lower than the historical average of 38%. The market-to-face value ratio for defaulted loans was 65%, two percentage points higher than the historical average of 62%. Market-to-face value ratios are potentially an important indication of trading opportunities, especially if one believes in "regression to the mean."

Year-End	Number of Facilities	Number of Firms	Face Value ($ Billions)	Market Value ($ Billions)	Market/Face Ratio
1995	17	14	2.9	2.0	0.69
1996	23	22	4.2	3.3	0.79
1997	18	15	3.4	2.4	0.71
1998	15	13	3.0	1.9	0.63
1999	45	23	12.9	6.8	0.53
2000	100	39	26.9	13.6	0.51
2001	141	56	44.7	23.8	0.53
2002	64	51	37.7	17.4	0.46
2003	76	43	39.0	23.9	0.61
2004	45	26	22.9	18.2	0.80
2005	41	21	18.7	16.2	0.86
2006	27	23	11.2	10.0	0.89
2007	31	13	13.0	10.4	0.79
2008	71	31	27.5	10.7	0.39
2009	67	27	57.6	34.1	0.59
2010	20	12	11.3	5.9	0.52
2011	28	15	9.1	4.7	0.52
2012	34	21	10.5	5.8	0.55
2013	22	13	6.9	4.6	0.67
2014	13	8	4.2	2.0	0.48
2015	26	12	12.5	4.9	0.39
2016	36	19	17.6	13.2	0.75
2017	**30**	**18**	**9.8**	**6.3**	**0.65**
Average Annual	**43**	**23**	**17.7**	**10.5**	**0.62**

FIGURE 15.2 Size of the Altman-Kuehne Defaulted Bank Loan Index, 1995–2017
Source: NYU Salomon Center.

The return history shows that the seniority of the bond issue is an extremely important characteristic of the performance of defaulted securities over specific periods, whether from issuance to emergence or from default to emergence. On the other hand, bank loans in default, which are primarily senior-secured, do not do well during reorganization, as our analysis below will show, although their returns are far less volatile.

The performance measure is based on average quotes from market makers and a fully invested, long-only strategy. Returns are calculated from individual bond and bank loan price movements. Returns are gross returns and do not reflect manager fees and expenses. There are, however, several distressed debt hedge fund indexes that reflect samples of investment funds' average performance and do include management fees (see Altman and Kuehne 2018b for their performances).

Figure 15.4 shows the time series returns on Defaulted Bonds compared to common stocks and high-yield bonds from 1987–2017. The arithmetic average returns on all three asset classes are quite similar, averaging between 9.25% for

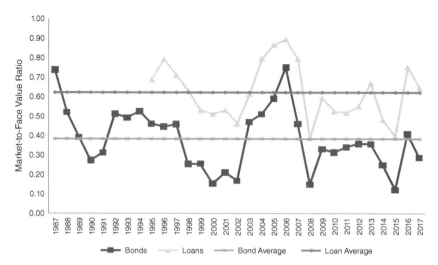

FIGURE 15.3 Altman-Kuehne Default Debt Indexes: Market-to-Face Value Ratios, Annual 1987–2017
Note: The loans' median market-to-face value is 0.61 and average market-to-face value is 0.62. Bonds' median market-to-face value is 0.36 and the average market-to-face value is 0.38.
Source: Altman and Kuehne (2018b).

HY bonds to 11.94% for common stocks, with Defaulted Bonds in the middle at 10.90%. However, the compounded average annual rate of return (+5.82%) for Defaulted Bonds is considerably lower, reflecting its time series negative performance in 13 of the 31 years, compared to just five negative performance years for the S&P 500 and six for high-yield bonds. Using the time series' compounded annual returns as a basis of comparison, the stock market outperformed high-yield bonds, which outperformed defaulted bonds, over the past 31 years.

The volatility of the defaulted bond index is considerably greater than either high-yield bonds or common stocks when measured on an annual basis, but only slightly greater than common stocks, when measured on a monthly basis. No doubt, the "calming" influence of coupon payments on high-yield bonds is a major reason why that index's volatility measure (both annual and monthly) is considerably below those of defaulted bonds and common stocks. Indeed, defaulted bonds are "no-yield" bonds since they trade "flat." Still, as we have shown earlier, this high relative volatility of defaulted bonds is somewhat mitigated by its low correlation with most other asset classes, and spectacular returns in selected years, for example, 1991, 2003, and 2009.

From a return/risk standpoint, the average annual return to standard deviation ratio favored the high-yield bond market and the stock market. Using arithmetic average returns, the ratios are 0.63 for High-Yield Bonds, 0.70 for the S&P

Year	Altman-Kuehne Defaulted Bond Index (%)	S&P 500 Stock Index (%)	Citigroup High-Yield Market Index (%)
1987	37.85	5.26	3.63
1988	26.49	16.61	13.47
1989	−22.78	31.68	2.75
1990	−17.08	−3.12	−7.04
1991	43.11	30.48	39.93
1992	15.39	7.62	17.8
1993	27.91	10.08	17.36
1994	6.66	1.32	−1.25
1995	11.26	37.56	19.71
1996	10.21	22.96	11.29
1997	−1.58	34.36	13.18
1998	−26.91	28.58	3.60
1999	11.34	20.98	1.74
2000	−33.09	−9.11	−5.68
2001	17.47	−11.87	5.44
2002	−5.98	−22.08	−1.53
2003	84.87	28.70	30.62
2004	18.93	10.88	10.79
2005	−1.78	4.92	2.08
2006	35.62	15.80	11.85
2007	−11.53	5.50	1.84
2008	−55.09	−37.00	−25.91
2009	96.42	26.46	55.19
2010	25.76	15.06	14.32
2011	−3.66	2.11	5.52
2012	2.63	15.99	15.17
2013	29.25	32.39	7.22
2014	−12.98	13.69	1.83
2015	−39.54	1.38	−5.56
2016	75.39	11.96	17.82
2017	**−6.66**	**21.83**	**7.05**
Arithmetic Average (Annual) Rate, 1987–2017	**10.90**	**11.94**	**9.25**
Standard Deviation	**34.16**	**17.00**	**14.66**
Compounded Average (Annual) Rate, 1987–2017	**5.82**	**10.52**	**8.32**
Return (Arithmetic)/Standard Deviation Ratio	**0.32**	**0.70**	**0.63**
Arithmetic Average (Monthly) Rate, 1987–2017	**0.59**	**0.93**	**0.69**
Standard Deviation	**4.85**	**4.27**	**2.40**
Compounded Average (Monthly) Rate, 1987–2017	**0.47**	**0.84**	**0.67**
Return (Arithmetic)/Standard Deviation Ratio	**0.12**	**0.22**	**0.29**

FIGURE 15.4   Altman-Kuehne Defaulted Bond Index Comparison of Returns, 1987–2017
Source: NYU Salomon Center, Standard & Poor's, and Citi, now FTSE, High-Yield Index.

500, and 0.32 for Defaulted Bonds. On a monthly return basis, the Defaulted Bond Index performs relatively better, as does the High-Yield Bond Index (which performs best).

Figure 15.5 shows that from 1996 to 2017, despite their seniority, Defaulted Bank Loans perform very poorly in comparison to the S&P 500 and High-Yield bonds. Figure 15.6 shows the combined Bond and Loan Index absolute and relative performance. This index does well, but not as well as common stocks or HY bonds.

Performance of "Distressed" Corporate Bonds

A related subasset class is those bonds that are distressed, but not defaulted, defined earlier as trading at least 1,000bp over comparable duration U.S. government bonds. Indeed, this definition is now a standard in the distressed debt "industry," with a number of financial institutions and rating agencies maintaining indexes on this group of securities. We will utilize a prominent example of these indexes, one that has been maintained since 1990 and is published by Bank of America, Merrill-Lynch (ICE BofAML US High Yield Master II).

Figure 15.7 shows the past 10-year, five-year, and three-year return performance of Distressed Bonds and the Altman-Kuehne Defaulted Bond Index, as well as the S&P 500 and Citi's HY Index (now FTSE). Over the period 2008–2017 Distressed Bond's arithmetic average return was 12.94%, compared to 11.5% for Defaulted Bonds, 9.27% for HY Bonds, and 10.30% for common stocks. Due to the sizeable negative returns of 20.18% in 2014 and 37.99% in 2015, Distressed Debt's return underperformed both high-yield bonds and stocks based on a geometric-mean calculation basis for this same 10-year period; the geometric-mean return is considerably lower at 4.70% – similar lower geometric-means (assumes reinvestment each year) can be observed for all asset classes due to the impact of negative years' returns. In the case of Distressed Bonds, there were negative returns in four of the 10 years, five for Defaulted Bonds, four for an Index composed of both Distressed and Defaulted Bonds, two for the High-Yield Index, and one for the S&P 500 index. The 10-year Sharpe ratio is highest for the S&P 500 (0.42), followed by High-Yield Bonds (0.35), Distressed Bonds (0.09), Distressed and Defaulted Bonds (0.07), and Defaulted Bonds (0.03). Similar comparative results can be observed for the past five-year and three-year periods, although both the three- and five-year returns on the S&P 500 far outrank the other asset classes.

post-Default Experience

It is also meaningful to analyze the performance of a large sample of defaulted securities during the post-Default period. We cover the period 1987–2Q 2016 and involve a sample including 1,189 issues with bond prices at default and emergence from 803 firms, and the period 1996–2Q 2016 for defaulted loans involving

Year	Altman-Kuehne Defaulted Bank Loan Index (%)	S&P 500 Stock Index (%)	Citigroup High Yield Market Index (%)
1996	19.56	22.96	11.29
1997	1.75	34.36	13.18
1998	−10.22	28.58	3.60
1999	0.65	20.98	1.74
2000	−6.59	−9.11	−5.68
2001	13.94	−11.87	5.44
2002	3.03	−22.08	−1.53
2003	27.48	28.70	30.62
2004	11.70	10.88	10.79
2005	7.19	4.92	2.08
2006	4.35	15.80	11.85
2007	2.27	5.50	1.84
2008	−43.11	−37.00	−25.91
2009	32.80	26.46	55.19
2010	9.98	15.06	14.32
2011	−2.31	2.11	5.52
2012	17.24	15.99	15.17
2013	5.78	32.39	7.22
2014	5.53	13.69	1.83
2015	−25.63	1.38	−5.56
2016	22.39	11.96	17.82
2017	**10.78**	**21.83**	**7.05**
Arithmetic Average (Annual) Rate, 1996–2017	4.93	10.57	8.09
Standard Deviation	16.68	18.01	15.09
Compounded Average (Annual) Rate, 1996–2017	3.44	8.94	7.12
Return (Arithmetic)/Standard Deviation Ratio	0.30	0.59	0.54
Arithmetic Average (Monthly) Rate, 1996–2017	0.33	0.81	0.61
Standard Deviation	3.19	4.27	2.60
Compounded Average (Monthly) Rate, 1996–2017	0.28	0.72	0.58
Return (Arithmetic)/Standard Deviation Ratio	0.10	0.19	0.23

FIGURE 15.5 Altman-NYU Salomon Center Defaulted Bank Loan Index Versus S&P 500 and Citigroup High-Yield Market Index: Comparison of Returns, 1996–2017
Source: NYU Salomon Center Index of Defaulted Bank Loans, Standard & Poor's, and Citi.

Year	Altman-Kuehne Defaulted Public Bond and Bank Loan Index (%)	S&P 500 (%)	Citigroup High Yield Market Index (%)
1996	15.62	22.96	11.29
1997	0.44	34.36	13.18
1998	−17.55	28.58	3.60
1999	4.45	20.98	1.74
2000	−15.84	−9.11	−5.68
2001	15.53	−11.87	5.44
2002	−0.53	−22.08	−1.53
2003	49.30	28.70	30.62
2004	15.40	10.88	10.79
2005	1.84	4.92	2.08
2006	23.40	15.80	11.85
2007	−3.30	5.58	1.84
2008	−47.52	−37.00	−25.91
2009	55.99	26.46	55.19
2010	17.70	15.06	14.32
2011	−3.02	2.11	5.52
2012	7.63	15.99	15.17
2013	19.37	32.39	7.22
2014	−6.45	13.69	1.83
2015	−30.94	1.38	−5.56
2016	43.90	11.96	17.82
2017	**1.50**	**21.83**	**7.05**
Arithmetic Average (Annual) Rate, 1996–2017	6.78	10.57	8.09
Standard Deviation	24.24	18.01	15.09
Compounded Average (Annual) Rate, 1996–2017	3.95	8.94	7.12
Return (Arithmetic)/Standard Deviation Ratio	0.28	0.59	0.54
Arithmetic Average (Monthly) Rate, 1996–2017	0.38	0.81	0.61
Standard Deviation	3.70	4.27	2.60
Compounded Average (Monthly) Rate, 1996–2017	0.31	0.72	0.58
Return (Arithmetic)/Standard Deviation Ratio	0.10	0.19	0.23

FIGURE 15.6 Combined Altman-NYU Salomon Center Defaulted Public Bond and Bank Loan Index Comparison of Returns, 1996–2017
Source: NYU–Salomon Center, Standard & Poor's, and Citi.

Year	BofAML Distressed Index	Altman-Kuehne Defaulted Bond Index	Citi HY Bond Index	S&P 500 Index
2008	−44.91%	−55.09%	−25.91%	−37.00%
2009	116.67%	96.42%	55.19%	26.46%
2010	25.41%	25.76%	14.32%	15.06%
2011	−6.61%	−3.66%	5.52%	2.11%
2012	24.10%	2.63%	15.17%	15.99%
2013	11.66%	29.25%	7.22%	32.39%
2014	−20.18%	−12.98%	1.83%	13.69%
2015	−37.99%	−39.54%	−5.56%	1.38%
2016	54.25%	75.39%	17.82%	11.96%
2017	7.01%	−6.66%	7.05%	21.83%
2008–2017 (10 year)[a]	12.94%	11.15%	9.27%	10.39%
Arithmetic avg. return	4.70%	2.02%	7.59%	8.50%
Geometric avg. return				
2013–2017 (5 year)	2.95%	9.09%	5.67%	16.25%
Arithmetic avg. return	−1.82%	2.17%	5.40%	15.79%
Geometric avg. return				
2015–2017 (3 year)	7.75%	9.73%	6.44%	11.73%
Arithmetic avg. return	0.78%	−0.34%	6.01%	11.41%
Geometric avg. return				
Sharpe Ratio (10 year)	0.090	0.034	0.350	0.420

[a]Return is based upon arithmetic average of annual return in each year from the BofAML Distressed Index and Altman-Kuehne Defaulted Bond Index.

FIGURE 15.7 Comparison of Various Indexes Total Annual Returns, 2008–2017
Source: BofAML, Citi, Standard & Poor's, and NYU Salomon Center.

a sample with 730 loan facilities from 398 firms. We also analyze the monthly performance 1–24 months after default, as well as the performance from before and after the major changes to the U.S. Bankruptcy Code in late 2005. Finally, we analyze the six-month prior-to-default experience.

The default-to-emergence period can last from as little as less than one month to more than 60 months. The average emergence period was about 28 months for defaulted bonds for the period 1987–2016, 34 months for 1987–2005, and much less, only 16 months, for 2006–2016. It will come as no surprise to most practitioners in the bankruptcy industry that the 2006–2016 period witnessed a considerably shorter average reorganization period due to a number of changes in the Bankruptcy Code, most notably the limit of 18 months on plan-exclusivity for the debtor and the popular use of late of prepackaged Chapter 11s. Many prepackaged or prearranged cases take less than 12 months, and some even less than six months, to conclude.

In terms of return performance, Figure 15.8 shows the annualized change in average bond prices for four intervals for our entire sample of defaulted bonds:

Default to Emergence						
Avg. Price at Default	# Issues	Avg. Price at Emergence	# Issues	Avg. # Months in Default	Avg. Annual Price Change	St. Dev.
1987–2Q16	36.92	1,727	47.18	1,189	28	11.08%	70.76%
2006–2Q16	34.85	635	47.09	452	16	25.34%	106.64%

	Avg. Price at Default	# Issues	Avg. Price at Emergence	# Issues	Avg. # Months in Default	Avg. Annual Price Change	St. Dev.
1987–2Q16	36.92	1,727	47.18	1,189	28	11.08%	70.76%
2006–2Q16	34.85	635	47.09	452	16	25.34%	106.64%

Default to Other Time Periods			
	Avg. Price 12 mo. post-Default	# Issues	Avg. Price Change 1–12 mo.	St. Dev.
1987–2Q16	40.05	954	8.49%	81.46%
2006–2Q16	43.98	259	26.20%	147.45%

	Avg. Price 24 mo. post-Default	# Issues	Avg. Price Change 12–24 mo.	St. Dev.	Avg. Price Change 1–24 mo.	St. Dev.
1987–2Q16	47.62	493	18.91%	52.84%	13.58%	45.54%
2006–2Q16	50.12	103	13.96%	109.10%	19.92%	81.26%

FIGURE 15.8   post-Default Average Price Changes on Defaulted Bonds, 1987–2Q 2016
Source: Altman/Kuehne NYU Salomon Center Defaulted Bond Pricing Database.

(1) from default to emergence, (2) from default to 12 months post-Default, (3) from default to 24 months post-Default, and (4) from the 12–24 months period. These results are shown for both the entire 30-year sample period and for the most recent 10 years. We can observe that the annualized average rate of return for all bonds from default to emergence (1987–2016) was 11.08% per year. This is decent but much lower than the 25.34% per year for the more recent 10-year period (2006-2016). If an investor purchased our entire sample of defaulted bonds at default and held them for 12 or 24 months, the average annualized returns were 8.49% for 12 months and 13.58% for 24 months on just those bonds that lasted for those intervals. For the most recent 10-year sample period, the annualized returns were much greater; 26.20% for 12 months, and 19.92% for 24 months.

Volatility of Defaulted Bond Returns

In general, returns appear to be quite impressive for investing in a broad, diversified portfolio of defaulted bonds. However, while average annual returns are high, there is a considerably high "cost" in terms of volatility. In our standard deviation calculations, we truncated the sample by eliminating the top 5% of issues that had exceptionally high returns (mainly "penny" bonds). Even with our truncation of outlier positive returns, the standard errors are quite high, ranging from about 70.8% for the default-to-emergence interval for the period 1987–2016 to 106.6% for the 2006-2016 period. We also looked at alternative ways to compute returns and volatilities. First, instead of annualizing the returns for those cases where the default-to-emergence period is less than 12 months, we assume that the proceeds at emergence are reinvested at the prevailing six-month U.S. Treasury bill rate for the remaining time up to 12 months. This lowered the above standard error of 70.8% to 56.5%. Second, rather than using the traditional standard deviation to measure volatility, we calculated the standard deviation of returns below the mean, the semideviation. The resulting volatility is 34.1% in the case when annualizing returns for those bonds with a default-to-emergence period less than 12 months and is 31.6% in the case when the proceeds at emergence in these bonds are reinvested at the six-month Treasury bill rate.

Performance by Seniority

Despite reasonably good average return performance for defaulted bonds, especially in the last 10-year period, there are some stark differences between bonds with different seniorities. Figure 15.9 shows the changes in price performance for the 1–12, 12–24, and 1–24 month periods, and from default to emergence for the entire 30-year sample period, stratified by the four seniority classes.

Senior unsecured bonds demonstrated excellent average annual returns for all four interval periods, ranging from 17.4% for default to emergence, to 28.5% for the 12–24 months period, while the senior secured sample showed only modest high-single-digit returns. The results for the subordinated class were very disappointing, with the average annual return from default to emergence actually negative, at –2.4%, and even lower, at –10.9%, during months 1–12. These results are consistent with Altman and Eberhart (1994).

Defaulted Loan Performance

The defaulted loan performance statistics for all periods indicate much poorer performance than for defaulted bonds (not shown in a figure). Indeed, all average annual returns were in the low to medium single digits for the interval default to emergence. Specifically, the annualized average rate of return for all loans (1996–2016) was 2.76% per year. This compares with 4.45% per year for the more recent 10-year period (2006–2016). The volatility of these loan returns, as measured by the standard deviation for the 492-issue sample in the 1996–2016 20-year period, was considerably lower than the 30-year period for bonds, but still quite high.

Default to Emergence						
Avg. Price at Default	# Issues	Avg. Price at Emergence	# Issues	Avg. # Months in Default	Avg. Annual Price Change	St. Dev.
Senior secured	48.81	335	55.26	236	22	6.92%	66.93%
Senior unsecured	37.07	925	53.96	631	28	17.39%	68.87%
Subordinated	28.46	402	26.71	287	32	-2.35%	75.61%
Discounted bonds	23.06	59	29.91	33	22	14.90%	74.22%

Default to Other Time Periods				
Avg. Price 12-mo. Postdefault	# Issues	Avg. Price Change, 1-12 mo.	St. Dev.
Senior secured	53.09	176	8.76%	61.27%
Senior unsecured	43.87	496	18.34%	93.42%
Subordinated	25.37	241	-10.85%	66.49%
Discounted bonds	19.53	32	-15.29%	55.65%

	Avg. Price 24 mo. Postdefault	# Issues	Avg. Price Change 12–24 mo.	St. Dev.	Avg. Price Change 1–24 mo.	St. Dev.
Senior secured	56.38	85	6.21%	39.13%	7.48%	40.67%
Senior unsecured	56.35	259	28.46%	52.03%	23.30%	39.20%
Subordinated	25.59	130	0.86%	58.28%	-5.17%	44.52%
Discounted bonds	27.22	13	39.33%	101.22%	8.64%	58.37%

FIGURE 15.9   Defaulted Corporate Bonds Return and Volatility Performance by Seniority, 1987–2Q 2016
Source: Altman/Kuehne NYU Salomon Center Defaulted Bond Pricing Database.

Analysis of Price Movements in Corporate Bonds Prior to Default

Finally, we examine the price behavior of defaulted bonds for a period of 6 months to 1 month *before* the default date. During a holding period from 1 month to 6 months prior to default, as shown in Figure 15.10, using a large sample of bonds with available prices, the percent change in the average price shows a steady monotonic decline across all periods from –38.52% at six months prior to default to –16.81% at one month prior to default. The volatility of the returns, as measured by the standard deviation, is fairly stable across all periods, ranging from a low of 33.11% to a high of 39.32%. So, despite the fact that distressed debt markets have become increasingly competitive and analyzed by numerous investors, defaulted bond prices still fall significantly from six months prior to default and even from just one month prior.

Holding Period (Months) Prior to Default (*t*)						
	t − 1	*t* − 2	*t* − 3	*t* − 4	*t* − 5	*t* − 6
Number of bonds	513	527	523	522	517	510
Average change	−16.81%	−23.03%	−24.77%	−30.85%	−36.06%	−38.52%
Median change	−27.03%	−40.13%	−42.39%	−45.19%	−52.77%	−54.24%
Standard deviation	34.62%	35.18%	39.32%	34.71%	33.11%	33.52%

FIGURE 15.10   Price Changes on Defaulted Corporate Bonds Return and Volatility
Performance Prior to Default, 2002–2Q 2016
Sources: Altman/Kuehne NYU Salomon Center Defaulted Bond Pricing Database.

CONCLUSION

In looking at our long-term empirical results, one might conclude that distressed
investing, except for the senior unsecured seniority class, is not a particularly
attractive asset class, at least not in comparison to assets like common stocks and
high-yield bonds. On closer observation, however, a number of unique aspects
makes this asset class very attractive, especially to hedge fund managers who can
move in and out of the securities depending upon the stage of the credit cycle.
Also, hedging can mitigate the devastating impact of particularly difficult periods, as well as the ability to switch from debt to equity. And, as mentioned earlier, distressed and defaulted debt is compelling to institutional investors, due to
its relatively low correlation to other asset classes. Finally, big-bet hedge funds
will usually find assets that combine hedging possibilities with the opportunity to
exploit periods of spectacular returns as incredibly appealing. Not all asset managers have exploited those challenges in the past, however, or have the staying
power to survive large negative periods.

Distressed debt as an asset class has matured greatly in the past 30 years or
so, that we have been following its development. Still, its popularity and general
interest ebbs and flows over the credit cycle. As for the near-term future of this
market, based on the record-long benign credit cycle of late and the enormous
increase in corporate debt in the past 10 years, we expect dramatically heightened
interest and importance once the next stressed credit cycle commences.

NOTES

1. The NYU Salomon Center maintains indexes on monthly returns on portfolios of
   defaulted bonds (1987–present) and defaulted loans (1996–present). Subscribers
   receive a newsletter discussing the monthly results as well as quarterly reports.
2. This index, originally developed in "Investing in Distressed Securities," E. Altman, The
   Foothill Group (1990) is maintained and published on a monthly basis by the NYU
   Salomon Center.

CHAPTER 16

Modeling and Estimating Recovery Rates

Conceptually, three main variables affect the credit risk of a financial asset: (1) the probability of default (PD); (2) the "loss given default" (LGD), which is equal to one minus the recovery rate in the event of default (RR); and (3) the exposure at default (EAD). Earlier in this book, we discussed models and procedures useful for estimating PD of a counterparty in a credit transaction. Of equal importance is estimation of LGD or RR on the defaulted bond or loan (Kalotay and Altman 2017). LGD and RR were among the most important variables underlying the efforts of the Basel Committee when they completed their recommendations in 2004 for specifying capital requirements on credit assets held by banks throughout the world (Basel Committee, 2003).

The RR, usually defined as the market price of the security just after default, is one of the two key variables analyzed by market practitioners in the pricing and other variables in the hugely important credit market. RR is also measured as of the end of the reorganization period, usually Chapter 11, where the rate or amount expected is referred to as the ultimate recovery. Standard & Poor's and Fitch both launched a special category of ratings on RR in the early 2000s, referred to as the recovery rating. This is essentially an estimate of post-Default ultimate recovery of nominal principal on large commercial and institutional loans. In doing so, both agencies recognized the market's need for bifurcated information on the two main elements of credit risk (PD and RR). Moody's, on the other hand, has argued for many years that their credit ratings incorporate both PD and RR estimates.[1]

The traditional focus of the credit risk literature had been on the estimation of the first component, PD. Much less attention was dedicated to the estimation of RR and to the relationship between PD and RR. This is mainly the consequence of two related factors. First, credit pricing models and risk management applications tend to focus on the systematic risk components of credit risk, as these are the only ones that attract risk-premium. Second, credit risk models traditionally assumed RR to be dependent on individual features (e.g., collateral or seniority) that do not respond to systematic factors, and to be independent of PD. However, the traditional focus on default analysis was partly reversed in the early 2000s, potentially because of the parallel increase in default rates and decrease of recovery rates since the 2001–2002 recession and the large magnitude of LGD in some of the largest defaults in history during the 2008–2009 crisis. More generally, evidence from many countries suggests that collateral values and recovery rates can be volatile and they tend to go down just when the number of defaults goes up in economic or industry downturns (Altman, Resti, and Sironi 2004, 2005).

This chapter first presents a detailed review of the way credit risk models, developed during the past 40 years, have treated the recovery rate and, more specifically, its relationship with the probability of default of an obligor. We discuss these approaches together with their basic assumptions, advantages, drawbacks, and empirical performance. We review recent studies that explicitly model and empirically investigate the relationship between PD and RR. We also present and assess some recent empirical evidence on recovery rates on both defaulted bonds and loans.

CREDIT RISK MODELS AND THEIR TREATMENT OF RR

Credit risk models can be generally divided into two main categories: (1) credit pricing models, and (2) portfolio credit value-at-risk (VaR) models. Credit pricing models can in turn be divided into three main approaches: (1) "first-generation" structural-form models, (2) "second generation" structural-form models, and (3) reduced-form models. We provide a broad overview of the evolvement of these models with an emphasis on how recovery rates are treated by these models.

First-Generation Structural-Form Models

The first category of credit risk models are the ones based on the original framework developed by Merton (1974) using the principles of option pricing (Black and Scholes 1973). In such a framework, the default process of a company is driven by the value of the company's assets and its liabilities, and the risk of a firm's default is therefore explicitly linked to the variability of the firm's asset value to liability value. The basic intuition behind the Merton model is relatively simple: default occurs when the value of a firm's assets (the market value of the firm) is lower than that of its liabilities. The payment to the debtholders at the maturity of the debt is therefore the smaller of two quantities: the face value of the debt or the market value of the firm's assets. Assuming that the company's debt is entirely represented by a zero-coupon bond, if the value of the firm at maturity is greater than the face value of the bond, then the bondholder gets back the face value of the bond. However, if the value of the firm is less than the face value of the bond, the shareholders get nothing and the bondholder gets back the market value of the firm. The payoff at maturity to the bondholder is therefore equivalent to the face value of the bond minus a put option on the value of the firm, with a strike price equal to the face value of the bond and a maturity equal to the maturity of the bond. Following this basic intuition, Merton derived an explicit formula for risky bonds, which can be used both to estimate the PD of a firm and to estimate the yield differential between a risky bond and a default-free bond.

In addition to Merton, first-generation structural-form models include Black and Cox (1976), Geske (1977), and Vasicek (1984). Each of these models tries to refine the original Merton framework by removing one or more of the unrealistic assumptions. Black and Cox introduce the possibility of more complex capital structures, with subordinated debt; Geske introduces interest-paying debt; Vasicek introduces the distinction between short- and long-term liabilities, which now represents a distinctive feature of the KMV model.[2]

Under these models, all the relevant credit risk elements, including default and recovery at default, are a function of the structural characteristics of the firm: asset levels, asset volatility (business risk) and leverage (financial risk). The RR is therefore an endogenous variable, as the creditors' payoff is a function of the residual value of the defaulted company's assets. More precisely, under Merton's theoretical framework, PD and RR tend to be inversely related. If, for example, the firm's value increases, then its PD tends to decrease while the expected RR at default increases (ceteris paribus). On the other side, if the firm's debt increases, its PD increases while the expected RR at default decreases. Finally, if the firm's asset volatility increases, its PD increases while the expected RR at default decreases, since the possible asset values can be quite low relative to liability levels.

Although the line of research that followed the Merton approach has proven very useful in addressing the qualitatively important aspects of pricing credit risks, it has been less successful in practical applications.[3] This lack of success has been attributed to different reasons. First, under Merton's model the firm defaults only at maturity of the debt, a scenario that is at odds with reality (see Altman and Keuhne 2018a, 2018b for empirical evidence). Second, for the model to be used in valuing default-risky debts of a firm with more than one class of debt in its capital structure (complex capital structures), the priority/seniority structures of various debts have to be specified. In addition, this framework assumes that the absolute-priority rules are actually adhered to upon default in that debts are paid off in the order of their seniority. However, empirical evidence, such as in Weiss (1990) and in Franks and Torous (1994), indicates that the absolute-priority rule is often violated. Moreover, the use of a lognormal distribution in the basic Merton model (instead of a more fat-tailed distribution) tends to overstate recovery rates in the event of default (Altman, Resti, and Sironi 2004).

Second-Generation Structural-Form Models

In response to such difficulties, an alternative approach has been developed that still adopts the original Merton framework as far as the default process is concerned but, at the same time, removes one of the unrealistic assumptions of the model; namely, that default can occur only at maturity of the debt when the firm's assets are no longer sufficient to cover debt obligations. Instead, it is assumed that default may occur anytime between the issuance and maturity of the debt and that default is triggered when the value of the firm's assets reaches an exogenous or endogenous threshold level.[4] These models include Kim, Ramaswamy, and Sundaresan (1993); Nielsen, Saà-Requejo, and Santa Clara (1993); Leland (1994); Hull and White (1995); Longstaff and Schwartz (1995); Leland and Toft (1996); and Collin-Dufresne and Goldstein (2001), and others.

Under many of these models, the RR in the event of default is exogenous and independent from the firm's asset value (exogenous barrier models). It is generally defined as a fixed ratio of the outstanding debt value and is therefore independent from the PD. For example, Longstaff and Schwartz (1995) argue that, by looking at the history of defaults and the recovery rates for various classes of debt of comparable firms, one can form a reliable estimate of the RR. In their model, they allow for a stochastic term structure of interest rates and for some correlation between defaults and interest rates. They find that this correlation between default risk and the interest rate has a significant effect on the properties of the credit spread.[5] This approach simplifies the first class of models by both exogenously specifying the cash flows to risky debt in the event of bankruptcy and simplifying the bankruptcy process. The latter occurs when the value of the firm's underlying assets hits some exogenously specified boundary. Collin-Dufresne and Goldstein (2001) consider a more general model that assumes the default threshold follows a stochastic process. Their model is able to generate mean-reverting process for the leverage ratio (i.e., the leverage ratio eventually moves back toward the mean), consistent with empirical observations.

In endogenous bankruptcy-barrier models such as those in Leland (1994), Leland and Toft (1996), and Fan and Sundaresan (2000) assume that a firm defaults when its asset value reaches an endogenous default boundary. The optimal default boundary is determined by shareholders to maximize the value of equity. Shareholders default when the value of continuation of their call option on assets is below the required new equity payment. Under these assumptions the value of coupon-paying bond and the optimal default boundary can be calculated analytically. The recovery rates are the ratio of the default boundary subtracting expected deadweight costs of default, scaled by the total face value of debt.[6]

Despite these improvements with respect to Merton's original framework, second generation structural-form models still suffer from three main drawbacks, which represent the main reasons behind their relatively poor empirical performance.[7] First, they still require estimates for the parameters of the firm's asset process, which is nonobservable. Indeed, unlike the stock price in the Black and Scholes formula for valuing equity options, the current underlying value of a firm is not easily observable. Second, structural-form models cannot incorporate credit-rating changes. This is at odds with empirical evidence that almost all corporate bonds undergo credit downgrades before they actually default. Finally, most structural-form models assume that the value of the firm is continuous in time. As a result, the time of default can be predicted just before it happens and hence, as argued by Duffie and Lando (2000), there are no "sudden surprises." In other words, without recurring to a "jump process," the PD of a firm is almost known with certainty immediately before defaults.

Reduced-Form Models

The attempt to overcome the previously mentioned shortcomings of structural-form models gave rise to reduced-form models. These include Litterman and Iben (1991); Madan and Unal (1995); Jarrow and Turnbull (1995); Jarrow, Lando, and Turnbull (1997); Lando (1998); Duffie (1998); and Duffie and Singleton (1999). Unlike structural-form models, reduced-form models do not condition default on the value of the firm, and parameters related to the firm's value need not be estimated to implement them. In addition to that, reduced-form models introduce separate explicit assumptions on the dynamic of both PD and RR. These variables are modeled independently from the structural features of the firm, its asset volatility and leverage. Generally speaking, reduced-form models assume an exogenous RR that is independent from the PD and specify a stochastic process for default intensity (i.e., the arrival rate of default). At each instant, there is some probability that a firm defaults on its obligations. Both this probability and the RR in the event of default may vary stochastically through time. Those stochastic processes determine the price of credit risk. Although these processes are not formally linked to the firm's asset value, there is presumably some underlying relation. Thus, Duffie and Singleton (1999) describe these alternative approaches as reduced-form models.

Reduced-form models fundamentally differ from typical structural-form models in the degree of predictability of default as they can accommodate defaults that are sudden surprises. A typical reduced-form model assumes that an exogenous random variable drives default and that the probability of default over any time interval is nonzero. Default occurs when the random variable undergoes a discrete shift in its level. The time at which the discrete shift will occur cannot be foretold on the basis of information available today. The simplest version of these models defines default as the first arrival time of a Poisson process with some constant arrival rate, referred to the default intensity (Duffie and Singleton 1999).

Reduced-form models somewhat differ by the manner in which the RR is parameterized. For example, Jarrow and Turnbull (1995) assumed that, at default, a bond would have a market value equal to an exogenously specified fraction of an otherwise equivalent default-free bond. Duffie and Singleton (1999) followed with a model that, when market value at default (i.e., RR) is exogenously specified, allows for closed-form solutions for the term-structure of credit spreads. Their model also allows for a random RR that depends on the predefault value of the bond. While this model assumes an exogenous process for the expected loss at default, meaning that the RR does not depend on the value of the defaultable claim, it allows for correlation between the default hazard-rate process and RR. Indeed, in this model, the behavior of both PD and RR may be allowed to depend on firm-specific, or macroeconomic variables, and therefore to be correlated with each other.

Other models assume that bonds of the same issuer, seniority, and face value have the same RR at default, regardless of the remaining maturity. For example, Duffie (1998) assumes that, at default, the holder of a bond of given face value receives a fixed payment, irrespective of the coupon level or maturity, and the same fraction of face value as any other bond of the same seniority. This allows him to use recovery parameters based on statistics provided by rating agencies such as Moody's. Jarrow, Lando, and Turnbull (1997) also allow for different debt seniorities to translate into different RRs for a given firm. Both Lando (1998) and Jarrow, Lando, and Turnbull (1997) use transition matrices (historical probabilities of credit rating changes) to price defaultable bonds.

Empirical evidence concerning reduced-form models is rather limited. Using the Duffie and Singleton (1999) framework, Duffee (1999) finds that these models have difficulty in explaining the observed term structure of credit spreads across firms of different credit risk qualities. In particular, such models have difficulty generating both relatively flat yield spreads when firms have low credit risk and steeper yield spreads when firms have higher credit risk.

An attempt to combine the advantages of structural-form models – a clear economic mechanism behind the default process – and the ones of reduced-form models – unpredictability of default – can be found in Zhou (2001). This is done by modeling the evolution of firm value as a jump-diffusion process (i.e., a stochastic process that involves both jumps and diffusion). This model links RRs to the firm value at default so that the variation in RRs is endogenously generated and the correlation between RRs and credit ratings, reported first in Gupton, Gates, and Carty (2000) and also Altman, Brady, Resti, and Sironi (2005), is justified.

Credit Value-at-Risk Models

During the second half of 1990s, banks and consultants started developing credit risk models aimed at measuring the potential loss, with a predetermined confidence level, that a portfolio of credit exposures could suffer within a specified time horizon (generally one year). These were mostly motivated by the growing importance of credit risk management especially since the now completed Basel II was anticipated to be proposed by the BIS. These value-at-risk (VaR) models include J.P. Morgan's CreditMetrics (Gupton, Finger, and Bhatia 1997), Credit Suisse Financial Products' CreditRisk+ (1997), McKinsey's CreditPortfolioView (Wilson 1998), KMV's CreditPortfolioManager, and Kamakura's Risk Manager.

Credit VaR models can be gathered in two main categories: (1) default mode (DM) models and (2) mark-to-market (MTM) models. In the former, credit risk is identified with default risk and a binomial approach is adopted. Therefore, only two possible events are taken into account: default and survival. The latter includes all possible changes of the borrower creditworthiness, technically called credit migrations. In DM models, credit losses only arise when a default occurs. On the other hand, MTM models are multinomial, as losses arise also when negative credit migrations occur. The two approaches basically differ in the amount of data necessary to feed them: limited in the case of DM models, much more plentiful in the case of MTM ones.

The main output of a credit VaR model is the probability density function (PDF) of the future losses on a credit portfolio. From the analysis of such a loss distribution, a financial institution can estimate both the expected loss and the unexpected loss on its credit portfolio. The expected loss equals the (unconditional) mean of the loss distribution; it represents the amount the investor can expect to lose within a specific period of time (usually one year). On the other side, the unexpected loss represents the deviation from expected loss and measures the actual portfolio risk. This can in turn be measured as the standard deviation of the loss distribution. Such a measure is relevant only in the case of a normal distribution and is therefore hardly useful for credit risk measurement: indeed, the distribution of credit losses is usually highly asymmetrical and fat-tailed. This implies that the probability of large losses is higher than the one associated with a normal distribution. Financial institutions typically apply credit risk models to evaluate the economic capital necessary to face the risk associated with their credit portfolios. In such a framework, provisions for credit losses should cover expected losses,[8] while economic capital is seen as a cushion for unexpected losses. Indeed, Basel II in its final iteration (June 2004) separated these two types of losses.

Credit VaR models can largely be seen as reduced-form models, where the RR is typically taken as an exogenous constant parameter or a stochastic variable independent from PD. Some of these models, such as CreditMetrics, treat the RR in the event of default as a stochastic variable – generally modeled through a beta distribution – independent from the PD. Others, such as CreditRisk+, treat it as a constant parameter that must be specified as an input for each single credit exposure. While a comprehensive analysis of these models goes beyond the aim of this review,[9] it is important to highlight that all credit VaR models treat RR and PD as two independent variables.

STATISTICAL MODELS AND EMPIRICAL EVIDENCE

This section focuses on measurements and empirical evidence of recovery rates. We start by describing a series of studies that document the strong negative relation between the probability of default and the recovery rates, and then briefly discuss implication of the procyclicality of the capital requirement. Finally, we provide the most up-to-date statistical models and empirical evidence on both bond and loan recovery rates.

The PD-RR Relationship

During the last two decades, new approaches explicitly modeling and empirically investigating the relationship between PD and RR have been developed. These typical models include Frye (2000a and 2000b); Jarrow (2001); Jokivuolle and Peura (2003); Carey and Gordy (2003); Bakshi, Madan, and Zhang (2001); Altman, Brady, Resti, and Sironi (2005); and others.

The model proposed by Frye draws from the conditional approach suggested by Finger (1999) and Gordy (2000). In these models, defaults are driven by a single systematic factor – the state of the economy – rather than by a multitude of correlation parameters. These models are based on the assumption that the same economic conditions that cause defaults to rise might cause RRs to decline, that is, the distribution of recovery is different in high-default periods from low-default ones. In Frye's model, both PD and RR depend on the state of the systematic factor. The correlation between these two variables therefore derives from their mutual dependence on the systematic factor. The intuition behind Frye's theoretical model is relatively simple: if a borrower defaults on a loan, a bank's recovery may depend on the value of the loan collateral. The value of the collateral, like the value of other assets, depends on economic conditions. If the economy experiences a recession, RRs may decrease just as default rates tend to increase. This gives rise to a negative correlation between default rates and RRs.

While the model originally developed by Frye (2000a) implied recovery to be taken from an equation that determines collateral, Frye (2000b) modeled recovery directly. This allowed him to empirically test his model using data on defaults and recoveries from Moody's Default Risk Service database for the 1982–1997 period. Results show a strong negative correlation between default rates and RRs for corporate bonds. This evidence is consistent with the most recent U.S. bond market data, indicating a simultaneous increase in default rates and LGDs for the 2008–2009 period (Figure 16.1). Frye's (2000b and 2000c) empirical analysis allows him to conclude that in a severe economic downturn, bond recoveries might decline 20–25 percentage points from their normal-year average. Loan recoveries may decline by a similar amount, but from a higher level.

Year	Par Value Outstanding[a] ($)	Par Value of Default ($)	Default Rate (%)	Weighted Price after Default ($)	Weighted Coupon (%)	Default Loss (%)[b]
2017	1,622,365	29,301	1.81	56.7	8.38	0.86
2016	1,656,176	68,066	4.11	30.0	8.31	3.05
2015	1,595,839	45,122	2.83	33.9	9.28	2.00
2014	1,496,814	31,589	2.11	63.2	10.44	0.89
2013	1,392,212	14,539	1.04	53.6	10.04	0.54
2012	1,212,362	19,647	1.62	57.8	8.97	0.76
2011	1,354,649	17,963	1.33	60.3	9.10	0.59
2010	1,221,569	13,809	1.13	46.6	10.59	0.66
2009	1,152,952	123,878	10.74	36.1	8.16	7.30
2008	1,091,000	50,763	4.65	42.5	8.23	2.83
2007	1,075,400	5,473	0.51	66.6	9.64	0.19
2006	993,600	7,559	0.76	65.3	9.33	0.30
2005	1,073,000	36,209	3.37	61.1	8.61	1.46
2004	933,100	11,657	1.25	57.7	10.30	0.59
2003	825,000	38,451	4.66	45.5	9.55	2.76
2002	757,000	96,858	12.79	25.3	9.37	10.15
2001	649,000	63,609	9.80	25.5	9.18	7.76
2000	597,200	30,295	5.07	26.4	8.54	3.95
1999	567,400	23,532	4.15	27.9	10.55	3.21
1998	465,500	7,464	1.60	35.9	9.46	1.10
1997	335,400	4,200	1.25	54.2	11.87	0.65
1996	271,000	3,336	1.23	51.9	8.92	0.65
1995	240,000	4,551	1.90	40.6	11.83	1.24
1994	235,000	3,418	1.45	39.4	10.25	0.96
1993	206,907	2,287	1.11	56.6	12.98	0.56
1992	163,000	5,545	3.40	50.1	12.32	1.91
1991	183,600	18,862	10.27	36.0	11.59	7.16
1990	181,000	18,354	10.14	23.4	12.94	8.42
1989	189,258	8,110	4.29	38.3	13.40	2.93
1988	148,187	3,944	2.66	43.6	11.91	1.66
1987	129,557	7,486	5.78	75.9	12.07	1.74
1986	90,243	3,156	3.50	34.5	10.61	2.48
1985	58,088	992	1.71	45.9	13.69	1.04
1984	40,939	344	0.84	48.6	12.23	0.48
1983	27,492	301	1.09	55.7	10.11	0.54
1982	18,109	577	3.19	38.6	9.61	2.11
1981	17,115	27	0.16	72.0	15.75	0.15
1980	14,935	224	1.50	21.1	8.43	1.25
1979	10,356	20	0.19	31.0	10.63	0.14
1978	8,946	119	1.33	60.0	8.38	0.59
Arithmetic Average 1978–2017			**3.31**	**45.88**	**10.39**	**2.19**
Weighted Average 1978–2017			**3.38**	**38.55**		**2.21**

[a]Excludes defaulted issues.

[b]Default loss rate adjusted for fallen angels is 9.3% in 2002, 1.82% in 2003, 0.59% in 2004, 1.56% in 2005, 0.04% in 2006, 0.20% in 2007, 3.42% in 2008, 7.38% in 2009, 0.66% in 2010, 0.58% in 2011, 0.86% in 2012, 0.54% in 2013, 0.91% in 2014, 1.99% in 2015, 3.11% in 2016 and 0.87% in 2017.

FIGURE 16.1 Annual Default Rates and Losses 1978–2017 (Dollars in Millions)
Source: NYU Salomon Center.

Jarrow (2001)'s methodology is similar to Frye in that RRs and PDs are correlated and depend on the state of the economy. However, Jarrow's methodology explicitly incorporates equity prices in the estimation procedure, allowing the separate identification of RRs and PDs and the use of an expanded and relevant dataset. In addition, the methodology explicitly incorporates a liquidity premium in the estimation procedure, which is considered essential in light of the high variability in the yield spreads between risky debt and U.S. Treasury securities.

Using four different datasets ranging from 1970 to 1999, Carey and Gordy (2003) analyze LGD measures and their correlation with default rates. Their results contrast with the findings of Frye (2000b): estimates of simple default rate-LGD correlation are close to zero. They find, however, that limiting the sample period to 1988–1998, estimated correlations are more in line with Frye's results (0.45 for senior debt and 0.8 for subordinated debt). The authors postulate that during this short period the correlation rises not because LGDs are low during the low-default years 1993–1996, but rather because LGDs are relatively high during the high-default years 1990 and 1991. They conclude that the basic intuition behind Frye's model may not adequately characterize the relationship between default rates and LGDs. Indeed, a weak or asymmetric relationship suggests that default rates and LGDs may be influenced by different components of the economic cycle.

Using defaulted bonds' data for the sample period 1982–2002, which includes the relatively high-default years of 2000–2002, Altman, Resti, and Sironi (2005) find empirical results consistent with Frye's intuition. However, they find that the single systematic risk factor – the performance of the economy – is less predictive than Frye's model would suggest. Their econometric models assign a key role to the supply of defaulted bonds (the default rate) and show that this variable, together with variables that proxy the size of the high-yield bond market and the economic cycle, explain a substantial proportion (close to 90%) of the variance in bond recovery rates aggregated across all seniority and collateral levels. They conclude that a simple microeconomic mechanism based on supply and demand drives aggregate recovery rates more than a macroeconomic model based on the common dependence of default and recovery on the state of the cycle. In high default years, the supply of defaulted securities tends to exceed demand,[10] thereby driving secondary market prices down. This in turn negatively affects RR estimates, as these are generally measured using bond prices shortly after default. In fact, Figure 9.10 presents the simple linear and nonlinear regression relationships between default rates and recovery rates. The graph shows that 58–62% of the variations in recovery rates can be explained simply by just one explanatory variable – the default rate.

Bakshi, Madan, and Zhang (2001) enhance the reduced-form models to allow for a flexible correlation between the risk-free rate, the default probability and the recovery rate. Based on some evidence published by rating agencies, they force recovery rates to be negatively associated with default probability. They find some strong support for this hypothesis through the analysis of a sample of BBB-rated corporate bonds: more precisely, their empirical results show that, on average, a 4% worsening in the (risk-neutral) hazard rate (i.e., arrival rate of default at a given time) is associated with a 1% decline in (risk-neutral) recovery rates.

Gupton and Stein (2002, 2005) analyze the recovery rate on over 3,000 corporate bond, loan and preferred stock defaults, from more than 1,400 companies, from 1981 to 2004, in order to specify and test Moody's LossCalc model for predicting LGD. Their model estimates LGD at two points in time – immediately after default and in one year after default – adding a holding period dimension to the analysis. The authors find that their multifactor model, incorporating micro variables (e.g., debt type, seniority), industry, and some macroeconomics factors (e.g., default rates, changes in leading indicators) outperforms traditional historic recovery average methods in predicting LGD.

Jokivuolle and Peura (2003) present a rather different approach for bank loans in which collateral value is correlated with the PD. They use the option-pricing framework for modeling risky debt: the borrowing firm's total asset value triggers the event of default. However, the firm's asset value does not determine the RR. Rather, the collateral value is in turn assumed to be the only stochastic element determining recovery. Because of this assumption, the model can be implemented using an exogenous PD, so that the firm's asset value parameters need not be estimated. In this respect, the model combines features of both structural-form and reduced-form models. Assuming a positive correlation between a firm's asset value and collateral value, the authors obtain a similar result as Frye (2000b), that realized default rates and recovery rates have an inverse relationship.

Pykhtin (2003) and Dullmann and Trapp (2004) extend the single factor model proposed by Gordy (2000), by assuming that the recovery rate follows a lognormal distribution (Pykhtin, 2003) or a logit-normal (Dullmann and Trapp, 2004) pattern. The latter study empirically compares the results obtained using the three alternative models, using time series of default rates and recovery rates from 1982 to 1999 from *Standard and Poor's Credit Pro* database. They find that estimates of RR based on market prices at default are significantly higher than the ones obtained using recovery rates at emergence from restructuring (usually bankruptcy). The findings are in line with previous ones: systematic risk is an important factor that influences recovery rates. The authors show that ignoring this risk component may lead to downward biased estimates of economic capital.

Altman et al. (2001, 2005) also highlight the implications of their results for credit risk modelling and for the issue of procyclicality of capital requirements.[11] In order to assess the impact of a negative correlation between default rates and recovery rates on credit risk models, they run Monte Carlo simulations on a sample portfolio of bank loans and compare the key risk measures (expected and unexpected losses). They show that both the expected loss and the unexpected loss are vastly understated if one assumes that PDs and RRs are uncorrelated. Therefore, credit models that do not carefully factor in the negative correlation between PDs and RRs might lead to insufficient bank reserves and cause unnecessary shocks to financial markets. They show that procyclicality effect tends to be exacerbated by the correlation between DRs and RRs: low recovery rates when defaults are high would amplify cyclical effects. This would especially be true under the so-called advanced internal rating based approach, where banks are free to estimate their own recovery rates and might tend to revise them downwards when defaults increase and ratings worsen.

Recent empirical studies document the strong relation between recovery rates and industry specific conditions, which affect industrywide default rates. For example, Acharya, Bharath, and Srinivasan (2007) find that industry conditions at the time of default are found to be robust and important determinants of recovery rates. This result is consistent with those of Altman, Resti, and Sironi (2001, 2005) in that there is little effect of macroeconomic conditions over and above the industry conditions. They suggest that the linkage between bond market aggregate variables and recoveries arising due to supply-side effects in segmented bond markets may be a manifestation of Shleifer and Vishny's (1992) industry equilibrium effect: macroeconomic variables and bond market conditions appear to be picking up the effect of omitted industry conditions. James and Kizilaslan (2014) further document that firm's exposure to industry dowturns have strong explanatory power for both the recovery rates in bankrutpcy and the likelihood of the firm experiecing financial distress when its peers are also in distress. The importance of the "industry" factor in determining LGD is also highlighted in a survey by Schuermann (2004).

Recent Empirical Evidence

Bond and Loan Recovery Rates Figure 16.1 presents annual par value outstanding of straight bonds, default rates, RR (based on market prices just after default), and default losses (the product of default rates and LGD) across all seniority and industry classifications for more than 3,200 bond defaults during 1971–2017 (Altman and Kuehne 2018a, 2018b). The arithmetic average annual recovery rates on defaulted bonds was 45.88% (38.55% on a weighted average basis). The figure shows strong cyclicality of recovery rates. During recession periods, the recovery rates were mostly in the 20s and 30s percentage range. Earlier studies (Keisman 2004) also find that during the most "extreme stress" default years of 1998 to 2002, the recovery rates on all seniorities declined compared to their longer 1988–2002 sample period. Figure 16.2 presents the frequency of bond recovery. Clearly, the vast majority falls in the 0–50% range.

The frequency distribution of default recovery rates for corporate loans are presented in Figure 16.3. The recovery rates distribution are based on the price one month after default of a smaller, but still relevant, sample of 766 loan

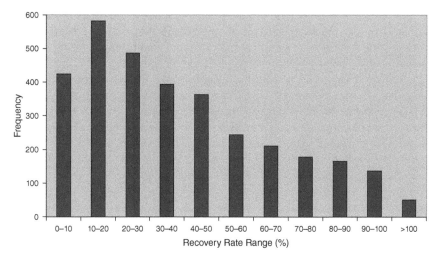

FIGURE 16.2   Corporate Bond Default Recovery Rate Frequency Based on Number of Issues 1971–2017[a]
[a]Number of Observations = 3,236.
Source: NYU Salomon Center Default Database.

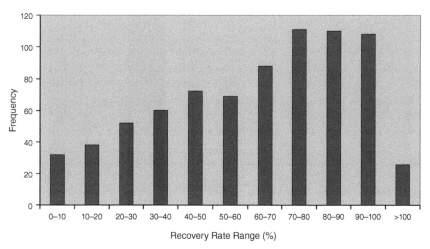

FIGURE 16.3   Loan Default Recovery Rate Frequency Based on Number of Issues 1996–2017[a]
[a]Number of Observations = 766.
Source: NYU Salomon Center.

defaults during 1996–2017 (Altman and Kuehne 2018a, 2018b). Not surprisingly, the recovery rates on loans are higher, on average, than those for bonds. The distribution is skewed more toward the higher end, with the bulk in the 60–100% range; the opposite is the case with bonds. The higher average recovery rates on defaulted loans compared to bonds reflects its senior, and sometimes secured, status. Indeed, earlier studies such as Emery, Cantor, Ou, Soloman, and Stumpp (2004) find that average loss rates on bonds are greater than similarly rated loans. The standard deviation of loan recoveries was about 33%, somewhat higher than the same for bonds (28%). Relative to the means, however, the standard deviation divided by mean recoveries for loans was 0.52, compared to 0.74 for bonds, indicative of the higher variability of bond default recoveries.[12]

Recovery by Seniority, Ratings, and Industry Figure 16.4 shows the average recovery rate by bond seniority (senior secured, senior unsecured, senior subordinated, subordinated, and discount and zero coupon bond) for 1978–2017. Senior unsecured bonds account for the largest fraction of the bonds in the sample. There is a clear monotonic relation between recovery rates and bond seniority with the mean recovery rates of 59% for senior secured bonds and 19% for discount bonds.

Varma, Cantor, and Hamilton (2003) and Altman and Fanjul (2004) are among earlier studies that report bond recoveries by the original rating (fallen angels vs. original rating noninvestment bonds) of different seniorities. Figure 16.5 provides updated statistics of recovery rates by seniority and original rating of corporate bond defaults from 1971 to 2017. We observe that senior-secured bonds, that were originally rated investment grade, recovered a median rate of 50.5% versus 45% (54% vs. 49% for mean recoveries) for the same seniority bonds that were noninvestment grade when issued. The median differential was even greater (39.5% vs. 30.2%) for senior unsecured bonds. Note that for senior-subordinated bonds, however, the rating at issuance is of no consequence, although the sample sizes for

Bond Seniority	Number of Issues	Median %	Mean %	Standard Deviations[*] %
Senior secured	596	58.51	59.05	17.44
Senior unsecured	1,763	37.13	45.64	13.61
Senior subordinated	513	32.13	35.59	16.27
subordinated	267	30.58	31.96	17.95
Discount and zero coupon	165	25.80	19.41	23.48
Total sample bonds	3,304	39.09	42.98	13.67

[*] Standard deviations are calculated based on the annual averages.

FIGURE 16.4 Weighted Average Recovery Rates on Defaulted Debt by Seniority per $100 Face Amount, 1978–2017
Source: NYU Salomon Center and Altman and Kuehne (2018b).

Seniority	Original Rating	No. of Issues	Mean Price ($)	Weighted Price ($)	Median Price ($)	Std. Dev.	Mini-mum Price ($)	Maxi-mum Price ($)
Senior secured	Investment Grade	150	54.01	58.83	50.50	27.72	3.00	111.00
	Non-Investment Grade	447	49.30	53.81	45.00	31.21	0.10	119.69
	All	670	50.07	53.83	46.88	30.06	0.05	119.69
Senior unsecured	Investment Grade	532	42.47	40.16	39.50	25.18	2.00	100.50
	Non-Investment Grade	1024	37.05	33.66	30.17	26.07	0.02	116.63
	All	1673	38.83	36.00	32.75	25.84	0.02	116.63
Senior sub-ordinated	Investment Grade	16	37.10	34.29	27.31	27.48	1.00	83.75
	Non-Investment Grade	472	33.84	31.23	28.08	25.49	0.13	107.75
	All	516	33.61	31.08	28.00	25.40	0.13	107.75
Sub-ordinated	Investment Grade	27	21.95	7.03	4.00	28.91	0.50	103.00
	Non-Investment Grade	206	32.41	29.32	28.42	22.60	1.00	112.00
	All	246	31.08	21.78	27.50	23.33	0.50	112.00
Discount	Investment Grade	1	13.63	13.63			13.63	13.63
	Non-Investment Grade	104	28.03	26.32	17.65	25.65	0.42	102.50
	All	131	27.28	26.33	18.00	23.95	0.42	102.50

FIGURE 16.5 Recovery Rates by Seniority and Original Bond Rating, 1971–2017
Source: NYU Salomon Center and Altman and Kuehne (2018b).

investment grade, low seniority bonds was very small. It seems that the quality of assets and the structure of the defaulting company's balance sheets favors higher recoveries for higher quality original issue bonds.

Figure 16.6 reports recoveries broken down by 13 different industry sectors. There is a high variation across industrial sectors. The overall weighted-average recovery rates are highest for utilities (63.6%), miscellaneous industries (50.0%), and financial services (48.2%), and lowest for communications and media (30.7%), real estate/construction (29.8%) and auto/motor carrier (22.6%). The rest vary between 35–45%. Altman and Kishore (1996) report similar evidence using earlier data. It is of interest to observe that the historic average recovery rate on energy defaults was considerably higher based upon data through 2014 (51.5%), than through 2017 (36.7%).

Other Economic Factors Recent empirical studies document other important economic determinants beyond macroeconomic factors, industry conditions, and debt specific characteristics. For example, Jankowitsch, Nagler, and Subrahmanyam

Industry	No. of Issues	Mean Price ($)	Weighted Price ($)	Median Price ($)	Std. Dev.	Minimum Price ($)	Maximum Price ($)
Auto/Motor carrier	123	30.03	26.65	25.75	22.53	2.71	93.60
Conglomerates	6	38.13	40.93	40.69	25.10	5.00	71.00
Energy	454	39.25	36.69	32.00	27.21	0.05	116.63
Financial services	244	42.86	48.20	38.25	29.02	0.02	103.00
Leisure & entertainment	179	40.93	40.34	34.50	28.67	3.00	112.00
General manufacturing	553	34.90	34.28	30.00	24.78	0.10	108.75
Healthcare	71	34.44	34.71	29.75	26.49	0.13	99.00
Misc. industries	178	45.93	49.96	41.00	28.16	0.05	105.16
Real estate & construction	132	38.12	29.84	31.93	25.77	1.21	100.50
Retailing	327	40.30	39.51	40.00	23.84	0.23	101.38
Communications & media	590	34.62	30.66	25.40	26.78	0.13	104.06
Transport (nonauto)	233	40.82	44.67	33.00	26.00	2.00	106.13
Utilities	146	64.24	63.60	70.81	28.39	1.30	119.69

FIGURE 16.6 Recovery Rates by Industry, 1971–2017
Source: NYU Salomon Center and Altman and Kuehne (2018).

(2014) provide a microstructure analysis of defaulted bonds from 2002 to 2010 using a multitude of data sources. Specifically, they examine the trading volume and prices of the defaulted bonds from 90 days before to 90 days after the observed default event. They find that trading volume is relatively high within a short window of the default event, suggesting temporary sell-side pressure as prices drop. They find that in addition to the default event types (e.g., distressed exchanges vs. bankruptcy), industry, ratings, and seniority, bond specific liquidity measures have strong explanatory power – illiquid bonds with high transaction costs recover less following default. They further document that bond covenants affect RR. One potential explanation is that covenants allow lenders to closely monitor the borrower and call default before it is too late.

Carey and Gordy (2016) find that the composition of debt has a strong effect on debt recovery. They build an extended Black and Cox (1976) model that allows private debt, referred to as loan lenders, to set the endogenous bankruptcy threshold value of assets. Banks have an incentive to foreclose (and the ability to do so due to covenants set in the contracts) only when the borrower's asset value drops to the neighborhood of the loan's value. The implication of their model is that the lower the bank loan share in total debt, the lower the asset value of the borrower at bankruptcy and the lower the recovery to debt as a whole. Their empirical evidence shows that the proportion of bank loan in a firm's debt structure has strong explanatory power for firm-level recovery rates.

Ultimate Recovery and Bond Returns during Restructuring Recovery rates that we have reported and discussed so far are measured at the time of default, not the *ultimate recovery* at the time of the completion of the restructuring period (usually at emergence from bankruptcy). A number of empirical studies pay attention to modelling and explaining ultimate recoveries. Standard and Poors (Van de Castle, Keisman, and Yang 2000 and Keisman 2004) document that an important determinant of ultimate recovery rates is the amount that a given seniority has junior liabilities below its level; the greater the proportion of junior securities, the higher the recovery rate on the senior tranches. The theory is that the greater the equity cushion, the more likely there will be assets of value, which, under absolute priority, go first in liquidation or reorganization to the more senior tranches. Altman and Kalotay (2014) propose a semi-parametric approach based on mixtures of distributions to forecast ultimate recoveries. Their study sample includes 4,720 defaulted loans and bonds from 1987 to 2011, obtained from Moody's Ultimate Recovery Database. Their results suggest that the approach based on a mixture of Gaussian (normal) distributions wherein the mixing probabilities are explicitly conditioned on borrower characteristics, debt features, and the economic conditions prevailing at the time of default outperforms both paramedic regression-based approaches and nonparametric approaches used in prior empirical research at forecasting RRs.

Figure 16.7 presents the summary statistics of ultimate recoveries, both the nominal and discounted (by the loan's predefault interest rate) values at the end of the restructuring period, for well over 2,239 defaulted loans and 3,342 defaulted bonds over the period 1987–2018, provided by Moody's LGD. Several items are of interest. First, the recovery on senior bank debt, which are mainly secured, was quite high at 87% and 81% for nominal and discounted values respectively. Senior secured and senior unsecured notes had lower recoveries and the more junior notes (subordinated bonds) had, not surprisingly, the lowest recoveries. Note, the differential between the nominal and discounted recovery rates diminish somewhat at the lower seniority levels.

Comparing ultimate recovery based on either the prices of bonds or loans at the end of reorganization period (usually Chapter 11) or the value of the package of cash or new securities issued upon emergence from restructuring and RR based on trading prices immediately after default/bankruptcy allows one to estimate the return performance of defaulted bonds. Altman and Eberhart (1994),

Debt Instrument	Average Nominal Recovery Rate	Average Discounted Recovery Rate	Counts	Standard Deviation
Revolver	94%	86%	1,149	26%
Term loan	81%	74%	1,090	34%
Senior secured bonds	72%	62%	743	35%
Senior unsecured bonds	55%	48%	1597	38%
Senior subordinated bonds	33%	29%	535	33%
Subordinated bonds	32%	27%	467	34%
Total term loans/ revolvers	87%	81%	2,239	
Total bonds	51%	45%	3,342	

Summary statistics are provided by David Keisman using Moody's Ultimate Recovery Database: the sample includes 5,581 defaulted loans and bond issues 1987–2018. Recoveries are discounted at each instruments' predefault interest rate.

FIGURE 16.7   Ultimate Recovery Rates on Bank Loan and Bond Defaults Nominal and Discounted Values (1987–2018)

being first to investigate returns of defaulted bonds in bankruptcy restructuring, conclude that the most senior bonds in the capital structure (senior secured and senior unsecured) perform well in the post-Default period (20–30% per annum returns) but the junior bonds (senior subordinated and subordinated) perform poorly. These junior bonds barely break even on a nominal basis and losing money on a discounted basis. Using an updated sample of 115 defaulted bonds from 1992 to 2000, Fridson and Gao (2002) find that senior bonds realize an internal rate of return of 11.3% and subordinated bonds realize a more moderate return of 8.3%. Employing a much larger sample of 424 defaulted bonds of 148 nonprepackaged Chapter 11 filings by large U.S. public firms from 1996 to 2007, Wang (2011) finds that senior bonds, on average, achieve about 20% annualized gains during bankruptcy restructuring. However, in contrast to finding in prior studies he finds that subordinated bonds realize more than 20% annualized losses in bankruptcy. He further finds that distressed funds' active involvement on the unsecured creditors' side brings higher returns to bondholders while they realize lower returns in lengthy cases and liquidated cases.

Distressed Exchanges and the Bankruptcy Code   Because distressed exchanges (DE) and other similar forms of out-of-court restructurings (see Chapter 4 herein) are not as dramatic a reflection of a firm's distressed status as a bankruptcy, one might expect the recovery rate on DE defaults to be higher than other in-court proceedings. Jankowitsch, Nagler, and Subrahmanyam (2014) also show that distressed exchanges have the highest recovery rates and bankruptcy filings show significantly lower recoveries. That is exactly what Altman and Kuehne (2018a, 2018b) observe. They show that the arithmetic-average recovery rate on all distressed exchanges was 57.0% for 1984–2017, compared to 44.4% for all defaults, and 37.1% for all non-distressed-exchange defaults. Their differences in means test on the arithmetic-average recovery rates show that differences in recovery rates are statistically significant at the 1% confidence level. One reason for the larger recoveries in distressed exchanges is that bondholders need to be offered a "premium" in order to be persuaded to participate in the exchange. It is not surprising that bondholders will choose, in many instances, to accept a recovery with certainty from a distressed exchange, rather than take the chance of holding out for an uncertain — and likely lower – recovery in bankruptcy. However, restructuring out of court may not be the best form of restructuring for many, as Altman and Kuehne (2018a, 2018b) suggest that 35% of successful DEs eventually end up in Chapter 11.

Two notable studies compare creditors' recoveries across different Chapters of the U.S. Bankruptcy Code as well as across bankruptcy codes in different countries. Bris, Welch, and Zhu (2006) compare recoveries between firms that file for Chapter 11 reorganization and Chapter 7 liquidation of the bankruptcy code, and find that creditors fare better in Chapter 11. Davydenko and Franks (2008) provide evidence on how lenders fare differently in defaulted borrowers in France, Germany, and the UK.

NOTES

1. Empirical evidence presented in Jankowitsch, Nagler, and Subrahmanyam (2014) suggests that the rating frameworks of Moody's and Fitch seem to incorporate recovery rate information to a greater extent than that of Standard & Poor's.
2. In the KMV model (see Crosbie and Bohn 2002), default occurs when the firm's asset value goes below a threshold represented by the sum of the total amount of short-term liabilities and half of the amount of long-term liabilities.
3. The standard reference is Jones, Mason, and Rosenfeld (1984), who find that, even for firms with very simple capital structures, a Merton-type model is unable to price investment-grade corporate bonds better than a naive model that assumes no risk of default.
4. One of the earliest studies based on this framework is Black and Cox (1976). However, this model is not included in the second-generation models in terms of the treatment of the recovery rate.
5. Using Moody's corporate bond yield data, they find that credit spreads are negatively related to interest rates and that durations of risky bonds depend on the correlation with interest rates.
6. Suo, Wang, and Zhang (2013) show that the endogeneously determined default threshold of the Leland and Toft model, estimated from equity prices, has strong explanatory power for debt recovery observed in the market.
7. See Eom, Helwege, and Huang (2001) and Huang and Huang (2012) for an empirical analysis of the performance of structural-form models.
8. As discussed in Jones and Mingo (1998), reserves are used to cover expected losses.

9. For a comprehensive analysis of these models, see Crouhy, Galai, and Mark (2000) and Gordy (2000).

10. Demand mostly comes from niche distress investors, who intentionally purchase bonds in default. These investors represent a specialized segment of the fixed income market.

11. Procyclicality involves the sensitivity of regulatory capital requirements to economic and financial market cycles. Since ratings and default rates respond to the cycle, the internal ratings-based (IRB) approach proposed by the Basel Committee in 2004 requires increasing capital charges and limiting credit supply, when the economy is slowing (the reverse being true when the economy is growing).

12. This statistic is known as the coefficient of variation and is a relevant comparative statistic for populations with different mean values.

References

Acharya, V.V., and K. Subramanian. 2009. "Bankruptcy Codes and Innovation." *Review of Financial Studies* **22**, 4949–4988.

Acharya, V.V., R.K. Sundaram, and K. John. 2011. "Cross-country Variation in Capital Structures: The Role of Bankruptcy Codes." *Journal of Financial Intermediation* **20**, 25–54.

Acharya, V.V., S.T. Bharath, and A. Srinivasan. 2007. "Does Industry-Wide Distress Affect Defaulted Firms? Evidence from Creditor Recoveries." *Journal of Financial Economics* **85**, 787–821.

Acharya, V.V., Y. Amihud, and L. Litov. 2011. "Creditor Rights and Corporate Risk-taking." *Journal of Financial Economics* **102**, 150–166.

Aghion, P., O. Hart, and J. Moore. 1992. "The Economics of Bankruptcy Reform." *Journal of Law, Economics, and Organization* **8**(3), 523–546.

Akins, B., J. Bitting, D. De Angelis, and M. Gaulin. 2018. "The Salience of Creditors' Interests and CEO Compensation." Working paper.

Alderson, M.J., and B.L. Betker. 1995. "Liquidation Costs and Capital Structure." *Journal of Financial Economics* **39**, 45–69.

Aldersen, M.J., and B.L. Betker. 1999. "Assessing Postbankruptcy Performance: An Analysis of Reorganized Firms' Cash Flows." *Financial Management* **28**, 68–82.

Altman, E.I. 1967. *The Prediction of Corporate Bankruptcy*. UCLA dissertation, Michigan University Press.

Altman, E.I. 1968. "Financial Ratios Discriminant Analysis and the Prediction of Corporate Bankruptcy." *Journal of Finance* **23**(4), 189–209.

Altman, E.I. 1970. "Corporate Bankruptcy Prediction and Its Implications for Commercial Loan Evaluation." *Journal of Commercial Bank Lending* (December), 1–18.

Altman, E.I. 1977. "The Cost of Lending Errors for Commercial Banks: Some Conceptual and Empirical Issues." *Journal of Commercial Bank Lending* (October), 16–32.

Altman, E.I. 1983. *Corporate Financial Distress and Bankruptcy*. New York: Wiley.

Altman, E.I. 1984a. "A Further Empirical Investigation of the Bankruptcy Cost Question." *Journal of Finance* **39**(4), 1067–1089.

Altman, E.I. (editor). 1984b. "Special Issues on International Bankruptcy Prediction Models." *Journal of Banking and Finance* **8**(2).

Altman, E.I. 1987. "The Anatomy of the High Yield Bond Market." *Financial Analysts Journal* (July/August), 12–25.

Altman, E.I. 1988. "Special Issues on International Bankruptcy Prediction Models." *Journal of Banking and Finance* **12**(7).

Altman, E.I. 1989. "Measuring Corporate Bond Mortality and Performance." *Journal of Finance* **39**(4), 909–922.

Altman, E.I. 1990. *Investing in Distressed Securities*. Los Angeles: Foothill Corporation.

315

Altman, E.I. 1992. "Revisiting the High Yield Bond Market." *Financial Management Journal*, 78–92.

Altman, E.I. 1992. *The Market for Distressed Securities and Bank Loans.* Los Angeles: Foothill Corporation.

Altman, E.I. 2007. "Global Debt Markets in 2007, New Paradigm or Great Credit Bubble." *Journal of Applied Corporate Finance* 10(3) (Summer), 17–31.

Altman, E.I. 2011. "Italy: The Hero or Villain of the Euro, Insight." *Financial Times* (June 21).

Altman, E.I. 2014. "The Role of Distressed Debt Markets and Trends in Bankruptcy." *Institute Law Review* 22(1), 75–267.

Altman, E.I., 1973. "Predicting Railroad Bankruptcies in America." *Bell Journal of Economics & Management Science* (Spring), 372–395.

Altman, E.I., 1976. "A Financial Early Warning System for Over-the-Counter Broker Dealers." *Journal of Finance* (September,) 460–476.

Altman, E.I., 1991. *Distressed Securities.* Chicago: Probus. Reprint, Frederick, MD: Beard Books, 1999.

Altman, E.I., 1993. *Corporate Financial Distress & Bankruptcy*, 2nd ed. New York: Wiley.

Altman, E.I., 2001, Altman High Yield Bond and Default Study, Salomon Smith Barney, U.S. Fixed Income High Yield Report, July.

Altman, E.I., A. Resti, A., and A. Sironi. 2004. "Default Recovery Rates in Credit Risk Modeling: A Review of the Literature and Recent Evidence." *Economic Notes* 33(2), 183–208.

Altman, E.I., A. Resti, and A. Sironi. 2001. "Analyzing and Explaining Default Recovery Rates." ISDA Research Report, London, December.

Altman, E.I., A. Resti, and A. Sironi. 2005. *Recovery Risk: The Challenge in Credit Risk Management.* London: Risk Books.

Altman, E.I., and A.C. Eberhart. 1994. "Do Seniority Provisions Protect Bondholders' Investments?" *Journal of Portfolio Management* (Summer), 67–75.

Altman, E.I., and B. Karlin. 2009. "The Re-emergence of Distressed Exchanges in Corporate Restructuring." *Journal of Credit Risk* 5(2) (Summer), 43–45.

Altman, E.I., and B. Kuehne. 2017. "Defaults and Returns in the High-Yield Bond & Distressed Debt Markets." NYU Salomon Center Special Report, February.

Altman, E.I., and B.J. Kuehne. 2017. "Special Report on Defaults and Returns in the High-Yield Bond Market: First-Half 2017 Review." NYU Salomon Center.

Altman, E.I. and B.J. Kuehne. 2018a. "Defaults & Returns in the High Yield Bond and Distressed Debt Markets." NYU Salomon Center. Special Report. February 2018.

Altman, E.I., and B.J. Kuehne. 2018b. "The Investment Performance and Market Dynamics of Defaulted and Distressed Corporate Bonds and Bank Loans: 2017 Review and 2018 Outlook." NYU Salomon Center Special Report, March 1.

Altman, E.I., and D.L. Kao. 1992. "The Implications of Corporate Bond Rating Drift." *Financial Analysis Journal* (May/June), 64–75.

Altman, E.I., and E. Hotchkiss. 2006. *Corporate Financial Distress and Bankruptcy*, 3rd ed. Hoboken, NJ: Wiley.

Altman, E.I., and E.A. Kalotay. 2014. "Ultimate Recovery Mixtures." *Journal of Banking and Finance* 40, 116–129.

Altman, E.I., and G. Fanjul. 2004. "Defaults and Returns in the High Yield Bond Market: Analysis Through 2003 and Through Second Quarter 2004." NYU Salomon Center Special Reports, January and July.

Altman, E.I., and G. Sabato. 2007. "Modeling Credit Risk of SMEs: Evidence from the U.S. Market." *ABACUS* **43**(2) (September), 332–357.

Altman, E.I., and H. Rijken. 2004. "How Rating Agencies Achieve Rating Stability." *Journal of Banking & Finance* **28**, 2679–2714.

Altman, E.I., and H. Rijken. 2006. "A Point-in-Time Perspective on Through-the-Cycle Ratings." *Financial Analysts Journal* **62**(1) (January/February), 54–70.

Altman, E.I., and H. Rijken. 2011. "A Bottom-Up Approach to Assessing Sovereign Default Risk." *Journal of Applied Corporate Finance* **23**(3) (Winter), 20–31.

Altman, E.I., and H.Y. Izan. 1982. "Identifying Corporate Distress in Australia." *AGSM, W. P.* 83–103, Sydney.

Altman, E.I., and J. Bencivenga. 1995. "A Yield Premium Model for the High Yield Debt Market." *Financial Analysts Journal* (September/October), 26–41.

Altman, E.I., and J. LaFleur. 1981. "Managing a Return to Financial Health." *Journal of Business Strategy* (Summer), 31–38.

Altman, E.I., and L. Goodman. 1981. "An Economic and Statistical Analysis of the Falling Company Doctrine." Working paper, New York University.

Altman, E.I., and L. Goodman. 2002. "An Economic & Statistical Analysis of the Failing Company Doctrine," in E. Altman, ed., *Bankruptcy, Credit Risk and High Yield Bonds*. Malden, MA: Blackwell.

Altman, E.I., and M. Brenner. 1981. "Information Effects and Stock Market Response to Signs of Firm Deterioration." *Journal of Financial Qualitative Analysis* (March XVI, #1), 35–51.

Altman, E.I., and M. Lavallee. 1980. "Business Failure Classification in Canada." *Journal of Business Administration* (Fall), 79–91.

Altman, E.I., and R. Benhenni. 2017. "The Anatomy of Investing in Defaulted Bonds and Loans." *Journal of Corporate Renewal* **30**(6), 5–11.

Altman, E.I., and R.D. Gritta. 1984. "Airline Bankruptcies Propensity: A ZETA Analysis." Transportation Research Forum. Washington, DC: Harmony Press.

Altman, E.I., and S. Nammacher. 1987. *Investing in Junk Bonds: Inside the High Yield Debt Market*. New York: Wiley.

Altman, E.I., and T.P. McGough. 1974. "Evaluation of a Company as a Going Concern." *Journal of Accountancy* (December), 50–57.

Altman, E.I., and V.M. Kishore. 1996. "Almost Everything You Wanted to Know About Recoveries on Defaulted Bonds." *Financial Analysts Journal* (November/December), 57–64.

Altman, E.I., B. Brady, A. Resti, and A. Sironi. 2005. "The Link between Default and Recovery Rates: Theory, Empirical Evidence and Implications." *Journal of Business* **78**(6), 2203–2228.

Altman, E.I., G. Sabato, and M. Esentato. 2016. "Assessing the Credit Worthiness of Italian SMEs and Mini-Bond Issuers." *Borsa Italiana*. Newsletter. April.

Altman, E.I., G. Sabato, and N. Wilson. 2010. "The Value of Nonfinancial Information in Small & Median-Sized Enterprise Risk Management." *Journal of Credit Risk* **6**(2), 1–33.

Altman, E.I., G. Sabato, and N. Wilson. 2017. "Assessing SME Default Risk in the U.K." Working paper, Wiserfunding, London.

Altman, E.I., H. Rijken, M. Watt, D. Balan, J. Forero, and J. Mina. 2010. The Z-Metrics™ Methodology for Estimating Company Credit Ratings and Default Risk Probabilities, RiskMetrics Group, NY, June, available from Prof. Altman's website.

Altman, E.I., J. Cziel, and H. Rijken. 2016. "Anatomy of Bank Distress: The Information Content of Accounting Fundamentals." IRMC Conference, Jerusalem, Israel, June 13.

Altman, E.I., J. Hartzell, and M. Peck. 1995. "A Scoring System for Emerging Market Corporate Bonds." Salomon Brothers (May), and in *Emerging Market Review 6* (December 2005).

Altman, E.I., J.F. Gonzalez-Heres, P. Chen, and S.S. Shin. 2014. "The Return/Volatility Trade-Off of Distressed Corporate Debt Portfolios." *Journal of Portfolio Management* (Winter), 69–85.

Altman, E.I., L. Zhang, and J. Yen. 2010. "Corporate Financial Distress Diagnosis Model and Application in Credit Rating for Listing Firms in China." *Frontier of Computer Science in China* **4**(2), 220–236.

Altman, E.I., M. Iwanicz-Drozdowska, E.K. Laitinen, and A. Suvas. 2016. "Financial Distress Prediction in an International Context: A Review and Empirical Analysis of Altman's Z-Score Model." *Journal of International Financial Management and Accounting* **28**(2), 131–171.

Altman, E.I., M. Iwanicz-Drozdowska, E.K. Laitinen, and A. Suvas. 2017. "Financial and Non-Financial Variables as Long-Horizon Predictors of Bankruptcy." *The Journal of Credit Risk* **12**(4), 49–78.

Altman, E.I., M. Margaine, M. Schlosser, and P. Vernimmen. 1974. "Financial and Statistical Analysis for Commercial Loan Evaluations: A French Experience." *Journal of Finance & Quantitative Analysis* **9**(2) (March), 195–211.

Altman, E.I., R.G. Haldeman, and P. Narayanan. 1977. "ZETA Analysis: A New Model to Identify Bankruptcy Risk of Corporations." *Journal of Banking & Finance* **1**(1) (June), 29–54.

Altman, E.I., T.K.N. Baidya, and L.M. Riberio-Dias. 1979. "Assessing Potential Financial Problems of Firms in Brazil." *Journal of International Business Studies* (Fall), 9–24.

Altman, E.I., Y.H. Eom, and D.K. Kim. 1995. "Failure Prediction: Evidence from Korea." *Journal of International Financial Management & Accounting* (Winter), 230–249.

Andrade, G., and S. Kaplan. 1998. "How Costly Is Financial (not economic) Distress? Evidence from Highly Leveraged Transactions That Became Distressed." *Journal of Finance* **53**, 1443–1493.

Ang, J.S., J.H. Chua, and J.J. McConnell. 1982. "The Administrative Costs of Corporate Bankruptcy: A Note." *Journal of Finance* **37**, 219–226.

Appenzeller, P., and R. Parker. 2005. "Deepening Insolvency is a Liability Trap for the Unwary." *Journal of Corporate Renewal*, TMA, Chicago, July.

Asquith, P., R. Gertner, and D. Scharfstein. 1994. "Anatomy of Financial Distress: An Examination of Junk-Bond Issuers." *The Quarterly Journal of Economics* **109**(3), 625–658.

Ayotte, K.M., and E.R. Morrison. 2009. "Creditor Control and Conflict in Chapter 11." *Journal of Legal Analysis* **1**(2), 511–551.

Ayotte, K.M., and E.R. Morrison. 2018. "Valuation Disputes in Corporate Bankruptcy." Working paper.

Ayotte, K.M., E.S. Hotchkiss, and K.S. Thorburn. 2012. "Governance in Financial Distress and Bankruptcy." In M. Wright, D.S. Siegel, K. Keasey, and I. Filatotchev, eds., *The Oxford Handbook of Corporate Governance*. Oxford, UK: Oxford University Press.

Babbel, D. F. 1996. "Insuring Sovereign Debt against Default." World Bank Discussion Papers, #328.

Baek, I.M., A. Bandopadhyaya, and C. Du. 2005. "Determinants of Market-Assessed Sovereign Risk: Economic Fundamentals or Market Risk Appetite?" *Journal of International Money and Finance* **24**(4), 533–48.

Baghai, R., R. Silva, and L. Ye. 2017. "Bankruptcy, Team-Specific Human Capital, and Productivity: Evidence from U.S. Inventors." Working paper.

Baird, D.G. 2010. *The Elements of Bankruptcy*, 5th ed. New York: Thomson Reuters/ Foundation Press.

Baird, D.G., and R.K. Rasmussen. 2002. "The End of Bankruptcy." *Stanford Law Review* **55**, 751–789.

Baird, D.G., and R.K. Rasmussen. 2006. "Private Debt and the Missing Lever of Corporate Governance." *University of Pennsylvania Law Review* **154**, 1209–1252.

Bakshi, G., D.B. Madan, and F. Zhang. 2001. "Understanding the Role of Recovery in Default Risk Models: Empirical Comparisons and Implied Recovery Rates." Finance and Economics Discussion Series, 2001–37, Federal Reserve Board of Governors, Washington D.C.

Balsam, S., Y. Gu, and X. Mao. 2018. "Creditor Influence and CEO Compensation: Evidence from Debt Covenant Violation." *The Accounting Review*, forthcoming.

Barboza, F., H Kumar, and E.I. Altman. 2017. "Machine Learning Models & Bankruptcy Prediction." *Expert Systems with Applications* **83**, 405–417.

Bary, A. 2010. "Post-Bankrupts: What Does the Next Chapter Hold?" *Barron's* (July 24).

Basel Committee on Banking Supervision. 2003. "The New Basel Capital Accord." Consultative Document, Bank for International Settlements, April.

Beaver, W. 1966. "Financial Ratios as Predictors of Failures." *Journal of Accounting Research* **4**, 71–111.

Bebchuk, L. 1988. "A New Approach to Corporate Reorganizations." *Harvard Law Review* **101**, 775–804.

Bebchuk, L. 2002. "Ex Ante Costs of Violating Absolute Priority in Bankruptcy." *Journal of Finance* **57**(1), 445–460.

Becker, B., and P. Strömberg. 2012. "Fiduciary Duties and Equity-Debtholder Conflicts." *Review of Financial Studies* **25**(6), 1931–1969.

Becker, B., and V. Ivashina. 2016. "Covenant-Light Contracts and Creditor Coordination." Sveriges Riksbank Working Paper Series.

Bedendo, M., L. Cathcart, and L. EL-Jahel. 2016. "Distressed Debt Restructuring in the Presence of Credit Default Swaps." *Journal of Money, Credit and Banking* **48**(1), 165–201.

Beers, D., M. Cavanaugh, and O. Takahira. 2002. "Sovereign Credit Ratings: A Primer." New York: Standard & Poor's Corp., April.

Bellovary, J., D. Giacomino, and M. Akers. 2007. "A Review of Bankruptcy Prediction Studies: 1930 to Present." *Journal of Financial Education* (Winter), 1–42.

Bellucci, M., and J. McCluskey. 2016. *The LSTA's Complete Credit Agreement Guide*, 2nd ed. New York: McGraw-Hill.

Benmelech, E., J. Dlugosz, and V. Ivashina. 2012. "Securitization without Adverse Selection: The Case of CLOs." *Journal of Financial Economics* **106**, 91–113.

Benmelech, E., N.K. Bergman, and R. Enriquez. 2012. "Negotiating with Labor under Financial Distress." *Review of Corporate Finance Studies* **1**(1), 28–67.

Bennett, B., J.C. Bettis, R. Gopalan, and T. Milbourn. 2017. "Compensation Goals and Firm Performance." *Journal of Financial Economics* **124**(2), 307–330.

Berk, J., and P. DeMarzo. 2017. *Corporate Finance*, 4th ed. Pearson Series in Finance. Boston: Pearson Education.

Berlin, M., G. Nini, and E. Yu. 2016. "Concentration of Control Rights in Leveraged Loan Syndicates." Working paper, Federal Reserve Bank of Philadelphia and Drexel University.

Bernstein, R. 1990. "The Decaying Financial Infrastructure – Revisited." *Quantitative Viewpoint,* Merrill Lynch (December 4), 1–7.

Bernstein, S., E. Colonnelli, and B. Iverson. 2018. "Asset Allocation in Bankruptcy." *Journal of Finance*, forthcoming.

Betker, B.L. 1997. "The Administrative Costs of Debt Restructurings: Some Recent Evidence." *Financial Management* **26**, 56–68.

Betker, B.L., S.P. Ferris, and R.M. Lawless. 1999. "Warm with Sunny Skies: Disclosure Statement Forecasts." *American Bankruptcy Law Journal* **73**, 809–836.

Bharath, S.T., and T. Shumway. 2008. "Forecasting Default with the Merton Distance to Default Model." *Review of Financial Studies* **21**, 1339–1369.

Bharath, S.T., and V. Panchapegesan, and I. Werner. 2010. "The Changing Nature of Chapter 11." Working paper, Arizona State University.

Bhattacharjee, A., and J. Han. 2014. "Financial Distress of Chinese Firms: Microeconomic, Macroeconomic and Institutional Influences." *China Economic Review* **30**, 244–262.

Billett, M.T., R. Elkamhi, L. Popov, and R.S. Pungaliya. 2016. "Bank Skin in the Game and Loan Contract Design: Evidence from Covenant-Lite Loans." *Journal of Financial and Quantitative Analysis* **51**, 839–873.

Black, F., and J.C. Cox. 1976. "Valuing Corporate Securities: Some Effects of Bond Indenture Provisions." *Journal of Finance* **31**, 351–367.

Black, F., and M. Scholes. 1973. "The Pricing of Options and Corporate Liabilities." *Journal of Political Economics* (May), 637–659.

Bolton, P., and D. Scharfstein. 1996. "Optimal Debt Structure and the Number of Creditors." *Journal of Political Economy* **104**(1), 1–25.

Bolton, P., and M. Oehmke. 2011. "Credit Default Swaps and the Empty Creditor Problem." *Review of Financial Studies* **24**(8), 2617–2655.

Borenstein, S., and N.L. Rose. 1995. "Bankruptcy and Pricing Behavior in U.S. Airline Markets." *The American Economic Review* **85**(2), 397–402.

Bradley, M., and M. Rosenzweig. 1992. "The Untenable Case for Chapter 11." *Yale Law Journal* **101**, 1043–1089

Branch, B., and H. Ray. 1992. *2007 Bankruptcy Investing*. Chicago: Dearborn Press.

Brav, A., H. Kim, and W. Jiang., 2010, Hedge Fund Activism: A Review, Foundation and Trends in Finance, Vol. 4, No. 3, 185–246 (IN CHAPTER 6)

Bris, A., I. Welch, and N. Zhu. 2006. "The Costs of Bankruptcy: Chapter 7 Liquidation versus Chapter 11 Reorganization." *Journal of Finance* **61**, 1253–1303.

Brown, D.T., C.M. James, and R.M. Mooradian. 1993. "The Information Content of Distressed Restructuring Involving public and Private Debt Claims." *Journal of Financial Economics* **33**, 93–118

Brown, J., and A.D. Matsa. 2016. "Boarding a Sinking Ship? An Investigation of Job Applications to Distressed Firms." *Journal of Finance* **71**, 507–550.

Buttwill, K., and C. Wihlborg. 2004. "The Efficiency of the Bankruptcy Process. An International Comparison." Working paper.

Campbell, J.Y., J. Hilcher, and J. Szilagi. 2008. "In Search of Distress Risk." *Journal of Finance* **63**(6), 2899–2939.

Campello, M., J. Gao, J. Qiu, and Y. Zhang. 2018. "Bankruptcy and the Cost of Organized Labor: Evidence from Union Elections." *Review of Financial Studies* **31**(3), 980–1013.

Caouette, J.B., E.I. Altman, P. Narayanan, and R. Nimmo. 1998. *Managing Credit Risk: The Next Great Financial Challenge*, 2nd ed. New York: Wiley.

Capkun, V., and E. Ors. 2016. "When the Congress Says PIP Your KERP: Performance Incentive Plans, Key Employee Retention Plans, Chapter 11 Bankruptcy, and Regulatory Arbitrage." Working paper, HEC Paris.

Carey, M., and M. Gordy. 2003. "Systematic Risk in Recoveries on Defaulted Debt." Mimeo, Federal Reserve Board, Washington.

Carey, M., and M. Gordy. 2016. "The Bank as Grim Reaper: Debt Composition and Bankruptcy Thresholds." Mimeo, Federal Reserve Board, Washington.

Carter, M.E., E.S. Hotchkiss, and M. Mohseni. 2018. "Payday Before Mayday: CEO Compensation Contracting for Distressed Firms." Working paper.

Chambers, W.J. 1997. "Understanding Sovereign Risk." *Credit Week,* Standard & Poor's (January 1).

Chang, T., and A. Schoar. 2013. Judge Specific Differences in Chapter 11 and Firm Outcomes. Working paper.

Chava, S., and M. Roberts. 2008. "How Does Financing Impact Investment? The Role of Debt Covenant Violations." *Journal of Finance* **63**, 2085–2121.

Chava, S., and R. Jarrow. 2004. "Bankruptcy Prediction with Industry Effects." *Review of Finance* **8**, 537–569.

Cho, S-Y., L. Fu, and Y. Yu. 2012. "New Risk Analysis Tools with Accounting Changes: Adjusted Z-Score." *The Journal of Credit Risk* **8**(1) (Spring), 89–108.

Choi, F. 1997. *International Accounting & Finance Handbook,* 2nd ed. (Ch. 35). New York: Wiley.

Chu, Y.Q., H.D. Nguyen, J. Wang, W. Wang, and W.Y. Wang. 2018. "Debt-Equity Simultaneous Holdings and Distress Resolution." Working paper.

Ciliberto, F., and C. Schenone. 2012. "Bankruptcy and Product-market Competition: Evidence from the Airline Industry." *International Journal of Industrial Organization* **30**(6), 564–577.

Claessens, S., and L.F. Klapper. 2005. "Bankruptcy Around the World: Explanations of Its Relative Use." *American Law and Economics Review* **7**, 253–283.

Claessens, S., S. Djankov, and L.F. Klapper. 2003. "Resolution of Corporate Distress in East Asia." *Journal of Empirical Finance* **10**, 199–216.

Cline, W. 1983. "A Logit Model of Debt Restructuring, 1963–1982." Institute for International Economics, WP, June.

Collin-Dufresne, P., and R.S. Goldstein. 2001. "Do Credit Spreads Reflect Stationary Leverage Ratios?" *Journal of Finance* **56**, 1929–1957.

Credit Suisse Financial Products. 1997. "CreditRisk+. A Credit Risk Management Framework." Technical Document.

Crosbie, P.J., and J.R. Bohn. 2002. "Modelling Default Risk." *Moody's KMV.*

Crouhy, M., D. Galai, and R. Mark. 2000. "A Comparative Analysis of Current Credit Risk Models." *Journal of Banking & Finance* 24, 59–117.

Cutler, D.M., and L.H. Summers. 1988. "The Costs of Conflict Resolution and Financial Distress: Evidence from the Texaco-Pennzoil Litigation." *Rand Journal of Economics* **19**, 157–172.

Dahiya, S., K. John, M. Puri, and G. Ramırez. 2003. "Debtor-in-Possession Financing and Bankruptcy Resolution: Empirical Evidence." *Journal of Financial Economics* **69**, 259–280.

Damodaran, A. 1996. *Investment Valuation: Tools and Techniques for Determining the Value of Any Asset.* New York: Wiley.

Danis, A. 2016. "Do Empty Creditors Matter? Evidence from Distressed Exchange Offers." *Management Science*, 1–17.

Das, S.R., and S. Kim. 2014. "Going for Broke: Restructuring Distressed Debt Portfolios." *Journal of Fixed Income* **24**(1), 5–27.

Das, S.R., P. Hanouna, and A. Sarin. 2009. "Accounting-Based vs. Market-Based Cross Sectional Models of CDS Spreads." *Journal of Banking & Finance* **33**, 719–730.

Davydenko, S.A., and J.R. Franks. 2008. "Do Bankruptcy Code Matter? A Study of Defaults in France, Germany, and the U.K." *Journal of Finance* **63**, 565–608.

Deakin, E. 1972. "A Discriminant Analysis of Predictors of Business Failure." *Journal of Accounting Research* **10**(1, March), 167–179.

Demiroglu, C., and James, C. 2015. "Bank Loans and Troubled Debt Restructurings." *Journal of Financial Economics* **118**, 192–210.

Demiroglu, C., J. Franks, and R. Lewis. 2018. "Do Market Prices Improve the Accuracy of Court Valuation in Chapter 11?" Working paper.

Denis, D.J., and J. Wang. 2014. "Debt Covenant Renegotiation and Creditor Control Rights." *Journal of Financial Economics* **113**, 348–367.

Denis, D.K. and K.J. Rodgers. 2007. "Chapter 11: Duration, Outcome, and Post-Reorganization Performance." *Journal of Financial and Quantitative Analysis* **42**, 101–118.

Dichev, I., and D.J. Skinner. 2002. "Large-Sample Evidence on the Debt Covenant Hypothesis." *Journal of Accounting Research* **40**, 1091–1123.

Djankov, S., C. McLiesh, and A. Shleifer. 2007. "Private Credit in 129 Countries." *Journal of Financial Economics* **12**, 77–99.

Djankov, S., O. Hart, C. McLiesh, and A. Shleifer. 2008. "Debt Enforcement Around the World." *Journal of Political Economy* **116**, 1105–1150.

Do Prado, J.W., V. de Castro Alcantara, F. de Melo Carvalho, K.C. Vieira, L.K.C. Machado, and D.F. Tonelli. 2016. "Multivariate Analysis of Credit Risk and Bankruptcy Research Data: A Bibliometric Study Involving Different Knowledge Fields (1984–2014)." *Scientometrics* **106**, 1007–1029.

Duan, Y., E.S. Hotchkiss, and Y. Jiao. 2015. "Corporate Pensions and Financial Distress." Working paper.

Duffee, G.R. 1999. "Estimating the Price of Default Risk." *Review of Financial Studies* **12**(1, Spring), 197–225.

Duffie, D. 1998. "Defaultable Term Structure Models with Fractional Recovery of Par." Graduate School of Business, Stanford University.

Duffie, D., and D. Lando. 2000. "Term Structure of Credit Spreads with Incomplete Accounting Information." *Econometrica* **69**, 633–664.

Duffie, D., and K.J. Singleton. 1999. "Modeling the Term Structures of Defaultable Bonds." *Review of Financial Studies* **12**, 687–720.

Duffie, D., and K.J. Singleton. 2003. *Credit Risk: Pricing, Measurement, and Management.* Princeton: Princeton University Press.

Duffie, D., L. Saita, and K. Wang. 2007. "Multi-Period Default Prediction with Stochastic Covariates." *Journal of Financial Economics* **83**, 635–662.

Dullman, K., and M. Trapp. 2004. "Systematic Risk in Recovery Rates – An Empirical Analysis of U.S. Corporate Credit Exposures." Mimeo, University of Mannheim, and also updated in Altman, Resti, and Sironi (2005).

Eberhart, A., E.I. Altman, and R. Aggarwal. 1999. "The Equity Performance of Firms Emerging from Chapter 11." *Journal of Finance* **54** (October), 1855–1868.

Eckbo, E.B., and K.S. Thorburn. 2003. "Control Benefits and CEO Discipline in Automatic Bankruptcy Auctions." *Journal of Financial Economics* **69**(1), 227–258.

Eckbo, E.B., K. Li, and W. Wang. 2018. "Why Is Low-risk Bankruptcy Financing So Expensive?" Working paper.

Eckbo, E.B., K.S. Thorburn, and W. Wang. 2016. "How Costly Is Corporate Bankruptcy for the CEO?" *Journal of Financial Economics* **121**(1), 210–229.

Edmister, R. 1972. "An Empirical Test of Financial Ratio Analysis for Small Business: Failure Prediction." *Journal of Financial Education Analysis* **7**(2), 1477–1493.

Eibl, A. 2014. "STOXX Strong Balance Sheet Indices: A Family that Selects the Financially Fittest Companies." *STOXX PULSE* (Spring), 12–15.

Ellias, J.A. 2016. "Do Activist Investors Constrain Managerial Moral Hazard in Chapter 11?: Evidence from Junior Activist Investing." *Journal of Legal Analysis* **8**, 493–547.

Ellias, J.A. 2018. "How Not to Regulate Executive Compensation: An Empirical Study of Chapter 11 Bankruptcy Reform." Working paper.

Ellias, J.A., and R.J. Stark. 2018. "The Erosion of Creditor Protection." Working paper.

Emery, K., R. Cantor, S. Ou, R. Soloman, and P. Stumpp. 2004. "Credit Loss Rates on Similarly Rated Loans and Bonds." Moody's Special Comment, December.

Emmott, W. 1991. "Theories at the Bottom of Their Jargon: International Finance Survey." *The Economist* (April 27), 5–24.

Eom, Y.H., J. Helwege, and J.Z. Huang. 2001. "Structural Models of Corporate Bond Pricing: An Empirical Analysis." Mimeo.

Ersahin, N., R.M. Irani, and K. Waldock. 2017. "Creditor Rights and Entrepreneurship: Evidence from Fraudulent Transfer Law." Working paper.

Fama, E., and K. French. 1993. "Common Risk Factors in the Returns on Stocks and Bonds." *Journal of Financial Economics* **33**(1), 3–56.

Fama, E., and K. French. 2002. "The Equity Premium." *Journal of Finance* **57**, 637–659.

Fan, H., and S.M. Sundaresan. 2000. "Debt Valuation, Renegotiation, and Optimal Dividend Policy." *Review of Financial Studies* **13**, 1057–99.

Feder, G., and R.E. Just. 1977. "A Study of Debt Servicing Capacity Applying Logit Analysis." *Journal of Development Economics* **4**(1), 25–38.

Feder, G., R.E. Just, and K. Ross. 1981. "Projecting Debt Capacity of Developing Countries." *Journal of Financial & Qualitative Analysis* **16**(5), 651–659.

Feldhütter, P., E. Hotchkiss, and O. Karakas. 2016. "The Value of Creditor Control in Corporate Bonds." *Journal of Financial Economics* **121**, 1–27.

Ferreira, D., M.A. Ferreira, and B. Mariano. 2017. "Creditor Control Rights and Board Independence." *Journal of Financial Economics*, forthcoming.

Finger, C. 1999. "Conditional Approaches for CreditMetrics® Portfolio Distributions." *CreditMetrics® Monitor* (April).

Frank Jr., C.R., and W.R. Cline. 1971. "Measurement of Debt Servicing Capacity: An Application of Discriminant Analysis." *Journal of International Economics* **1**(3), 327–344.

Franks, J.R., and O. Sussman. 2005. "Financial Distress and Bank Restructuring of Small to Medium Size UK Companies." *Review of Finance* **9**, 65–96.

Franks, J.R., and W.N. Torous. 1989. "An Empirical Investigation of U.S. Firms in Reorganization." *Journal of Finance* **44**, 747–769.

Franks, J.R., and W.N. Torous. 1994. "A Comparison of Financial Recontracting in Distressed Exchanges and Chapter 11 Reorganizations." *Journal of Financial Economics* **35**, 349–370.

Frenkel, M., A. Karmann, and B. Scholtens, eds. 2004. *Sovereign Risk and Financial Crises*, xii, 258. Heidelberg and New York: Springer.

Fridson, M.S., and Y. Gao. 2002. "Defaulted Bond Returns by Seniority Class." *Journal of Fixed Income* (Summer), 50–57.

Frydman, H., E.I. Altman, and D.L. Kao. 1985. "Introducing Recursive Partitioning Analysis for Financial Classification: The Case of Financial Distress." *Journal of Finance* **50**(1), 269–291.

Frye, J. 2000a. "Collateral Damage." *Risk* (April), 91–94.

Frye, J. 2000b. "Collateral Damage Detected." Federal Reserve Bank of Chicago, Working Paper, Emerging Issues Series (October), 1–14.

Frye, J. 2000c. "Depressing Recoveries." *Risk* (November).

Gande, A., E.I. Altman, and A. Saunders. 2010. "Bank Debt versus Bond Debt: Evidence from Secondary Market Prices." *Journal of Money, Credit and Banking* **42**(4), 755–767.

Gennaioli, N., A. Martin, and S. Rossi. 2010. "Sovereign Default, Domestic Banks and Financial Institutions." *Journal of Finance* **69**, 819–866.

Gerlach, S., A. Schultz, and G. Wolff, G. 2010. "Banking and Sovereign Risk in the Euro Area." Deutsche Bundesbank, Research Centre, Discussion Paper Series 1: Economic Studies: 2010.

Gertner, R., and D. Scharfstein. 1991. "A Theory of Workouts and the Effects of Reorganization Law." *Journal of Finance* **46**(4), 1189–1222.

Geske, R. 1977. "The Valuation of Corporate Liabilities as Compound Options." *Journal of Financial and Quantitative Analysis* **12**(4), 541–552.

Giammarino, R.M. 1989. "The Resolution of Financial Distress." *The Review of Financial Studies* **2**(1), 25–47.

Gilje, E.P. 2016. "Do Firms Engage in Risk-Shifting? Empirical Evidence." *The Review of Financial Studies* **29**(11), 2925–2954.

Gilson, S.C. 1989. "Management Turnover and Financial Distress." *Journal of Financial Economics* **25**(2), 241–262.

Gilson, S.C. 1990. "Bankruptcy, Boards, Banks and Blockholders: Evidence on Changes in Corporate Ownership and Control When Firms Default." *Journal of Financial Economics* **27**(2), 355–387.

Gilson, S.C. 1997. "Transactions Costs and Capital Structure Choice: Evidence from Financially Distressed Firms." *Journal of Finance* **52**, 161–197.

Gilson, S.C. 2010. *Corporate Restructuring: Case Studies in Bankruptcies, Buyouts & Breakups,* 2nd ed. Hoboken, NJ: Wiley.

Gilson, S.C. 2012. "Coming through in a Crisis: How Chapter 11 and the Debt Restructuring Industry Are Helping Revive the U.S. Economy." *Journal of Applied Corporate Finance* **24**, 23–35.

Gilson, S.C. 2014. "School Specialty, Inc." Harvard Business School Case (9–214–084).

Gilson, S.C., and M.R. Vetsuypens. 1993. "CEO Compensation in Financially Distressed Firms: An Empirical Analysis." *Journal of Finance* **48**(2), 425–458.

Gilson, S.C., and S.L. Abbott. 2009. "Kmart and ESL Investments (A)." Harvard Business School Case (9-209-044).

Gilson, S.C., E.S. Hotchkiss, and M.G. Osborn. 2016. "Cashing Out: The Rise of M&A in Bankruptcy." Working paper.

Gilson, S.C., E.S. Hotchkiss, and R.S. Ruback. 2000. "Valuation of Bankrupt Firms." *Review of Financial Studies* **13**, 43–74.

Gilson, S.C., K. John, and H.P. Lang. 1990. "Troubled Debt Restructurings: An Empirical Study of Private Reorganization of Firms in Default." *Journal of Financial Economics* **27**(2), 315–353.

Gopalan, R., X. Martin, and K. Srinivasan. 2017. "Weak Creditor Rights and Insider Opportunism: Evidence from an Emerging Market." Working paper.

Gordy, M. 2000. "A Comparative Anatomy of Credit Risk Models." *Journal of Banking & Finance* **24**(1–2), 119–149.

Gordy, M. 2003. "A Risk-Factor Model Foundation for Ratings-Based Capital Rules." *Journal of Financial Intermediation* **12**(3), 199–232.

Gormley, T., N. Gupta, and A. Jha. 2018. "Quiet Life No More? Corporate Bankruptcy and Bank Competition." *Journal of Financial and Quantitative Analysis* **53**, 581–611.

Goyal, V.K., and W. Wang. 2017. "Provision of Management Incentives in Bankrupt Firms." *Journal of Law, Finance, and Accounting* **2**, 87–123.

Goyal, A., M. Kahl, and W. Torous. 2003. "The Long-Run Stock Performance of Financial Distressed Firms: An Empirical Investigation." Working paper.

Graham, J.R., H. Kim, S. Li, and J. Qiu. 2016. "Employee Costs of Corporate Bankruptcy." Working paper.

Gray, D.F., R. Merton, and Z. Bodie. 2006. "A New Framework for Analyzing and Managing Macrofinancial Risk of an Economy." IMF Working Paper, October.

Gray, D.F., R. Merton, and Z. Bodie. 2007. "Contingent Claims Approach to Measuring and Managing Sovereign Credit Risk." *Journal of Investment Management* **5**(4), 1.

Grinols, E. 1976. "International Debt Rescheduling and Discrimination Using Financial Variables." U.S. Treasury Department, Washington, DC.

Groh, A.P., and O. Gottschalg. 2011. "The Effect of Leverage on the Cost of Capital of US Buyouts." *Journal of Banking and Finance* **35** (8), 2099–2110.

Gropper, A.L. 2012. "The Arbitration of Cross-Border Insolvencies." *American Bankruptcy Law Journal* **86**, 201–242.

Gupton, G., and R.M. Stein. 2002. "LossCalc: Moody's Model for Predicting Loss Given Default (LGD)." Moody's/KMV, New York, updated in 2005, "LossCalc V2, Dynamic Prediction of LGD." Moody's/KMV (January).

Gupton, G., and R.M. Stein. 2005. "Dynamic Prediction of LGD Modelling Methodology.", Special report, Moody's KMV.

Gupton, G., C. Finger, and M. Bhatia. 1977a. "CreditMetrics: The Benchmark for Understanding Credit Risk." J.P. Morgan, Inc., New York, April.

Gupton, G., C. Finger, and M. Bhatia. 1997b. "CreditMetrics™: Technical Document." J.P.Morgan & Co., New York.

Gupton, G., D. Gates, and L.V. Carty. 2000. "Bank Loan Loss Given Default." Moody's Investors Service, Global Credit Research, November.

Gutzeit, G., and J. Yozzo. 2011. "Z-Score Performance amid Great Recession." *ABI Journal* **44**(80, March), 44–46.

Halford, J., M. Lemmon, Y.-Y. Ma, and E. Tashjian. 2017. "Bankruptcy Restructuring and Recidivism." Working paper.

Hart, O., 1999. "Different Approaches to Bankruptcy." Working paper, Harvard Institute of Economic Research.

Haugh, D., P. Ollivaud, and D. Turner. 2009. "What Drives Sovereign Risk Premiums? An Analysis of Recent Evidence from the Euro Areas." OECD, Economics Department, Working Paper 718.

Helwege J., and F. Packer. 2003. "Determinants of the Choice of Bankruptcy Procedure in Japan." *Journal of Financial Intermediation* **12**(1), 96–120.

Hertzel, M., Z. Li, M.S. Officer, and K.J. Rodgers. 2008. "Inter-firm Linkages and the Wealth effects of Financial Distress along the Supply Chain." *Journal of Financial Economics* **87**, 374–387.

Hillegeist, S., E. Keating, D. Kram, and K. Lundstedt. 2004. "Assessing the Probability of Bankruptcy." *Review of Accounting Studies* **9**(1), 5–34.

Hilscher, J., and Y. Nosbusch. 2010. "Determinants of Sovereign Risk: Macroeconomic Fundamentals and the Pricing of Sovereign Debt." *Review of Finance* **14**(2), 235–262.

Hochberg, Y.V., C.J. Serrano, and R.H. Ziedonis. 2016. "Patent Collateral, Investor commitment, and the Market for Venture Lending." *Journal of Financial Economics*, forthcoming.

Hortaçsu, A., G. Matvos, C. Syverson, and S. Venkataraman. 2013. "Indirect Costs of Financial Distress in Durable Goods Industries: The Case of Auto Manufacturers." *The Review of Financial Studies*, **26**(5), 1248–1290.

Hoshi, T., A. Kashyap, and D. Scharfstein. 1990. "The Role of Banks in Reducing the Costs of Financial Distress in Japan." *Journal of Financial Economics* **27**, 67–88.

Hotchkiss, E.S. 1995. "Postbankruptcy Performance and Management Turnover." *Journal of Finance* **50**, 3–21.

Hotchkiss, E.S., and R.M. Mooradian. 1997. "Vulture Investors and the Market for Control of Distressed Firms." *Journal of Financial Economics* **43**, 401–432.

Hotchkiss, E.S., and R.M. Mooradian. 1998. "Acquisitions as a Means of Restructuring Firms in Chapter 11." *Journal of Financial Intermediation* **7**, 240–262.

Hotchkiss, E.S., and R.M. Mooradian. 2004. "Postbankruptcy Performance of Public Companies." Working paper.

Hotchkiss, E.S., D.C. Smith, and P. Stromberg. 2016. "Private Equity and the Resolution of Financial Distress." Working paper.

Hotchkiss, E.S., K. John, R.M. Mooradian, and K.S. Thorburn. 2008. "Bankruptcy and the Resolution of Financial Distress," in B. Eckbo, ed., *Handbook of Empirical Corporate Finance*, 2nd ed. (Amsterdam: Elsevier/North-Holland).

Hu, H.T-C., and B.S. Black., 2008a. "Debt, Equity and Hybrid Decoupling: Governance and Systemic Risk Implications." *European Financial Management* **14**(4), 663–709.

Hu, H.T-C., and B.S. Black. 2008b. "Equity and Debt Decoupling and Empty Voting II: Importance and Extensions." *University of Pennsylvania Law Review* **156**, 625–739.

Hu, H.T-C., and J.L. Westbrook. 2007. "Abolition of the Corporate Duty to Creditors." *Columbia Law Review* **107**, 1321–1403.

Huang, J.Z., and M. Huang. 2012. "How Much of the Corporate-Treasury Yield Spread Is Due to Credit Risk?" *The Review of Asset Pricing Studies* **2**, 153–202.

Huang, Z., H. Chen, C.-J. Hsu, W.-H. Chen, and S. Wu. 2004. "Credit Rating Analysis with Support Vector Machines and Neural Networks: A Market Comparative Study." *Decision Support Systems* **37**(4), 543–558.

Hull, J., and A. White. 1995. "The Impact of Default Risk on the Prices of Options and Other Derivative Securities." *Journal of Banking and Finance* **19**, 299–322.

Ivashina, V., and B. Iverson. 2018. "Trade Creditors' Information Advantage." Working paper.

Ivashina, V., and Z. Sun. 2011. "Institutional Demand Pressure and the Cost of Corporate Loans." *Journal of Financial Economics* **99**, 500–522.

Ivashina, V., B. Iverson, and D.C. Smith. 2016. "The Ownership and Trading of Debt Claims in Chapter 11 Restructurings." *Journal of Financial Economics* **119**, 316–335.

Iverson, B., J. Madsen, W. Wang, and Q. Xu. 2018. "Practice Makes Perfect: Judge Experience and Bankruptcy Outcomes." Working paper.

James, C. 1995. "When Do Banks Take Equity? An Analysis of Bank Loan Restructurings and the Role of Public Debt." *Review of Financial Studies* **8**, 1209–1234.

James, C. 1996. "Bank Debt Restructurings and the Composition of Exchange Offers in Financial Distress." *Journal of Finance* **51**(2), 711–727.

James, C., and A. Kizilaslan. 2014. "Asset Specificity, Industry-Driven Recovery Risk, and Loan Pricing." *Journal of Financial and Quantitative Analysis* **49**, 599–631.

Jankowitsch, R., F. Nagler, M.G. Subrahmanyam. 2014. "The Determinants of Recovery Rates in the US Corporate Bond Market." *Journal of Financial Economics* **114**, 155–177.

Jarrow, R.A. 2001. "Default Parameter Estimation Using Market Prices." *Financial Analysts Journal* **57**(5), 75–92.

Jarrow, R.A., and S.M. Turnbull. 1995. "Pricing Derivatives on Financial Securities Subject to Credit Risk." *Journal of Finance* **50**, 53–86.

Jarrow, R.A., D. Lando, and S.M. Turnbull. 1997. "A Markov Model for the Term Structure of Credit Risk Spreads." *Review of Financial Studies* **10**, 481–523.

Jefferies & Company. 2003. "How to Perform a Post-Restructuring Equity Valuation." New York: Jeffries Recapitalization and Restructuring Group.

Jefferies & Company. 2006. "The Jefferies Re-Org Index SM, Special Situation." New York, December 21.

Jensen, M.C. 1986. "Agency Costs of Free Cash Flow, Corporate Finance, and Takeovers." *American Economic Review* **76**, 323–329.

Jensen, M.C., and W.H. Meckling. 1976. "Theory of the Firm: Managerial Behavior, Agency Costs and Ownership Structure." *Journal of Financial Economics* **3**, 305–360.

Jiang, W., and W. Wang. 2019. "The Performance of Post-Reorganization Equity." Working paper.

Jiang, W., K. Li, and W. Wang. 2012. "Hedge Funds and Chapter 11." *Journal of Finance* **67**, 513–560.

Jiang, Y. 2014. "The Curious Case of Inactive Bankruptcy Practice in China: A Comparative Study of U.S. and Chinese Bankruptcy Law." *Northwestern Journal of International Law & Business* **34**(3), 559–582.

Jokivuolle, E., and S. Peura. 2003. "A Model for Estimating Recovery Rates and Collateral Haircuts for Bank Loans." *European Financial Management*, forthcoming.

Jones D.S., and J. Mingo. 1998. "Industry Practices in Credit Risk Modeling and Internal Capital Allocations: Implications for a Models-Based Regulatory Capital Standard: Summary of Presentation." *Economic Policy Review* **4**(3, October), 1998.

Jones, E.P., S.P. Mason, and E. Rosenfeld. 1984. Contingent Claims Analysis of Corporate Capital Structures: An Empirical Investigation. *Journal of Finance* **39**, 611–627.

Kahan, M., and E.B. Rock. 2009. "Hedge Fund Activism in the Enforcement of Bondholder Rights." *Northwestern University Law Review* **103**, 281–322.

Kaiser, K.M. 1996. "European Bankruptcy Laws: Implications for Corporations Facing Financial Distress." *Financial Management* **25**, 67–85.

Kalotay, E.A., and E.I. Altman. 2017. "Intertemporal Forecasts of Defaulted Bond Recoveries and Portfolio Losses." *Review of Finance*, 433–463.

Kamakura Corp. 2002. "Kamakura Default Probabilities, Kamakura.com and D. van Dementer. 2002. A 10-Step Program to Replace Legacy Credit Ratings with Modern Default Probabilities for Counter-Party & Credit Risk Assessment." Kamakura Corp. (November 27).

Kaplan, S.N. 1989. "Campeau's Acquisition of Federated: Value Created or Value Destroyed?" *Journal of Financial Economics* **25**, 191–212

Kaplan, S.N. 1994. "Campeau's Acquisition of Federated: Postbankruptcy Results." *Journal of Financial Economics* **35**, 123–136

Kaplan, S.N., and R. Ruback. 1995. "The Valuation of Cash Flow Forecasts: An Empirical Analysis." *Journal of Finance* **50**, 1059–1093.

Keasey, K., and R. Watson. 1991. "Financial Distress Prediction Models: A Review of their Usefulness." *British Journal of Management* **2**(2, July), 89–102.

Keisman, D. 2004. "Ultimate Recovery Rates on Bank Loan and Bond Defaults." S&P Loss Stats.

Kim, I.J., K. Ramaswamy, and S. Sundaresan. 1993. Does Default Risk in Coupons Affect the Valuation of Corporate Bonds? A Contingent Claims Model." *Financial Management* **22**, (3), 117–131.

Kolay, M., M. Lemmon, and E. Tashjian. 2016. "Spreading the Misery? Sources of Bankruptcy Spillover in the Supply Chain." *Journal of Financial and Quantitative Analysis* **51**, 1955–1990.

Koller, T., M. Goedhart, and D. Wessels. 2010. *Valuation: Measuring and Managing the Value of Companies* (6th ed.). Hoboken, NJ: John Wiley & Sons.

Kostin, D., N. Fox, C. Maasry, and A. Sneider. 2008. "Strategy Baskets, US: Portfolio Strategy." Goldman Sachs (December 11), 17–22.

Kurth, M.H. 2004. "The Search for Accountability: The Emergence of 'Deepening Insolvency' As an Independent Cause of Action." *Corp. Corruption Lit. Reporter* (June), 196–210.

La Porta, R., F. Lopez-de-Silanes, A. Shleifer, and R. Vishny. 1998. Law and Finance, *Journal of Political Economy* **106**, 1113–1155.

Lando, D. 1998. "On Cox Processes and Credit Risky Securities." *Review of Derivatives Research* **2**, 99–120.

Lawless, R.M., and S.P. Ferris.P. 1997. "Professional Fees and Other Direct Costs in Chapter 7 Bankruptcies." *Washington University Law Quarterly* **75**, 1207–1236.

Lawless, R.M., Ferris S.P., Jayaraman, N., and Makhija, A.K. 1994. "A Glimpse of Professional Fees and Other Direct Costs in Small Firm Bankruptcies." *University of Illinois Law Review* **1994**(4), 847–888.

Leland, H.E. 1994. "Corporate Debt Value, Bond Covenants, and Optimal Capital Structure." *Journal of Finance* **49**, 1213–1252.

Leland, H.E., and K.B. Toft. 1996. "Optimal Capital Structure, Endogenous Bankruptcy, and the Term Structures of Credit Spreads." *Journal of Finance* **51**, 987–1019.

Lemmon, M., Y. Ma, and E. Tashjian. 2009. "Survival of the Fittest? Financial and Economic Distress and Restructuring Outcomes in Chapter 11." Working paper.

Li, K., and W. Wang. 2016. "Debtor-in-possession Financing, Loan-to-loan, and Loan-to-own." *Journal of Corporate Finance* **39**, 121–138.

Li, Y., and Z. Zhong. 2013. "Investing in Chapter 11 Stocks: Trading, Value, and Performance." *Journal of Financial Markets* **16**(1), 33–60.

Lim, J. 2015. "The Role of Activist Hedge Funds in Financial Distressed Firms." *Journal of Financial and Quantitative Analysis* **50**, 1321–1351.

Litterman, R., and T. Iben. 1991. "Corporate Bond Valuation and the Term Structure of Credit Spreads." *Financial Analysts Journal* (Spring), 52–64.

Loeffler, G. 2004. "An Anatomy of Rating through the Cycle." *Journal of Banking & Finance* **28**, 695–720.

Longstaff, F.A., and E.S. Schwartz. 1995. "A Simple Approach to Valuing Risky Fixed and Floating Rate Debt." *Journal of Finance* **50**, 789–819.

Longstaff, F.A., J. Pan, L. Pedersen, and K. Singleton. 2007. "How Sovereign Is Sovereign Credit Risk?" National Bureau of Economic Research, Inc. NBER Working Paper: 13658.

LoPucki, L.M. 2005. *Courting Failure: How Competition for Big Cases if Corrupting the Bankruptcy Courts.* Ann Arbor, MI: University of Michigan Press.

LoPucki, L.M., and J.W. Doherty. 2004. "The Determinants of Professional Fees in Large Bankruptcy Reorganization Cases." *Journal of Empirical Legal Studies* **1**, 111–141.

LoPucki, L.M., and J.W. Doherty. 2008. "Professional Overcharging in Large Bankruptcy Reorganization Cases." *Journal of Empirical Legal Studies* **5**, 983–1017.

LoPucki, L.M., and W.C. Whitford. 1993. "Corporate Governance in the Bankruptcy Reorganization of Large, Publicly Held Companies." *University of Pennsylvania Law Review* **141**, 669–800.

Lubben, S.J. 2000. "The Direct Costs of Corporate Reorganization: An Empirical Examination of Professional Fees in Large Chapter 11 Cases." *American Bankruptcy Law Journal* **74**, 509–552.

Ma, S., T. Tong, and W. Wang. 2017. "Selling Innovation in Bankruptcy." Working paper.

Madan, D.B., and H. Unal. 1995. "Pricing the Risks of Default." Working paper, University of Maryland.

Maksimovic, V., and G. Phillips. 1998. "Asset Efficiency and Reallocation Decisions of Bankrupt Firms." *Journal of Finance* **53**, 1495–1532.

Mann, W. 2016. "Creditor Rights and Innovation: Evidence from Patent Collateral." Working paper, UCLA.

Maxwell, W., and Shenkman, M., 2010. *Leveraged Financial Markets: A Comprehensive Guide to Loans, Bonds and Other High Yield Instruments*. New York: McGraw-–Hill.

McHugh, C., A. Michel, and I. Shaked. 1998. "After Bankruptcy: Can Ugly Ducklings Turn into Swans?" *Financial Analyst Journal* **54**, 31–40.

McLean, R.D., and J. Pontiff. 2016. "Does Academic Research Destroy Stock Return Predictability?" *Journal of Finance* **71**(1), 5–32

McQuown, J., 1993. *A Comment on Market vs. Accounting Based Measures of Default Risk*. San Francisco: KMV Corp.

Merton, R.C. 1974. "On the Pricing of Corporate Debt: The Risk Structure of Interest Rates." *Journal of Finance* **29**, (May), 449–470.

Moody's Default Report. 2017. Moody's Advisory Services. New York.

Moody's. 2008. "Moody's Comments on Debtor-in-Possession Lending, Special Comment." Moody's.

Moyer, S.G. 2005. *Distress Debt Analysis: Strategies for Speculative Investors*. Boca Raton, FL: Ross J. Publishing.

Myers, S.C. 1977. "Determinants of Corporate Borrowing." *Journal of Financial Economics* **5**, 147–175.

Nadauld, T.D., and M.S. Weisbach. 2012. "Did Securitization Affect the Cost of Corporate Debt?" *Journal of Financial Economics* **105**, 332–352.

Neziri, H. 2009. "Can Credit Default Swaps Predict Financial Crises?" *Journal of Applied Economic Sciences* IV/ **1**(7), 75.

Nielsen, L.T., J. Saà-Requejo, and P. Santa-Clara. 1993. "Default Risk and Interest Rate Risk: The Term Structure of Default Spreads." Working paper, INSEAD.

Nini, G., D.C. Smith, and A. Sufi. 2009. "Creditor Control Rights and Firm Investment Policy." *Journal of Financial Economics* **92**, 400–420.

Nini, G., D.C. Smith, and A. Sufi. 2012. "Creditor Control Rights, Corporate Governance, and Firm Value." *Review of Financial Studies* **25**, 1713–1761.

Ohlson, J. 1980. "Financial Ratios and the Probabilistic Prediction of Bankruptcy." *Journal of Accounting Research* **18**(1), 109–131.

Opler, T., and S. Titman. 1994. "Financial Distress and Corporate Performance." *Journal of Finance* **49**, 1015–1040.

Oshiro, N., and Y. Saruwatari. 2005. "Quantification of Sovereign Risk: Using the Information in Equity Market Prices." *Emerging Markets Review* **6**(4), 346–62.

Ostrower, M., and A. Calderon. 2001. "Introducing Our Tenant Credit Index, Retail REITS, Equity Research." Morgan Stanley Dean Witter (June 19), 1–5.

Phillips, G., and G. Sertsios. 2013. "How Do Firm Financial Conditions Affect Product Quality and Pricing?" *Management Science* **59**(8), 1764–1782.

Pinto, C., F. Serra, F., and M. Ferreira. 2014. A Bibliometric Study on Culture Research in International Business. *Brazilian Administrative Review,* **11**(3), 340–363.

Pomerleano, M. 1998. "Corporate Finance Lessons from the East Asian Crisis, Viewpoint Note 155." World Bank Group, Washington, DC, October.

Pomerleano, M. 1999. "The East-Asia Crisis and Corporate Finance – The Untold Micro Study." *Emerging Markets Quarterly*.

Pulvino, T. 1998. "Do Asset Fire-Sales Exist? An Empirical Investigation of Commercial Aircraft Transactions." *Journal of Finance* **53**, 939–978.

Pulvino, T. 1999. "Effects of Bankruptcy Court Protection on Asset Sales." *Journal of Financial Economics* **52**, 151–186.

Pykhtin, M. 2003. "Unexpected Recovery Risk." *Risk* **16**, 74–78.

Pynchon, T. 2013. *Bleeding Edge*. New York: Penguin Press.

Ramaswami, M., and S. Moeller. 1990. *Investing in Financially Distressed Firms*. New York: Quorum Books.

Ravid, S.A., and S. Sundgren. 1998. "The Comparative Efficiency of Small-firm Bankruptcies: A Study of the US and Finnish Bankruptcy Codes." *Financial Management* **27**(4), 28–40.

Reinhart, M., and K. Rogoff. 2010. *This Time Is Different*. Princeton, NJ: Princeton University Press.

Remolona, E.M., M. Scatigna, and E. Wu. 2008. A Ratings-Based Approach to Measuring Sovereign Risk. *International Journal of Finance and Economics* **13**(1), 26–39.

Resti, A., and A. Sironi. 2005. "Loss Given Default and Recovery Risk Under Basel II." In Altman E.I., et al., eds., *Recovery Risk*. London: Risk Books.

Roberts, M. 2015. "The Role of Dynamic Renegotiation and Asymmetric Information in Financial Contracting." *Journal of Financial Economics* **116**, 61–81.

Roberts, M., and A. Sufi. 2009a. "Renegotiation of Financial Contracts: Evidence from Private Credit Agreements." *Journal of Financial Economics* **93**, 159–184.

Roberts, M., and A. Sufi. 2009b. "Creditor Rights and Capital Structure: An Empirical Investigation." *Journal of Finance* **64**, 1657–1695.

Rodano, G., N. Serrano-Velarde, and E. Tarantino. 2016. Bankruptcy Law and Bank Financing. *Journal of Financial Economics* **120**, 363–382.

Rosenberg, H., 1992. *The Vulture Investors*. New York: HarperCollins. Revised ed., Wiley, 2000, New York.

Ruback, R. 2002. "Capital Cash Flows: A Simple Approach to Valuing Risky Cash Flows." *Financial Management* **31**(2), 85–103.

Sargen, H. 1977. "Economics Indicators and Country Risk Appraisal, Federal Reserve Bank of San Francisco." *Economic Review* (Fall).

Schmidt, R. 1984. "Early Warning of Debt Rescheduling." *Journal of Banking and Finance* **8**(2), 357–370.

Schoenherr, D. 2017. "Managers' Personal Bankruptcy Costs and Risk-taking." Working paper.

Schuermann, T. 2004. "What Do We Know about Loss Given Default?" Working paper, Federal Reserve Bank of New York, forthcoming in Shimko D. (ed.), *Credit Risk Models and Management* (2nd ed.). London: Risk Books.

Schultze, G., and J. Lewis. 2012. *The Art of Vulture Investing: Adventures in Distressed Securities Management*. Hoboken, NJ: Wiley.

Schumpeter, J. 1942. *Capitalism, Socialism, and Democracy*. New York: Harper & Row.

Scott, J., 1981. "The Probability of Bankruptcy: A Comparison of Empirical Predictions and Theoretical Models." *Journal of Banking & Finance* (September), 317–344.

Segoviano, M.A., C. Caceres, and V. Guzzo. 2010. "Sovereign Spreads: Global Risk Aversion, Contagion or Fundamentals?" IMF Working Paper: 10/120.

Sgherri, S., and E. Zoli. 2009. "Euro Area Sovereign Risk During the Crisis." International Monetary Fund Working Paper: 09/222.

Shleifer, A., and R. Vishny. 1992. "Liquidation Values and Debt Capacity: A Market Equilibrium Approach." *Journal of Finance* **47**, 1343–1366.

Shumway, T. 2001. "Forecasting Bankruptcy More Accurately: A Simple Hazard Model." *Journal of Business* **74**(1), 101–124.

Skeel, D.A. 2003a. "The Past, Present, and Future of Debtor-in-possession Financing." *Cardozo Law Review*, 1905–1934.

Skeel, D.A. 2003b. "Creditors' Ball: The New New Corporate Governance in Chapter 11." *University of Pennsylvania Law Review* **152**, 917–951.

Smith, R., and I. Walter, I. 2003. *Global Banking*. London: Oxford University Press.

Smith, R., and M. Dion. 2008. "Who Said Zed's Dead? Reviving the Altman Z-Score." JPMorgan: Asia Pacific Equity Research (September 25), 1–61,

SP Global. 2017. 2016 Annual Global Corporate Default Study & Rating Transitions. SP Global Ratings. New York (April 13).

Standard & Poor's. 2017. "2016 Annual Global Corporate Default Study and Rating Transitions." New York.

Stark, R.J., H.L. Siegel, and E.S. Weisfelner. 2011. *Contested Valuation in Corporate Bankruptcy*. LexisNexis, Matthew Bender & Co.

Subrahmanyam, M.G., Y. Tang, and Q. Wang. 2014. "Does the Tail Wag the Dog?: The Effect of Credit Default Swaps on Credit Risk." *The Review of Financial Studies* **27**(10), 2927–2960.

Suo, W., W. Wang, and Q. Zhang. 2013. "Explaining Debt Recovery Using an Endogenous Bankruptcy Model." *The Journal of Fixed Income* **23**, 114–131.

Taillard, J.P. 2013. "The Disciplinary Effects of Non-Debt Liabilities: Evidence from Asbestos Litigation." *Journal of Corporate Finance* **23**, 267–293.

Tashjian, E., R.C. Lease, and J.J. McConnell. 1996. "An Empirical Analysis of Prepackaged Bankruptcies." *Journal of Financial Economics* **40**, 135–162.

Thorburn, K.S. 2000. "Bankruptcy Auction: Costs, Debt Recovery, and Firm Survival." *Journal of Financial Economics* **58**, 337–368.

Tirole, J. 2001. "Corporate Governance." *Econometrica* **69**, 1–35.

Van de Castle, K., D. Keisman, and R. Yang. 2000. *Suddenly Structure Mattered: Insights into Recoveries of Defaulted Debt*. S&P Corporate Ratings (May 24).

Varma, P., R. Cantor, and D. Hamilton. 2003. *Recovery Rates on Defaulted Corporate Bonds and Preferred Stocks*. Moody's Investors Service (December).

Vasicek, O.A. 1984. *Credit Valuation*. KMV Corporation (March).

Vig, V. 2013. "Access to Collateral and Corporate Debt Structure: Evidence from A Natural Experiment." *Journal of Finance* **68**, 881–928.

Waldock, K.P. 2017. "Unsecured Creditor Control in Chapter 11." Working paper, Georgetown University.

Wang, W. 2011. "Recovery and Returns of Distressed Bonds in Bankruptcy." *Journal of Fixed Income* **21**, 21–31.

Warner, J.B. 1977. "Bankruptcy Costs: Some Evidence." *Journal of Finance* **32**, 337–347.

Weiss, L.A. 1990. "Bankruptcy Resolution: Direct Costs and Violation of Priority of Claims." *Journal of Financial Economics* **27**, 285–314.

Whitman, M., and F. Diz, F. 2009. *Distress Investing: Principles and Techniques*. Hoboken, NJ: Wiley.

Wilcox, J. 1971. "A Gamblers' Ruin Prediction of Business Failure Using Accounting Data." *Sloan Management Review* **12** (September), 84–96.

Williams, J.F., S. Bernstein, and S.H Seabury. 2008. "Squaring Bankruptcy Valuation Practice with Daubert Demands." *ABI Law Review* (Spring).

Wilson, T.C. 1998. "Portfolio Credit Risk, Federal Reserve Board of New York." *Economic Policy Review* (October), 71–82.

Wilton, J.M., and J.A. Wright. 2011. "Parsing and Complying with New Rule 2019." *American Bankruptcy Institute Journal* (October).

Zhang, Z. 2018. "Bank Interventions and Trade Credit: Evidence from Debt Covenant Violation, Journal of Financial and Quantitative Analysis." *Journal of Financial and Quantitative* Analysis, forthcoming.

Zhou, C. 2001. "The Term Structure of Credit Spreads with Jump Risk." *Journal of Banking & Finance* **25**, 2015–2040.

Author Index

Subject Index